Student Solutions Manual

for Olmsted/Williams

Chemistry
THE MOLECULAR SCIENCE

Larry Peck
and
Frank Kolar
Texas A & M University

 Mosby

St. Louis Baltimore Berlin Boston Carlsbad Chicago London Madrid
Naples New York Philadelphia Sydney Tokyo Toronto

Mosby

Dedicated to Publishing Excellence

Printed in the United States of America

Mosby–Year Book, Inc.
11830 Westline Industrial Drive
St. Louis, Missouri 63146

ISBN 0-8016-5071-2

PREFACE

This solutions manual has been prepared as a supplement to CHEMISTRY-THE MOLECULAR SCIENCE by John Olmsted III and Gregory M. Williams.

The textbook by Olmsted and Williams presents an excellent, modern approach to general chemistry. To thoroughly understand the molecular world as it is presented in the textbook, a student needs to develop the ability to solve problems and answer questions related to the principles and concepts presented. We recognize that many students have difficulty developing the necessary problem-solving ability to best utilize the knowledge gained from material presented in the textbook. Through the use of this book and the textbook, the authors hope that students will improve their problem-solving skills while learning a considerable amount of chemistry. Students are strongly encouraged to attempt to solve each problem and to consult the textbook before referring to the solutions in this manual. It is highly likely that the knowledge gained and the problem-solving skills acquired in a general chemistry course will prove valuable in subsequent science and technical courses as well as assisting students to prepare for a variety of professions.

Many problems can be solved in more than one way. We encourage students to develop methods that they understand. For consistency we have striven to follow the methods used in the textbook. Where practical, we have tried to include dimensions and to show the correct number of significant figures; in some problems an extra digit is carried in intermediate steps. Remember, authors can make mistakes, too.

We wish to thank our friends and family for their support in the production of this book. Special thanks are extended to Sandra Peck, typist.

Larry Peck and Frank Kolar
Chemistry Department
Texas A&M University
College Station, TX 77843

CONTENTS

CHAPTER 1: THE SCIENCE OF CHEMISTRY

1.1   You may think of other examples to elaborate on, but a few are:
        a) Setting any pollution (environmental) standards
        b) Setting any food, drug or cosmetics purity standards
        c) Passing any drug control laws

1.3   A pharmacist must know nomenclature for chemicals (especially complex organic molecules) in order to identify chemically equivalent commercial drugs, must know how to handle and protect the potency of drugs, must be able to weigh out drugs (in some cases small amounts) and thoroughly mix them, and must know enough about chemical reactions so that he does not mix chemically incompatible drugs.

1.5   a) hydrogen, H   b) helium, He   c) hafnium, Hf   d) nitrogen, N
        e) neon, Ne   f) niobium, Nb

1.7   a) As, arsenic   b) Ar, argon   c) Al, aluminum   d) Am, americium
        e) Ag, silver   f) Au, gold   g) At, astatine   h) Ac, actinium

1.9 CCl_4

1.11   a) Br_2 b) HCl c) C_2H_5I d) PCl_3 e) SF_4 f) N_2O_4

1.13   Cs, cesium

1.15   O, oxygen; Cl, chlorine; Se, selenium; and P, phosphorus. Oxygen and selenium have chemical properties similar to those of sulfur.

1.17   a) Li, lithium; Na, sodium; K, potassium; Rb, rubidium; and Cs, cesium are possible answers. (Select any one).
        b) Be, beryllium; Mg, magnesium; Ca, calcium; Sr, strontium; Ba, barium; and Ra, radium are possible answers.

1.19   lithium, Li; beryllium, Be; boron, B; carbon, C; oxygen, O; fluorine, F; neon, Ne

1.21   a) iron - pure substance   b) cup of coffee - solution
        c) glass of milk - heterogeneous mixture   d) dust free atmosphere - solution
        e) dusty atmosphere - heterogeneous mixture
        f) block of wood - heterogeneous mixture

1.23   a) gasoline - liquid   b) molasses - liquid   c) snow - solid   d) chewing gum - solid

1.25   a) formation of frost - physical change   b) drying of clothes - physical change
        c) burning of leaves - chemical change

1.27   a) lake water - mixture   b) distilled water- compound   c) mud - mixture
        d) helium - element   e) rubbing alcohol - compound   f) paint - mixture

1.29 a) $100,000 = 1.00000 \times 10^5$ b) $10,000 \pm 100 = 1.00 \times 10^4$

c) $0.000400 = 4.00 \times 10^{-4}$ d) $0.0003 = 3 \times 10^{-4}$ e) $275.3 = 2.753 \times 10^2$

1.31 a) $430\ kg = 4.30 \times 10^2\ kg$

b) $1.35\ \mu m \times \dfrac{1 \times 10^{-6}\ m}{\mu m} = 1.35 \times 10^{-6}\ m$

c) $624\ ps \times \dfrac{1 \times 10^{-12}\ s}{ps} = 6.24 \times 10^{-10}\ s$

d) $1024\ ng \times \dfrac{10^{-9}\ g}{ng} \times \dfrac{1\ kg}{10^3\ g} = 1.024 \times 10^{-9}\ kg$

e) $93,000\ km \times \dfrac{10^3\ m}{km} = 9.3000 \times 10^7\ m$

f) $1\ day \times \dfrac{24\ h}{d} \times \dfrac{60\ min}{h} \times \dfrac{60\ s}{min} = 9 \times 10^4\ s$

(assuming that one day is only one significant figure)

g) $0.0426\ in \times \dfrac{2.54\ cm}{in} \times \dfrac{10^{-2}\ m}{cm} = 1.08 \times 10^{-3}\ m$

1.33 $5.0 \times 10^{-1}\ carat \times \dfrac{3.168\ grains}{1\ carat} \times \dfrac{1\ g}{15.4\ grains} \times \dfrac{10^{-3}\ kg}{1\ g} = 1.0 \times 10^{-4}\ kg$

$7.00\ g\ of\ Au \times \dfrac{10^{-3}\ kg}{1\ g} = 7.00 \times 10^{-3}\ kg$

$1.0 \times 10^{-4} kg + 7.00 \times 10^{-3}\ kg = 7.10 \times 10^{-3}\ kg$

1.35 $1\ qt\ H_2O \times \dfrac{1\ L}{1.057\ qt} \times \dfrac{10^3\ mL}{1\ L} \times \dfrac{1\ cm^3}{1\ mL} \times \dfrac{1.00\ g}{1\ cm^3} = 946\ g$

(assuming exactly 1 qt of H_2O)

1.37 Volume $= 15.5\ cm \times 4.6\ cm \times 1.75\ cm = 125\ cm^3$

density $= \dfrac{98.456\ g}{125\ cm^3} = 0.79\ \dfrac{g}{cm^3}$

1.39 Volume $= \dfrac{mass}{density} = \dfrac{15.4\ g}{2.70\ g/cm^3} = 5.70\ cm^3$

1.41 $r = 0.875$ in x $\dfrac{2.54 \text{ cm}}{1 \text{ in}} = 2.2225$ cm;   height $= 4.500$ in x $\dfrac{2.54 \text{ cm}}{1 \text{ in}} = 11.43$ cm

$V = \pi r^2 h = \pi (2.2225)^2 (11.43) = 177.37 \text{ cm}^3$

mass of $H_2O = 270.064$ g - 93.054 g = 177.010 g

$d = \dfrac{177.010 \text{ g}}{177.37 \text{ cm}^3} = 0.998 \text{ g}/\text{cm}^3$

1.43   a) 176 kg   b) 2.54 s   c) 73 mi hr^{-1} d) 924 kg m^{2}s^{-2} e) 34 J

1.45   1 year x $\dfrac{365 \text{ days}}{\text{yr}}$ x $\dfrac{24 \text{ hr}}{1 \text{ day}}$ x $\dfrac{60 \text{ min}}{1 \text{ hr}}$ x $\dfrac{60 \text{ s}}{1 \text{ min}}$ x $\dfrac{186{,}000 \text{ mi}}{1 \text{ s}}$

x $\dfrac{1.61 \text{ km}}{1 \text{ mi}} = 9.44 \times 10^{12}$ km

1.47

1.49   Eight elements have symbols beginning with "T." They are: Ta, tantalum; Tc, technetium; Te, tellurium; Tb, terbium; Tl, thallium; Th, thorium; Tm, thulium; and Ti, titanium.

1.51   Many of these metals were found in pure or nearly pure form and were used without chemical treatment. Others were available as ores that could be easily reduced to the metal.

1.53   Lead, Pb

1.55   100 years x $\dfrac{365.24 \text{ day}}{\text{year}}$ x $\dfrac{24 \text{ hr}}{\text{day}}$ x $\dfrac{60 \text{ min}}{\text{hr}}$ x $\dfrac{60 \text{ s}}{\text{min}} = 3.1557 \times 10^9$ s

1.57   silver and copper

1.59   Some examples are given. You may be able to find other examples. Be and Ba, Pd and Pt, Ag and Au, S and Se, and Pr and Pa.

1.61   Try by starting with things that you are interested in. What happens chemically when we taste? What happens chemically when we smell? What chemical materials can serve as superconductors? Are there chemical substances that can stimulate the immune system? Develop a method of storing radioactive wastes. If you are interested in food preparation, you might include examples associated with (1) the preserving of vitamins during cooking, (2) better microwave ovens, (3) new food additives that are natural substances, (4) genetically engineered plants, (5) the effects of metal ions in foods, (6) or what essential nutrients really are.

1.63 $-11.5°C + 273.15 = 261.7\ K$

1.65 a) 4.52×10^{34} b) 1×10^{-4}

1.67 $5.00\ lb \times \dfrac{454\ g}{1\ lb} \times \dfrac{1\ cm^3}{8.92\ g} = 254\ cm^3$;

Diameter $= 0.0508\ in \times \dfrac{2.54\ cm}{1\ in} = 0.129\ cm$

$V = \pi r^2 h = 254\ cm^3 = \pi(0.129\ cm/2)^2 h$;

$h = \dfrac{254\ cm^3}{\pi(0.129\ cm/2)^2} = 1.94 \times 10^4\ cm \times \dfrac{10^{-2}\ m}{cm} = 1.94 \times 10^2\ m$

1.69 $3\ min\ 57\ s = 237\ s$; $1\ mile = 1.61 \times 10^3\ m$;

$1500\ m \times \dfrac{237\ s}{1.61 \times 10^3\ m} = 221\ s$

1.71 $12\ ft \times 9.5\ ft \times 10.5\ ft \times \left(\dfrac{12\ in}{ft}\right)^3 \times \left(\dfrac{2.54\ cm}{in}\right)^3 \times \left(\dfrac{1\ m}{10^2\ cm}\right)^3 = 33.9\ m^3$

$\dfrac{5.5\ mg}{1.000\ m^3} \times 33.9\ m^3 \times \dfrac{1\ g}{10^3\ mg} = 0.19\ g$

1.73 $\left((7\ ft \times 12\ in/ft) + 1\ in\right) \times \dfrac{2.54\ cm}{in} \times \dfrac{10\ mm}{cm} = 2160.\ mm$

$\pm 1/4\ in \times 2.54\ cm\ in^{-1} \times 10\ mm\ cm^{-1} = \pm 6.35\ mm$

Answer $= 2160 \pm 6\ mm$

1.75 $9.3 \times 10^7\ miles \times \dfrac{1.61\ km}{mile} \times \dfrac{1000\ m}{km} \times \dfrac{1\ s}{3.00 \times 10^8\ m} \times \dfrac{1\ min}{60\ s} = 8.3\ min$

1.77 Follow the notations used in Figures 1-4 and 1-5.

1.79 a) $8.97 \times 10^5\ acre\ ft \times \dfrac{43,560\ ft^3}{acre\ ft} \times (12\ in/ft)^3 \times (2.54\ cm/m)^3$

$\times (1\ m/10^2\ cm)^3 \times \dfrac{1\ L}{10^{-3}\ m^3} = 1.11 \times 10^{12}\ L$

b) $8.97 \times 10^5\ acre\ ft \times \dfrac{43,560\ ft^3}{acre\ ft} = 3.91 \times 10^{10}\ ft^3$

c) $8.97 \times 10^5 \times 43,560\ ft^3 \times (12\ in/ft)^3 \times (2.54\ cm/in)^3 \times (1\ m/10^2\ cm)^3$

$= 1.11 \times 10^9\ m^3$

4

CHAPTER 2: THE ATOMIC NATURE OF MATTER

2.1   a) 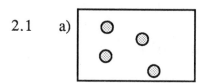   b) 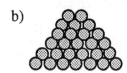   c)

   Helium, a gas   Tungsten, a solid   Mercury, a liquid

2.3   a) Oxygen gas contains oxygen atoms.
b) Oxygen atoms combine into diatomic oxygen molecules.
c) The diatomic oxygen molecules look and act differently than the solid carbon made from carbon atoms.
d) The molecules of oxygen and carbon react to give carbon monoxide; atoms of carbon combine with atoms of oxygen in a 1:1 ratio.
e) Atoms are rearranged in this chemical process, but the total number of each type of atom remains the same (4 atoms of oxygen and 13 atoms of carbon).

2.5   a) $25 \text{ atoms C } \times \dfrac{1 \text{ molecule } CH_4}{1 \text{ atom C}} = 25 \text{ molecules } CH_4$

   b) $25 \text{ atoms C} \times \dfrac{2 \text{ molecules } H_2}{1 \text{ atom C}} = 50 \text{ molecules } H_2$

   c) $25 \text{ atoms C} \times \dfrac{2 \text{ molecules } H_2}{1 \text{ atom C}} \times \dfrac{2 \text{ atoms H}}{\text{molecule } H_2} = 100 \text{ atoms H}$

2.7   In order for the odor to reach the olfactory nerves in the nose, molecules must move from the rose through the air to the nose.

2.9   Use Coulomb's Law to answer this question. $\text{Force} = k\dfrac{q_1 q_2}{r^2}$

   a) If one of the q values increases by a factor of 3, the force increases by a factor of 3.

   b) If both q values increase by a factor of 3, the force increases by a factor of 9.

   c) If the distance (r) doubles, the force changes by a factor of one-fourth.

2.11   By adjusting the amount of electrical force, the downward-acting gravitational force could be exactly counterbalanced by the upward-attracting electrical force. From the electrical force required to suspend the different droplets, calculation of the charges on the droplets is possible. The calculated values of charge would be multiples of the electron charge just like Millikan determines because the positively-charged oil droplets result from the removal of one or more electron.

2.13 (See Table 2-1) -1.00×10^{-6} C $x = \dfrac{9.1094 \times 10^{-31} \text{ kg}}{-1.6022 \times 10^{-19} \text{ C}} = 5.69 \times 10^{-18}$ kg

2.15 a) two electrons

b) Total mass of particles = $(2 \text{ e})\left(\dfrac{9.1094 \times 10^{-31} \text{ kg}}{1 \text{ e}}\right)$

$+(2 \text{ p})\left(\dfrac{1.6726 \times 10^{-27} \text{ kg}}{1 \text{ p}}\right) + (2 \text{ n})\left(\dfrac{1.6749 \times 10^{-27} \text{ kg}}{1 \text{ n}}\right)$

$= 6.6968 \times 10^{-27}$ kg;

$\dfrac{(2 \text{ e})\left(\dfrac{9.1094 \times 10^{-31} \text{ kg}}{1 \text{ e}}\right)}{6.69698 \times 10^{-27} \text{ kg}} \times 100\% = 0.027205\%$ or 0.00027205;

Using at. mass of He: $\dfrac{(2 \text{ e})\left(\dfrac{9.1094 \times 10^{-28} \text{ g}}{1 \text{ e}}\right)}{\dfrac{4.002602 \text{ g} / \text{mole}}{6.022 \times 10 \text{ atom}^{23} / \text{mole}}} \times 100\% = 0.027411\%$

or 0.00027411

2.17 $_{12}$Mg, magnesium; $_{25}$Mn, manganese; $_{29}$Cu, copper; $_{27}$Co, cobalt; $_{15}$P, phosphorus; $_{82}$Pb, lead

2.19 a) $^{56}_{26}$Fe   b) $^{38}_{18}$Ar   c) $^{236}_{92}$U   d) $^{19}_{9}$F

2.21

2.23 There are 3 peaks for Cl_2^+. They are 70, 72, and 74; 70 is the most intense; 74 is the least intense.

2.25 Cations: Cl_2^+, CO^+, Cr^{3+}; Anions: Cl^- and $Cr_2O_7^{2-}$; Neutral species: C, CCl_4, CO_2

2.27 a) $OH^- + H^+ \rightarrow H_2O$ b) $Na \rightarrow Na^+ + e^-$ c) $HCl \rightarrow H^+ + Cl^-$ d) $O + 2 \text{ e}^- \rightarrow O^{2-}$

2.29 Al^{3+} and O^{2-} yield Al_2O_3 as a neutral representation.

2.31 CsI and SrI_2

2.33 a) Some of the runner's fat is converted into energy for running.
b) As the apple falls, potential energy is converted to kinetic energy.
c) The kinetic energy is converted into thermal energy when the apple hits the ground.

6

2.35 $k.E. = 1/2\ mV^2$; $V = \sqrt{\dfrac{k.E.}{1/2\ m}} = \sqrt{\dfrac{3.75 \times 10^{-23}\ kg\ m^2/s^2}{1/2\left(\dfrac{4.002602 \times 10^{-3}\ kg}{6.022 \times 10^{23}}\right)}} = 106\ m/s$

2.37 a) radiant energy becomes thermal energy   b) potential energy becomes kinetic energy

c) potential energy (chemical energy) becomes thermal and radiant energy

2.39 X could be one of many elements such as Be, Mg, Ca, Sr, Ba, and Ra, but they must be metals if the compound formed is to be ionic as stated in the problem. OF_2, XeF_2 and other nonmetallic compounds are not ionic.

2.41 solid 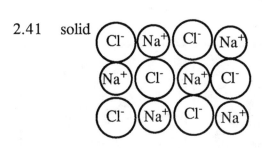   liquid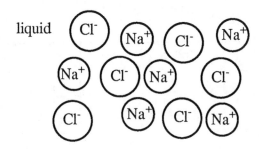

2.43 N_2^+, O_2^+, N^+ and O^+; N_2^-, O_2^-, N^- and O^-

2.45 a) 8 protons, 8 neutrons and 10 electrons
b) 7 protons, 8 neutrons and 7 electrons
c) 25 protons, 30 neutrons and 22 electrons
d) 17 protons, 18 neutrons and 18 electrons
e) 17 protons, 20 neutrons and 16 electrons

2.47 a) Co-60, 27 protons and 33 neutrons;   b) C-14, 6 protons and 8 neutrons;

c) U-235, 92 protons and 143 neutrons;   d) U-238, 92 protons and 146 neutrons

2.49 $^{14}_{7}N$, $^{14}_{6}C$

2.51 Read the amounts shown in the pie chart and from those values construct a mass spectrum similar to that at the end of the answer to question 2.46

2.53 57.5 mile hr^{-1} = velocity   2250 lb = mass   1 km = 0.6215 mi
454 g = 1 lb 10^3 m = km   1000 g = kg
60 min = 1 hr   1 $kg\ m^2\ s^{-2}$ = 1 J   60 s = 1 min
$k.E. = 1/2\ mV^2$

$\dfrac{57.5\ mi}{hr} \times \dfrac{1\ km}{0.6215\ mi} \times \dfrac{10^3\ m}{kg} \times \dfrac{1\ hr}{60\ min} \times \dfrac{1\ min}{60\ s} = 25.7\ m\ s^{-1}$ = velocity

$2250\ lb \times \dfrac{454\ g}{lb} \times \dfrac{kg}{10^3\ g} = 1.02 \times 10^3\ kg$

$k.E. = (1/2)(1.02 \times 10^3\ kg)(25.7\ m\ s^{-1})^2 = 3.37 \times 10^5\ J$

7

2.55 $^{40}_{18}Ar$, $^{40}_{19}K$, $^{40}_{20}Ca$

2.57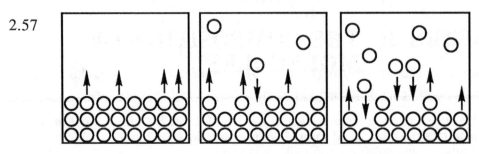

2.59   a) 30   b) 20   c) 20 N and 60 H

2.61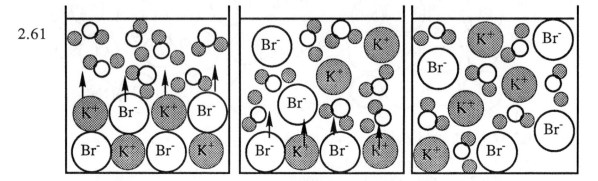

2.63 ^{98}Tc - 43 protons, 55 neutrons, 43 electrons;
        ^{99}Tc - 43 protons, 56 neutrons, 43 electrons

2.65   Two peaks will be obtained; one for F_2^+ and one for F^+.

2.67

	proton	electron

mass = 1.6726×10^{-27} kg   mass = 9.1094×10^{-31} kg
velocity = ?   velocity = 1.55×10^6 m s^{-1}
k.E. = $1/2\ mV^2$ k.E. = $1/2\ mV^2$
k.E.'s are equal or $1/2\ m_pV_p^2 = 1/2\ m_eV_e^2$

$(1.6726 \times 10^{-27}\ kg)V_p^2 = (9.1094 \times 10^{-31}\ kg)(1.55 \times 10^6\ m\ s^{-1})^2$
$V_p^2 = 1.308 \times 10^9\ m^2\ s^{-2}$
$V_p = 3.62 \times 10^4$ m s^{-1}

2.69   When magnesium metal is burned in air, the magnesium combines with oxygen from the air to produce MgO. The mass of MgO produced equals the mass of magnesium burned plus the mass of oxygen from the air that combined with it.

CHAPTER 3: THE COMPOSITION OF MOLECULES

3.1 a) methane, CH_4 b) ethylene, C_2H_4 c) dimethyl ether, C_2H_6O
d) hydrogen bromide, HBr e) phosphorus trichloride, PCl_3
f) urea, CH_4N_2O g) iodoethane, C_2H_5I h) hydrazine, N_2H_4

3.3 a) 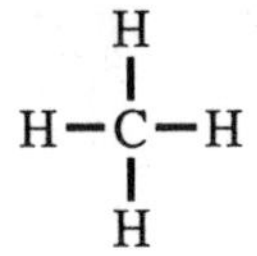

b) 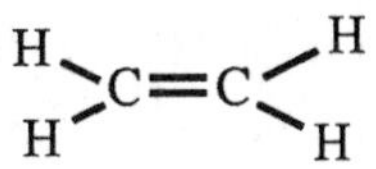

c) 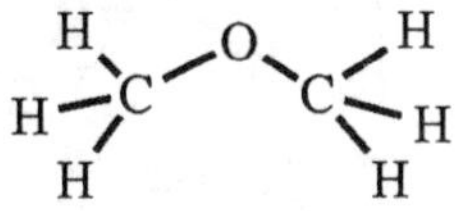

d) H—Br

e) 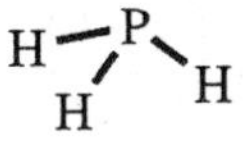

f) 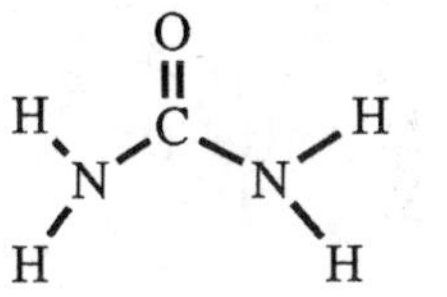

g) 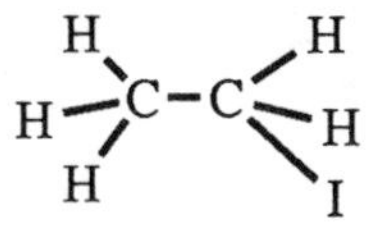

h) 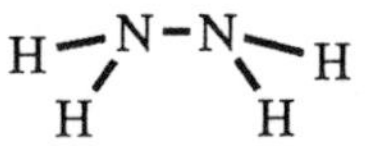

3.5 a) 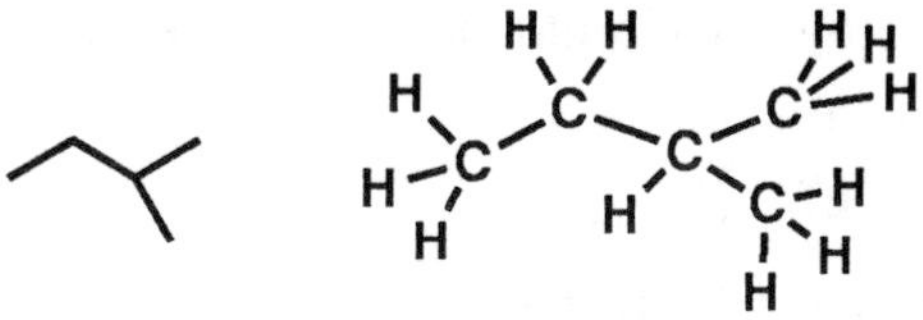

b) 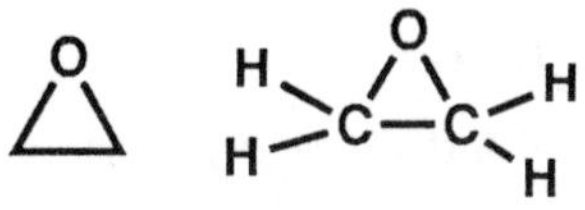

c) 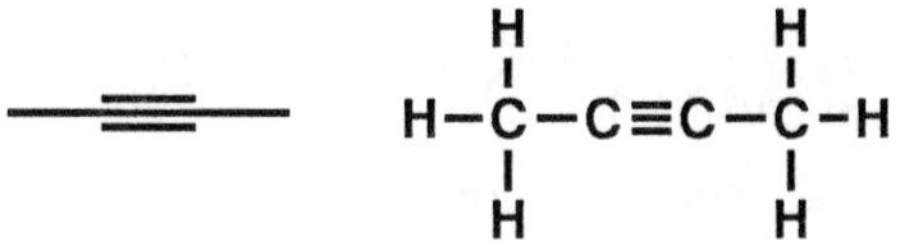

d) 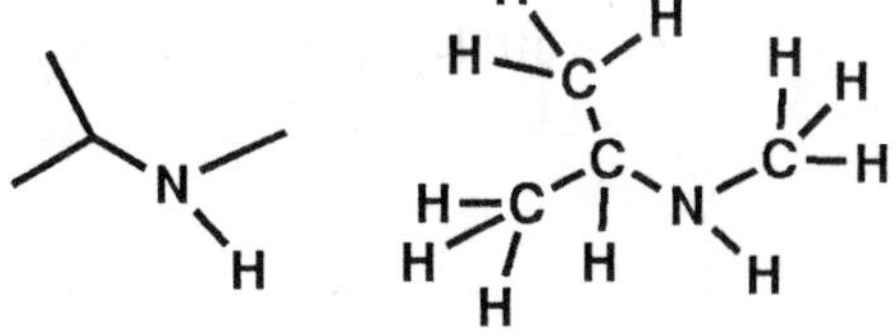

e) 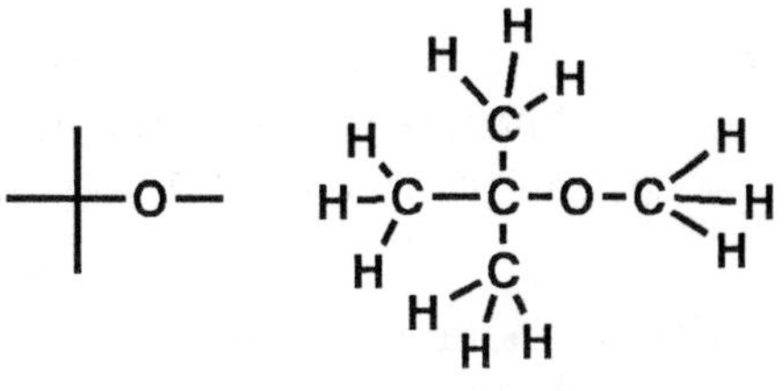

f) 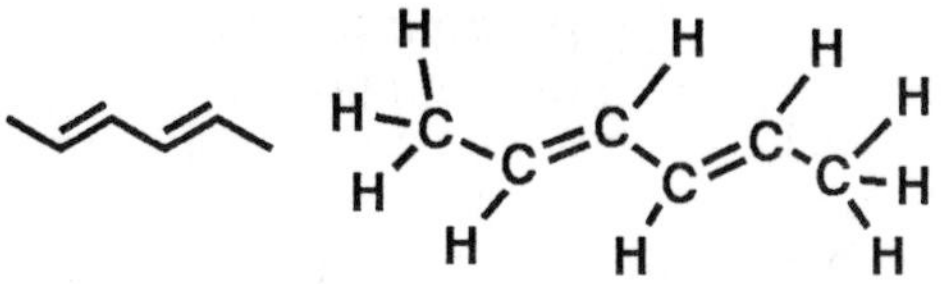

(continued)

(3.5 continued)

g)

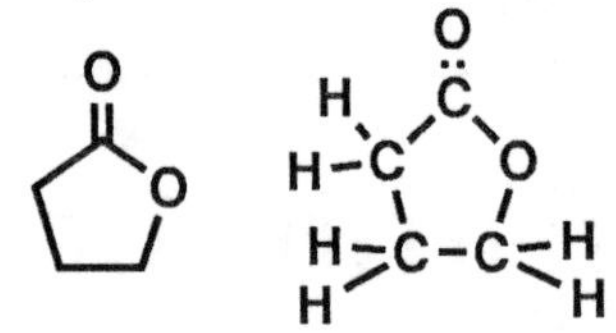

h)

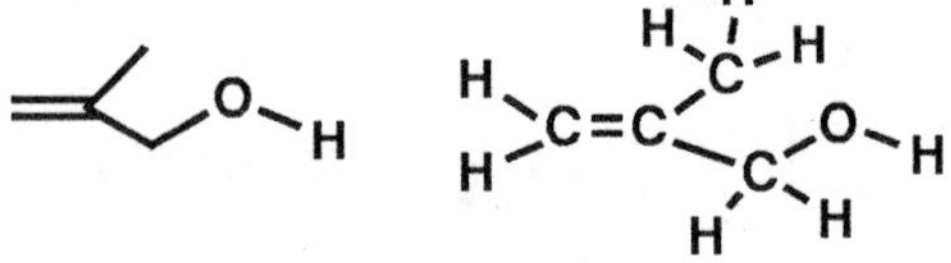

i)

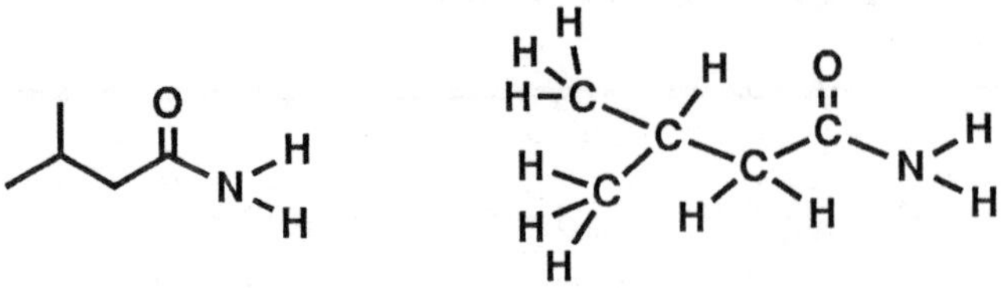

3.7   a) 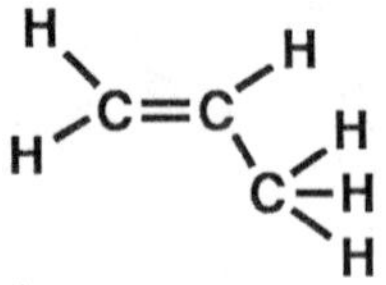   b) 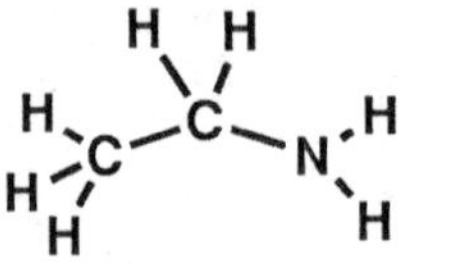   c) $H-C\equiv C-O^H$

d) 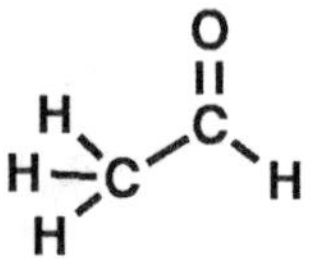   e)

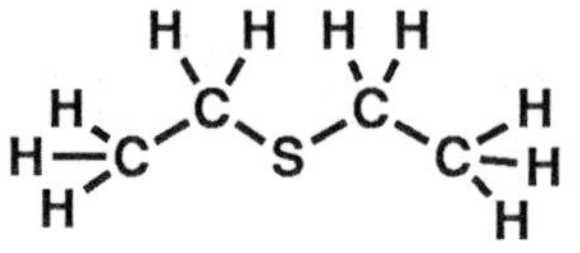

f) 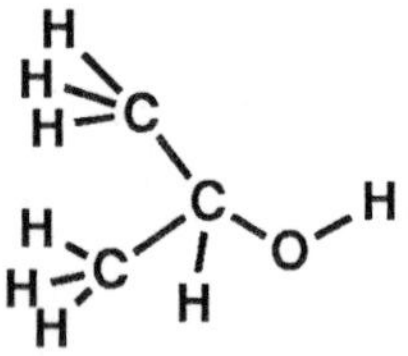   g)

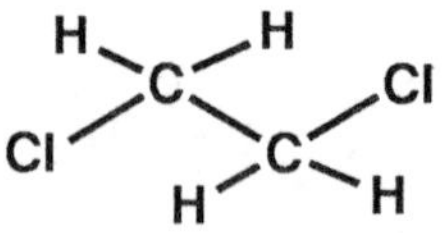

3.9   a) CO_2, carbon dioxide   b) HCl, hydrogen chloride   c) CCl_4, carbon tetrachloride
d) ClF_3, chlorine trifluoride   e) PH_3, phosphine
f) N_2O, dinitrogen oxide (nitrous oxide)

3.11   a) CH_4 b) HF   c) CaH_2 d) PCl_3 e) N_2O_5 f) SF_6
g) BF_3 h) $CH_3CH_2CH(OH)CH_2CH_3$

3.13   a) disulfur dichloride   b) iodine heptafluoride   c) hydrogen bromide
d) dinitrogen trioxide   e) silicon carbide
f) methanol (other names also possible for this compound)

3.15   a) HF   b) CaF_2, ionic   c) $Al_2(SO_4)_3$, ionic
d) $(NH_4)_2S$, ionic   e) SO_2 f) CCl_4

3.17   The ionic compounds are: CaO, K_2CO_3 and Na_2HPO_4.

3.19   a) calcium chloride hexahydrate   b) iron(II) ammonium sulfate
c) potassium carbonate   d) tin(II) chloride dihydrate   e) sodium hypochlorite
f) silver sulfate   g) copper(II) sulfate   h) potassium dihydrogen phosphate
i) sodium nitrate   j) calcium sulfite   k) potassium permanganate

10

3.21   a) MM of Fe = 55.85 $7.85 \text{ g Fe} \times \dfrac{1 \text{ mol}}{55.85 \text{ g}} = 0.141 \text{ mol}$

      b) $6.55 \times 10^{13} \text{ atoms} \times \dfrac{1 \text{ mol}}{6.022 \times 10^{23} \text{ atoms}} = 1.09 \times 10^{-10} \text{ mol}$

      c) 4.68 μg of Si,   MM of Si = 28.09,   1 μg = 10^{-6} g

      $4.68 \text{ μg} \times \dfrac{10^{-6} \text{ g}}{\text{μg}} \times \dfrac{1 \text{ mole}}{28.09 \text{ g}} = 1.67 \times 10^{-7} \text{ mol}$

      d) 1.46 met. ton of Al   1 met. ton = 10^3 kg   MM of Al = 26.98

      $1.46 \text{ met. ton} \times \dfrac{10^3 \text{ kg}}{\text{met. ton}} \times \dfrac{1 \text{ mole}}{26.98 \text{ g}} \times \dfrac{10^3 \text{ g}}{\text{kg}} = 5.41 \times 10^4 \text{ mol}$

3.23   Ar-36:   35.96755 x 0.00337   =   0.121
       Ar-38:   37.96272 x 0.00063   =   0.024
       Ar-40:   39.9624 x 0.99600   =   <u>39.803</u>
                Total   =   39.948 g/mol

3.25   a) MM of $(NH_4)_2CO_3$ = 2(14.01) + 8(1.008) + 12.01 + 3(16.00) = 96.09
      b) MM of K_2S = 2(39.10) + 32.07 = 110.27
      c) MM of $CaCO_2$ = 40.08 + 12.01 + 3(16.00) = 100.09
      d) MM of LiBr = 6.94 + 79.90 = 86.84
      e) MM of Na_2SO_4 = 2(22.99) + 32.07 + 4(16.00) = 142.05
      f) MM of $AgNO_3$ = 107.87 + 14.01 + 3(16.00) = 169.88

3.27   5.89 μg, 10^{-6} = 1 μg, and 6.022×10^{23} atom mol^{-1}
      a) Be, MM = 9.01

      $5.86 \text{ μg} \times \dfrac{10^{-6} \text{ g}}{\text{μg}} \times \dfrac{1 \text{ mol}}{9.01 \text{ g}} \times \dfrac{6.022 \times 10^{23} \text{ atoms}}{\text{mol}} = 3.92 \times 10^{17} \text{ atoms}$

      b) P, MM = 30.97

      $5.86 \text{ μg} \times \dfrac{10^{-6} \text{ g}}{\text{μg}} \times \dfrac{1 \text{ mol}}{30.97 \text{ g}} \times \dfrac{6.022 \times 10^{23} \text{ atoms}}{\text{mol}} = 1.14 \times 10^{17} \text{ atoms}$

      c) Zr, MM = 91.22

      $5.86 \text{ μg} \times \dfrac{10^{-6} \text{ g}}{\text{μg}} \times \dfrac{1 \text{ mol}}{91.22 \text{ g}} \times \dfrac{6.022 \times 10^{23} \text{ atoms}}{\text{mol}} = 3.87 \times 10^{16} \text{ atoms}$

      d) U, MM = 238.03

      $5.86 \text{ μg} \times \dfrac{10^{-6} \text{ g}}{\text{μg}} \times \dfrac{1 \text{ mol}}{238.03 \text{ g}} \times \dfrac{6.022 \times 10^{23} \text{ atoms}}{\text{mol}} = 1.48 \times 10^{16} \text{ atoms}$

3.29 a) MM of CH_4 = 12.01 + 4(1.008) = 16.04

$$375{,}000 \text{ molecules} \times \frac{1 \text{ mole}}{6.022 \times 10^{23} \text{ molecules}} \times \frac{16.04 \text{ g}}{\text{mole}} = 9.99 \times 10^{-18} \text{ g}$$

b) MM = 183.2

$$2.5 \times 10^9 \text{ molecules} \times \frac{1 \text{ mole}}{6.022 \times 10^{23} \text{ molecules}} \times \frac{183.2 \text{ g}}{\text{mole}} = 7.6 \times 10^{-13} \text{ g}$$

c) MM = 893.5

$$1 \text{ molecule} \times \frac{1 \text{ mole}}{6.022 \times 10^{23} \text{ molecules}} \times \frac{893.5 \text{ g}}{\text{mole}} = 1.484 \times 10^{-21} \text{ g}$$

3.31 MM of H_3PO_4 = 98.00, 454 g = 1 lb

$$2.47 \times 10^8 \text{ lb} \times \frac{454 \text{ g}}{\text{lb}} \times \frac{1 \text{ mole}}{98.00 \text{ g}} = 1.14 \times 10^9 \text{ moles of } H_3PO_4$$

1.14×10^9 mole of H_3PO_4 x 35% = 4.00×10^8 moles from P. Assume that for the H_3PO_4 made from P that one molecule P will produce one molecule of H_3PO_4. Therefore, 4.00×10^8 moles of H_3PO_4 produced from P will utilize 4.00×10^8 moles of P.

$$4.00 \times 10^8 \text{ mol of P} \times \frac{30.97 \text{ g}}{\text{mole}} \times \frac{\text{kg}}{10^3 \text{ g}} = 1.24 \times 10^7 \text{ kg of P}$$

3.33 CaO,
   MM = 40.08 + 16.00 = 56.08
   Ca = (40.08/56.08) x 100 = 71.47%
   O = (16.00/56.08) x 100 = 28.53%

   SiO_2,
   MM = 28.09 + 2(16.00) = 60.09
   Si = (28.09/60.09) x 100 = 46.75%
   O = (32.00/60.09) x 100 = 53.25%

   Al_2O_3,
   MM = 2(26.98) + 3(16.00) = 101.96
   Al = (53.96/101.96) x 100 = 52.92%
   O = (48.00/101.96) x 100 = 47.08%

   Fe_2O_3,
   MM = 2(55.85) + 3(16.00) = 159.70
   Fe = (111.70/159.70) x 100 = 69.94%
   O = (48.00/159.70) x 100 = 30.06%

3.35 $C: 74.0 \text{ g} \times \dfrac{1 \text{ mole}}{12.01 \text{ g}} = 6.1615 \text{ mole C};$ $\dfrac{6.1615 \text{ mole C}}{1.2384} = 4.975 \text{ mole C}$

  $H: 8.65 \text{ g} \times \dfrac{1 \text{ mole}}{1.008 \text{ g}} = 8.5813 \text{ mole H};$ $\dfrac{8.5813 \text{ mole H}}{1.2384} = 6.929 \text{ mole H}$

  $N: 17.35 \text{ g} \times \dfrac{1 \text{ mole}}{14.01 \text{ g}} = 1.2384 \text{ mole N};$ $\dfrac{1.2384 \text{ mole N}}{1.2384} = 1.000 \text{ mole N}$

Empirical Formula = C_5H_7N
Mass of Empirical Formula = 5(12.01) + 7(1.008) + 14.01 = 81.12
If MM is twice the mass of the Empirical Formula, then the Molecular Formula must be twice the Empirical Formula. Molecular Formula is $C_{10}H_{14}N_2$

3.37 $\%\ C = \left(\left(97.46\ \text{mg CO}_2 \times \left[\dfrac{12.01\ \text{mg C}}{44.01\ \text{mg CO}_2}\right]\right) \middle/ 38.7\ \text{mg}\right) \times 100 = 68.7\ \%$

$\%\ H = \left(\left(20.81\ \text{mg H}_2\text{O} \times \left[\dfrac{2 \times 1.008\ \text{mg H}}{18.02\ \text{mg H}_2\text{O}}\right]\right) \middle/ 38.7\ \text{mg}\right) \times 100 = 6.02\ \%$

$\%\ N = 3.8\%;\qquad \%\ O = 100 - 68.7 - 6.02 - 3.8 = 21.5\%$

C : $68.7\ \text{g} \times \dfrac{1\ \text{mole}}{12.01\ \text{g}} = 5.72\ \text{mole C};\qquad \dfrac{5.72\ \text{mole C}}{0.271} = 21.11\ \text{mole C}$

H : $6.02\ \text{g} \times \dfrac{1\ \text{mole}}{1.008\ \text{g}} = 5.97\ \text{mole H};\qquad \dfrac{5.97\ \text{mole H}}{0.271} = 22.0\ \text{mole H}$

N : $3.8\ \text{g} \times \dfrac{1\ \text{mole}}{14.01\ \text{g}} = 0.271\ \text{mole N};\qquad \dfrac{0.271\ \text{mole N}}{0.271} = 1.00\ \text{mole N}$

O : $21.5\ \text{g} \times \dfrac{1\ \text{mole}}{16.00\ \text{g}} = 1.34\ \text{mole O};\qquad \dfrac{1.34\ \text{mole O}}{0.271} = 4.94\ \text{mole O}$

These values are very indicative of heroin.

3.39 MM of KOH = 39.10 + 16.00 + 1.008 = 56.11

$4.75\ \text{g} \times \dfrac{1\ \text{mole}}{56.11\ \text{g}} = 0.0847\ \text{mole};$

$M = \dfrac{0.0847\ \text{mole}}{0.275\ \text{L}} = 0.308\ M = [\text{KOH}]$

Since: $\text{KOH} \rightarrow \text{K}^+ + \text{OH}^-$; $0.308\ M = [\text{K}^+] = [\text{OH}^-]$

3.41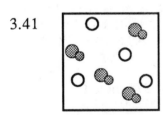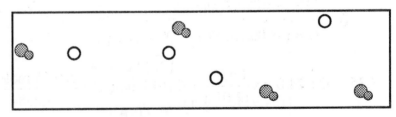

0.308 M = [K⁺] = [OH⁻]   0.077 M = [K⁺] = [OH⁻]

K⁺ = O   OH⁻ = ⬤⬤   The second solution is much more dilute.

3.43 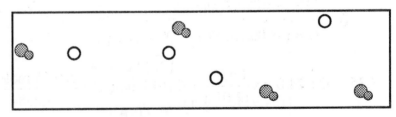 Na⁺ = O   CO₃²⁻ = ⬤⬤

3.45 $\% \text{ C} = \left(\left(4.34 \text{ g CO}_2 \text{ x} \left[\dfrac{12.01 \text{ g C}}{44.01 \text{ g CO}_2} \right] \right) \middle/ 4.00 \text{ g} \right) \text{x } 100 = 29.61 \%$

$\% \text{ Cl} = \left(\left(0.334 \text{ g AgCl x} \left[\dfrac{35.45 \text{ g Cl}}{143.32 \text{ g AgCl}} \right] \right) \middle/ 0.125 \text{ g} \right) \text{x } 100 = 66.09 \%$

$\% \text{ H} = 100.00 - 29.61 - 66.09 = 4.30$

$\text{C} : 29.61 \text{ g x} \dfrac{1 \text{ mole C}}{12.01 \text{ g}} = 2.47 \text{ mole C}; \quad \dfrac{2.47 \text{ mole C}}{1.86} = 1.33 \text{ mole C}$

$\text{Cl} : 66.09 \text{ g x} \dfrac{1 \text{ mole}}{35.45 \text{ g}} = 1.86 \text{ mole Cl}; \quad \dfrac{1.86 \text{ mole Cl}}{1.86} = 1.00 \text{ mole Cl}$

$\text{H} : 4.30 \text{ g x} \dfrac{1 \text{ mole}}{1.008 \text{ g}} = 4.27 \text{ mole H}; \quad \dfrac{4.27 \text{ mole H}}{1.86} = 2.30 \text{ mole H}$

Empirical Formula = $C_4H_7Cl_3$

3.47 a) NH_4Cl, ammonium chloride   b) XeF_4, xenon tetrafluoride
c) Fe_2O_3, iron(III) oxide   d) SO_2, sulfur dioxide
e) $KClO_4$, potassium perchlorate   f) $KClO_3$, potassium chlorate
g) $KClO_2$, potassium chlorite   h) $KClO$, potassium hypochlorite
i) KCl, potassium chloride   j) Na_2HPO_4, sodium hydrogen phosphate

3.49 Verapamil $C_{27}H_{38}O_4N_2$
a) $MM = (27 \text{ x } 12.01) + (38 \text{ x } 1.008) + (4 \text{ x } 16.00) + (2 \text{ x } 14.01) = 454.59$

b) $120.0 \text{ mg x} \dfrac{10^{-3} \text{ g}}{\text{mg}} \text{x} \dfrac{1 \text{ mole}}{454.6 \text{ g}} = 2.640 \text{ x } 10^{-4} \text{ mole of } C_{27}H_{38}O_4N_2$

c) $2.640 \text{ x } 10^{-4} \text{ mole of } C_{27}H_{38}O_4N_2 \text{ x} \dfrac{6.022 \text{ x } 10^{23} \text{ molecules}}{\text{mole}} \text{x} \dfrac{2 \text{ atoms N}}{\text{molecule}}$

$= 3.180 \text{ x } 10^{20} \text{ atoms N}$

3.51 $\% \text{ C} = \left[(\text{mass of CO}_2) \text{ x} \left(\dfrac{\text{At. Wt. C}}{\text{MM CO}_2} \right) \middle/ \text{mass of sample} \right] \text{x } 100 = ?\%$

$= \left[(54.9140 - 54.4375) \left(\dfrac{12.01}{44.01} \right) / (2.8954 - 2.7534) \right] \text{x } 100 = 91.57\%$

$\% \text{ H} = \left[(47.9961 - 47.8845) \left(\dfrac{2 \text{ x } 1.008}{18.02} \right) / (2.8954 - 2.7534) \right] \text{x } 100 = 8.79\%$

$\text{C} = 91.57 \text{ g x} \dfrac{1 \text{ mole}}{12.01 \text{ g}} = 7.62 \text{ mole C}; \quad \dfrac{7.62}{7.62} = 1.00; \ 1.00 \text{ x } 7 = 7.00 \text{ moles C}$

$\text{H} = 8.79 \text{ g x} \dfrac{1 \text{ mole}}{1.008 \text{ g}} = 8.72 \text{ mole H}; \quad \dfrac{8.72 \text{ mole}}{7.62} = 1.14; \ 1.14 \text{ x } 7 = 7.98 \text{ moles H}$

Empirical Formula = C_7H_8

3.53   a) $NaNO_2$ and $NaNO_3$
        b) K_2CO_3 and $KHCO_3$
        c) FeO and Fe_2O_3
        d) I_2 and I^-

3.55   a) $\dfrac{25.0 \text{ g Fe}_3\text{O}_4}{231.54 \text{ g/mol}} = 0.108 \text{ mol Fe}_3\text{O}_4$

        b) $\dfrac{25.0 \text{ g C}_2\text{H}_6\text{O}}{46.07 \text{ g/mol}} = 0.543 \text{ mol C}_2\text{H}_6\text{O}$

        c) $\dfrac{25.0 \text{ g C}_3\text{H}_5\text{(OH)}_3}{92.09 \text{ g/mol}} = 0.271 \text{ mol C}_3\text{H}_5\text{(OH)}_3$

        d) $\dfrac{25.0 \text{ g Al}_2\text{(SO}_4\text{)}_3}{342.17 \text{ g/mol}} = 0.0731 \text{ mol Al}_2\text{(SO}_4\text{)}_3$

        e) $\dfrac{25.0 \text{ g NiSO}_4 \cdot 6\text{H}_2\text{O}}{262.9 \text{ g/mol}} = 0.0951 \text{ mol NiSO}_4 \cdot 6\text{H}_2\text{O}$

3.57   a) $\dfrac{16.0 \text{ g O}}{1.00 \text{ mol O}} \times \dfrac{100.00 \text{ g Fe}_3\text{O}_4}{27.64 \text{ g O}} = 57.9 \text{ g Fe}_3\text{O}_4 / \text{mol O}$

        b) $\dfrac{16.0 \text{ g O}}{1.00 \text{ mol O}} \times \dfrac{100.00 \text{ g C}_2\text{H}_6\text{O}}{34.73 \text{ g O}} = 46.1 \text{ g C}_2\text{H}_6\text{O} / \text{mol O}$

        c) $\dfrac{16.0 \text{ g O}}{1.00 \text{ mol O}} \times \dfrac{100.00 \text{ g C}_3\text{H}_5\text{(OH)}_3}{52.12 \text{ g O}} = 30.7 \text{ g C}_3\text{H}_5\text{(OH)}_3 / \text{mol O}$

        d) $\dfrac{16.0 \text{ g O}}{1.00 \text{ mol O}} \times \dfrac{100.00 \text{ g Al}_2\text{(SO}_4\text{)}_3}{56.12 \text{ g O}} = 28.5 \text{ g Al}_2\text{(SO}_4\text{)}_3 / \text{mol O}$

        e) $\dfrac{16.0 \text{ g O}}{1.00 \text{ mol O}} \times \dfrac{100.00 \text{ g NiSO}_4 \cdot 6\text{H}_2\text{O}}{60.87 \text{ g O}} = 26.3 \text{ g NiSO}_4 \cdot 6\text{H}_2\text{O} / \text{mol O}$

3.59 $1 \text{ L sea water} \times \dfrac{2.2 \times 10^{-9} \text{ mol}}{1 \text{ L}} \times \dfrac{6.022 \times 10^{23} \text{ molecules}}{1 \text{ mol}} = 1.3 \times 10^{15} \text{ molecules}$

        $1 \text{ kg Rb} \times \dfrac{10^3 \text{ g}}{1 \text{ kg}} \times \dfrac{1 \text{ mol Rb}}{85.47 \text{ g Rb}} \times \dfrac{1 \text{ L}}{2.2 \times 10^{-9} \text{ mol}} = 5.3 \times 10^9 \text{ L} \cong 5 \times 10^9 \text{ L}$

3.61   a) IO_3^- b) IO_4^- c) PO_4^{3-} d) ClO_4^- e) SO_4^{2-} f) $CH_3CO_2^-$ g) HCO_3^- h) NO_3^-

3.63   MM of $C_3H_8 = (3 \times 12.01) + (8 \times 1.008) = 44.09$
        % C = $[(3 \times 12.01)/44.09] \times 100 = 81.72\%$

3.65 NH_4NO_3; MM = (2 x 14.01) + (4 x 1.008) + (3 x 16.00) = 80.05 g/mol

$(NH_4)_2SO_4$; MM = (2 x 14.01) + (8 x 1.008) + (1 x 32.07) + (4 x 16.00) =
$$132.15 \text{ g/mol}$$

$(NH_2)_2CO$; MM = (2 x 14.01) + (4 x 1.008) + 12.01 + 16.00 = 60.06 g/mol

$(NH_4)_2HPO_4$; MM = (2 x 14.01) + (9 x 1.008) + (30.97) + (4 x 16.00) =
$$132.06 \text{ g/mol}$$

NH_4NO_3 $1 \text{ kg N} \times \dfrac{10^3 \text{ g}}{\text{kg}} \times \dfrac{80.05 \text{ g } NH_4NO_3}{(2 \times 14.01) \text{ g N}} \times \dfrac{1 \text{ kg}}{10^3 \text{ g}} = 2.857 \text{ kg } NH_4NO_3$

$(NH_4)_2SO_4$ $\dfrac{132.15}{2 \times 14.01} = 4.716 \text{ kg } (NH_4)_2SO_4$

$(NH_2)_2CO$ $\dfrac{60.06}{2 \times 14.01} = 2.143 \text{ kg } (NH_2)_2CO$

$(NH_4)_2HPO_4$ $\dfrac{132.06}{2 \times 14.01} = 4.713 \text{ kg } (NH_4)_2HPO_4$

3.67 In addition to the atomic weights of carbon, oxygen and hydrogen and the weight of the sample burned, one would also need to know if any elements other than hydrogen, carbon and, possibly, oxygen are present and what the approximate molecular mass is before the molecular formula can be determined.

3.69 $\dfrac{8.3 \text{ mg } O_2}{L} \times \dfrac{10^{-3} \text{ g}}{\text{mg}} \times \dfrac{1 \text{ mole } O_2}{32.00 \text{ g}} = 2.6 \times 10^{-4} \dfrac{\text{mole } O_2}{L}$

2.6×10^{-4} M = molarity of oxygen as dissolved O_2

3.71 MM of $C_{16}H_{18}N_2O_4S$ =
(16 x 12.01) + (18 x 1.008) + (2 x 14.01) + (4 x 16.00) + 32.07 = 334.4

$50.0 \text{ mg} = \dfrac{10^{-3} \text{ g}}{\text{mg}} \times \dfrac{1 \text{ mole}}{334.4 \text{ g}} = 1.50 \times 10^{-4} \text{ moles}$

$50.0 \text{ mg} = \dfrac{10^{-3} \text{ g}}{\text{mg}} \times \dfrac{1 \text{ mole } C_{16}H_{18}N_2O_4S}{334.4 \text{ g}} \times \dfrac{6.022 \times 10^{23} \text{ molecules}}{\text{mole}}$

$\times \dfrac{16 \text{ C atoms}}{\text{molecule}} = 1.45 \times 10^{21} \text{ atoms C}$

$75.0 \text{ mg} = \dfrac{10^{-3} \text{ g}}{\text{mg}} \times \dfrac{32.07 \text{ g S}}{334.4 \text{ g penicillin}} = 7.19 \times 10^{-3} \text{ g S or } 7.19 \text{ mg S}$

3.73 $\% \text{ C} = \left(\left(9.1 \text{ g CO}_2\text{O} \times \left[\dfrac{12.01 \text{ g / mol C}}{44.01 \text{ g / mol CO}_2}\right]\right)\Big/ 5.00 \text{ g}\right) \times 100 = 49.8\% \text{ C}$

$\% \text{ H} = \left(\left(1.80 \text{ g H}_2\text{O} \times \left[\dfrac{2(1.008) \text{ g / mol H}}{18.02 \text{ g / mol H}_2\text{O}}\right]\right)\Big/ 5.00 \text{ g}\right) \times 100 = 4.03 \% \text{ H}$

$\% \text{ Fe} = 100.0 - 49.8 - 4.03 = 46.2$

C: 49.8 g × 1 mole/12.01 g = 4.15 mole C; 4.15/0.8272 = 5.02 moles C

H: 4.03 g × 1 mole/1.008 g = 3.998 mole H; 3.998/0.8272 = 4.83 moles H

Fe: 46.2 g × 1 mole/55.85 g = 0.8272 mole Fe; 0.8272/0.8272 = 1.00 mole Fe

C_5H_5Fe

3.75 $0.138 \text{ g CO}_2 \times \dfrac{12.01 \text{ g C}}{44.01 \text{ g CO}_2} = 0.0377 \text{ g C}$

$0.0566 \text{ g H}_2\text{O} \times \dfrac{2 \times 1.008 \text{ g H}}{18.02 \text{ g H}_2\text{O}} = 0.00633 \text{ g H}$

$0.0238 \text{ g NH}_3 \times \dfrac{14.01 \text{ g N}}{17.03 \text{ g NH}_3} = 0.0196 \text{ g N}$

$1.004 \text{ g AgCl} \times \dfrac{35.45 \text{ g Cl}}{143.32 \text{ g AgCl}} = 0.2483 \text{ g Cl}$

$\% \text{ C} = \dfrac{0.0377 \text{ g C}}{0.150 \text{ g sample}} \times 100\% = 25.1\%$

$\% \text{ H} = \dfrac{0.00633 \text{ g H}}{0.150 \text{ g sample}} \times 100\% = 4.22\%,\quad \% \text{ N} = \dfrac{0.0196 \text{ g N}}{0.200 \text{ g sample}} \times 100\% = 9.80\%$

$\% \text{ Cl} = \dfrac{0.2483 \text{ g Cl}}{0.500 \text{ g sample}} \times 100\% = 49.7\%$

$\% \text{ O} = 100.0\% - 25.1\% \text{ C} - 4.22\% \text{ H} - 9.80\% \text{ N} - 49.7\% \text{ Cl} = 11.2\%$

For 100 g sample;

$\dfrac{25.1 \text{ g C}}{12.01 \text{ g / mol}} = 2.09 \text{ mol C}\qquad \dfrac{4.22 \text{ g H}}{1.008 \text{ g / mol}} = 4.19 \text{ mol H}$

$\dfrac{49.7 \text{ g Cl}}{35.45 \text{ g / mol}} = 1.40 \text{ mol Cl}\qquad \dfrac{11.2 \text{ g O}}{16.00 \text{ g / mol}} = 0.700 \text{ mol O}$

$\dfrac{9.80 \text{ g N}}{14.01 \text{ g / mol}} = 0.700 \text{ mol N}$

$\dfrac{2.09}{0.700} = 2.99\qquad \dfrac{4.19}{0.700} = 5.99\qquad \dfrac{0.700}{0.700} = 1.00$

$\dfrac{1.40}{0.700} = 2.00\qquad \dfrac{0.700}{0.700} = 1.00\qquad C_3H_6Cl_2NO$

3.77   % Hg = (0.302 g Hg/0.350 g sample) x 100 = 86.3%

$$\% \text{ S} = \left(\left(0.0964 \text{ g} \times \left[\frac{32.07 \text{ g} / \text{mol S}}{64.07 \text{ g} / \text{mol SO}_2}\right]\right) \middle/ 0.350 \text{ g}\right) \times 100 = 13.8 \%$$

Hg: 86.3 g x (1 mole/200.6 g) = 0.430 mole; 0.430 mole/0.43 = 1.0
S: 13.8 g x (1 mole/32.07 g) = 0.430 mole; 0.430 mole/0.43 = 1.0
HgS

3.79   a) $C_{44}H_{69}O_{12}N$

MM = (44 x 12.01) + (69 x 1.008) + (12 x 16.00) + 14.01 = 804.0 g/mol

b) $5.0 \text{ mg} \times \dfrac{10^{-3} \text{ g}}{\text{mg}} \times \dfrac{1 \text{ mole}}{804.0 \text{ g}} = 6.2 \times 10^{-6}$ moles FK - 506

c) 6.2×10^{-6} moles FK - 506 $\times \dfrac{12 \text{ mole O}}{\text{mole}} \times \dfrac{6.022 \times 10^{23} \text{ atoms O}}{\text{mole O}}$

$= 4.5 \times 10^{19}$ atoms O

3.81 $8.50 \times 10^4 \text{ L} \times \dfrac{10^3 \text{ mL}}{\text{L}} \times \dfrac{1.0 \text{ g}}{\text{mL}} \times \dfrac{1 \text{ mol}}{18.02 \text{ g}} = 4.7 \times 10^6$ mol H_2O

3.83   a) carbon dioxide   b) potassium nitrate   c) sodium chloride
       d) sodium bicarbonate or sodium hydrogencarbonate   e) sodium carbonate
       f) sodium hydroxide   g) calcium oxide   h) magnesium hydroxide

3.85 $\dfrac{100 \text{ g B}_{12}}{4.34 \text{ g Co}} \times \dfrac{58.93 \text{ g Co}}{1 \text{ mole Co}} \times \dfrac{1 \text{ mole Co}}{1 \text{ mole B}_{12}} = 1.36 \times 10^3$ g B_{12} / mole B_{12}

3.87

       3-methylbutane thiol   trimethylamine   cadaverine   pyridine

3.89 $\dfrac{\$275}{2000 \text{ lb SO}_2} \times \dfrac{1 \text{ lb}}{454 \text{ g}} \times \dfrac{64.07 \text{ g SO}_2}{1 \text{ mole SO}_2} = \$0.0194 / \text{mol SO}_2 \approx 2\text{¢}$

$\$1.00 \times \dfrac{1 \text{ mol SO}_2}{\$0.0194} \times \dfrac{6.022 \times 10^{23} \text{ molecule}}{1 \text{ mole}} = 3.10 \times 10^{25}$ molecules of SO_2 / $1.00

3.91   CH$_3$–CH$_2$–CH$_2$–CH$_2$–CH$_2$–CH$_2$–OH   CH$_3$–CH$_2$–CH$_2$–CH$_2$–CH(OH)–CH$_3$
       1-hexanol   2-hexanol

       CH$_3$–CH$_2$–CH$_2$–CH(OH)–CH$_2$–CH$_3$
       3-hexanol
                                                             Any others would require that some
                                                             of the carbons not be in the largest
                                                             (straight) chain.

CHAPTER 4: CHEMICAL REACTIONS AND STOICHIOMETRY

4.1   Follow the methods and notations given in Section 4.1

a) $5H_2 + 2NO \rightarrow 2NH_3 + 2H_2O$

b) $2CO + 2NO \rightarrow N_2 + 2CO_2$

c) $2NH_3 + 2O_2 \rightarrow N_2O + 3H_2O$

d) $6NO + 4NH_3 \rightarrow 5N_2 + 6H_2O$

4.3   Follow the methods and notations given in Section 4.1

a) $N_2O_{5(g)} + H_2O_{(l)} \rightarrow 2HNO_{3(aq)}$

b) $2KClO_{3(s)} \rightarrow 2KCl_{(s)} + 3O_{2(g)}$

c) $2Fe_{(s)} + O_{2(g)} + 2H_2O_{(l)} \rightarrow 2Fe(OH)_{2(s)}$

d) $Au_2S_{3(s)} + 3H_{2(g)} \rightarrow 3H_2S_{(g)} + 2Au_{(s)}$

4.5   Follow the methods and notations given in Section 4.1

a) $4NH_3 + 5O_2 \rightarrow 4NO + 6H_2O$

b) $2NO + O_2 \rightarrow 2NO_2$

c) $3NO_2 + H_2O \rightarrow 2HNO_3 + NO$

d) $4NH_3 + 3O_2 \rightarrow 2N_2 + 6H_2O$

e) $4NH_3 + 6NO \rightarrow 5N_2 + 6H_2O$

4.7   For problems 4.6, 4.7, & 4.8 the molecular pictures need to show reactants and products in the indicated proportions. For a discussion of notations used, see chapter 2. In problem 4.5d, molecular ammonia ($NH_{3(g)}$) reacts with oxygen ($O_{2(g)}$) in the ratio of four molecules of ammonia with three molecules of oxygen to produce two molecules of nitrogen ($N_{2(g)}$) and six molecules of water ($H_2O_{(g)}$). The simplest molecular picture would illustrate that smallest number of reactants and products. Draw a picture showing that ratio of reactants and products.

4.9 In the following problems, obtain the number of moles of the first reactant by dividing 5.00 grams of that substance by its molecular mass. Then use the coefficients to convert to moles of the second reactant. Use the molecular mass of the second reactant to obtain the mass of that reactant.

a) $\dfrac{5.00 \text{ g H}_2}{2.016 \text{ g / mol}} \times \dfrac{2 \text{ mol NO}}{5 \text{ mol H}_2} \times \dfrac{30.01 \text{ g NO}}{\text{mol}} = 29.8 \text{ g NO}$

b) $\dfrac{5.00 \text{ g CO}}{28.01 \text{ g / mol}} \times \dfrac{2 \text{ mol NO}}{2 \text{ mol CO}} \times \dfrac{30.01 \text{ g NO}}{\text{mol}} = 5.36 \text{ g NO}$

c) $\dfrac{5.00 \text{ g NH}_3}{17.03 \text{ g / mol}} \times \dfrac{2 \text{ mol O}_2}{2 \text{ mol NH}_3} \times \dfrac{32.00 \text{ g O}_2}{\text{mol}} = 9.40 \text{ g O}_2$

d) $\dfrac{5.00 \text{ g NO}}{30.01 \text{ g / mol}} \times \dfrac{4 \text{ mol NH}_3}{6 \text{ mol NO}} \times \dfrac{17.03 \text{ g NH}_3}{\text{mol}} = 1.89 \text{ g NH}_3$

4.11 Perform three calculations similar to those in Problems 4.9 and 4.10.

$CCl_4 + 2 \text{ HF} \rightarrow CCl_2F_2 + 2 \text{ HCl}$

$175 \text{ kg CCl}_4 \times \dfrac{10^3 \text{ g}}{\text{kg CCl}_4} \times \dfrac{1 \text{ mol CCl}_4}{153.8 \text{ g}} \times \dfrac{2 \text{ mol HF}}{1 \text{ mol CCl}_4} \times \dfrac{20.01 \text{ g HF}}{\text{mol}} \times \dfrac{10^{-3} \text{ kg}}{\text{g}}$

$= 45.5 \text{ kg HF required}$

$175 \text{ kg CCl}_4 \times \dfrac{10^3 \text{ g}}{\text{kg CCl}_4} \times \dfrac{1 \text{ mol CCl}_4}{153.8 \text{ g}} \times \dfrac{1 \text{ mol CCl}_2F_2}{1 \text{ mol CCl}_4} \times \dfrac{120.9 \text{ g CCl}_2F_2}{\text{mol}}$

$\times \dfrac{10^{-3} \text{ kg}}{\text{g}} = 138 \text{ kg CCl}_2F_2 \text{ obtained}$

$175 \text{ kg CCl}_4 \times \dfrac{10^3 \text{ g}}{\text{kg}} \times \dfrac{1 \text{ mol CCl}_4}{153.8 \text{ g}} \times \dfrac{2 \text{ mol HCl}}{1 \text{ mol CCl}_4} \times \dfrac{36.46 \text{ g HCl}}{\text{mol}}$

$\times \dfrac{10^{-3} \text{ kg}}{\text{g}} = 83.0 \text{ kg HCl obtained}$

4.13 Convert metric tons to grams, grams to moles of ammonium sulfate, then to moles of ammonia, and then convert the moles of ammonia to grams of ammonia, etc.

$3.50 \text{ mton (NH}_4)_2\text{SO}_4 \times \dfrac{1000 \text{ kg}}{\text{mton}} \times \dfrac{10^3 \text{ g}}{\text{kg}} \times \dfrac{1 \text{ mol (NH}_4)_2\text{SO}_4}{132.2 \text{ g (NH}_4)_2\text{SO}_4}$

$\times \dfrac{2 \text{ mol NH}_3}{1 \text{ mol (NH}_4)_2\text{SO}_4} \times \dfrac{17.03 \text{ g NH}_3}{\text{mol}} \times \dfrac{10^{-3} \text{ kg}}{\text{g}} = 902 \text{ kg NH}_3$

4.15   Convert mass to moles, moles to moles, and back to mass.

$C_2H_5OH + 3\ O_2 \rightarrow 2\ CO_2 + 3\ H_2O$

$$\frac{5.75\ g\ C_2H_5OH}{46.07\ g/mol} \times \frac{2\ mol\ CO_2}{mol\ C_2H_5OH} \times \frac{44.01\ g\ CO_2}{mol\ CO_2} = 11.0\ g\ CO_2$$

$$\frac{5.75\ g\ C_2H_5OH}{46.07\ g/mol} \times \frac{3\ mol\ H_2O}{mol\ C_2H_5OH} \times \frac{18.02\ g\ H_2O}{mol\ H_2O} = 6.75\ g\ H_2O$$

4.17   Assuming the fluoroapatite is the limiting reagent: $Ca_5(PO_4)_3F \rightarrow 3\ H_3PO_4$

$$1.00\ kg\ Ca_3(PO_4)_3F \times \frac{10^3\ g}{kg} \times \frac{1\ mol\ Ca_3(PO_4)_3F}{504.3\ g} \times \frac{3\ mol\ H_3PO_4}{1\ mol}$$

$$\times \frac{97.99\ g\ H_3PO_4}{mol} = \text{theoretical yield of } H_3PO_4 = 583\ g$$

$$= \text{percent yield} = (400/583) \times 100 = 68.6\%$$

4.19 $C_7H_8 \rightarrow (1) \rightarrow (2) \rightarrow (3) \rightarrow (4) \rightarrow (5) \rightarrow (6) \rightarrow (7) \rightarrow C_{12}H_{12}N_2O_3$
Finish   107.6 mol
Working backwards one step at a time: if 25 kg of $C_{12}H_{12}N_2O_3$ is 107.6 moles,
then x moles or 107.9 divided by 90% of the intermediate in the last step must have
been available. Repeating this backward approach yields: 250.1 mol $\rightarrow$ 225.1 $\rightarrow$
202.6 $\rightarrow$ 182.3 $\rightarrow$ 164.1 $\rightarrow$ 147.7 $\rightarrow$ 132.9 $\rightarrow$119.6 mol
Check: 250.1 mol x $(.90)^8$ = 107.7
Answer: 250.1 mol C_7H_8 x (92.13 g/mol) x (1 kg/10^3 g) = 23 kg

4.21
	Reaction:	$C_6H_{12}O_6$	$\rightarrow$	$2\ C_2H_5OH$
MM		180.2		46.07
Th. Yield				4.58 kg
Finish				99.4 mol
Start		49.7 mole		
Start		8.92 kg		

4.23
	#8 [1/2"]	#8 [1"]	#10 [1/2"]	#10 [11/2"]	#12 [2"]
# per lb	38	21	31	15	10
# per assort	10	8	8	6	4
2 assort's	20	16	16	12	8

One could prepare 2 "handyman assortments" per pound of each type of screw.

4.25   a)
Reaction:	N_2	+	$3\ H_2$	$\rightarrow$	$2\ NH_3$
Start (10^3)	26.4 mol		367		0
Change	-26.4		-26.4 x 3		+26.4 x 2
Finish (10^3)	0		287.8 mol		52.8 mol
Mass	0		580 kg		<u>899 kg</u>

(continued)

(4.25 continued)

b) Reaction:

	$2 H_2$	+	CO	$\rightarrow$	CH_3OH
Start (10^3)	367		26.4		0
Change	-52.8		-26.4		+26.4
Finish (10^3)	314 mol		0		26.42 mol
Mass	633 kg		0		<u>847 kg</u>

c) Reaction:

	CaO	+	$3 C$	$\rightarrow$	CO	+	CaC_2
Start (10^3)	13.2 mol		61.6 mol		0		0
Change	-13.2		-39.6 mol		+13.2 mol		+13.2
Finish (10^3)	0		22.0 mol		13.2 mol		13.2
Mass	0		264 kg		<u>370 kg</u>		<u>846 kg</u>

d) Reaction:

	$2 C_2H_4$	+	O_2	+	$4 HCl$	$\rightarrow$	$2 C_2H_4Cl_2$	+	$2 H_2O$
Start (10^3)	26.4 mol		23.1 mol		20.3 mol		0		0
Change	-10.15		-5.08		-20.3		+10.15		+10.15
Finish (10^3)	16.25 mol		18.02 mol		0		10.15 mol		10.15
Mass	456 kg		577 kg		0		<u>1.01×10^3 kg</u>		<u>184 kg</u>

4.27 Reaction:

	P_4	+	$5 O_2$	$\rightarrow$	P_4O_{10}
Start	0.0303 mol		0.205 mol		0
Change	-0.0303 mol		-0.151 mol		+0.0303 mol
Finish	0		0.054 mol		0.0303 mol
Mass	0		<u>1.7 g</u>		<u>8.60 g</u>

4.29 Reaction:

	SiO_2	+	$2 C$	+	$2 Cl_2$	$\rightarrow$	$SiCl_4$	+	$2 CO$
Start	1.25 mol		6.24 mol		1.06		0		0
Change	-0.53 mol		-1.06 mol		-1.06 mol		+0.53 mol		+1.06 mol
Finish	0.72		5.18 mol		0 mol		0.53 mol		1.06 mol
Mass					0		90.0 g		29.7 g

Yield: 90.0 g of $SiCl_4$ x 95.7% = <u>86.1 g of $SiCl_4$</u>

4.31 Follow the solubility guidelines given in Section 4.5.
a) $H_2O_{(l)}$, $NH_4^+{}_{(aq)}$ and $Cl^-{}_{(aq)}$ b) $H_2O_{(l)}$, $Fe^{2+}{}_{(aq)}$ and $ClO_4^-{}_{(aq)}$
c) $H_2O_{(l)}$, $Na^+{}_{(aq)}$ and $SO_4^{2-}{}_{(aq)}$ d) $H_2O_{(l)}$, $K^+{}_{(aq)}$ and $Br^-{}_{(aq)}$

4.33 a) $AgNO_3$ will form a precipitate when mixed with NH_4Cl, Na_2SO_4 or KBr solutions. The precipitates will be $AgCl$, Ag_2SO_4 and $AgBr$, respectively.
b) Na_2CO_3 would be expected to form a precipitate when mixed with $Fe(ClO_4)_2$. The precipitate would be $FeCO_3$.
c) $Ba(OH)_2$ would be expected to form precipitates when mixed with $Fe(ClO_4)_2$ and Na_2SO_4 solutions. The precipitates would be $Fe(OH)_2$ and $BaSO_4$ respectively.

4.35 a) $Al^{3+}{}_{(aq)} + 3 OH^-{}_{(aq)} \rightarrow Al(OH)_{3(s)}$
b) $3Mg^{2+}{}_{(aq)} + 2 PO_4^{3-}{}_{(aq)} \rightarrow Mg_3(PO_4)_{2(s)}$
c) $Ba^{2+}{}_{(aq)} + SO_4^{2-}{}_{(aq)} \rightarrow BaSO_{4(s)}$

4.37

Reaction:	Pb^{2+}	$+$	$2\,Cl^-$	$\rightarrow$	$PbCl_{2(s)}$
Start	5.625×10^{-2} mol		0.1069 mol		0
Change	-5.344×10^{-2} mol		-0.1069 mol		$+5.344 \times 10^{-2}$
Finish	2.81×10^{-3} mol		0		5.344×10^{-2}
Mass					14.9 g

14.9 g of $PbCl_2$ will form. Remaining in solution will be ammonium ions, nitrate ions and excess lead(II) ions.

4.39

$$2\,H_3O^+ + 2\,NO_3^- + Ba^{2+} + 2\,OH^- \rightarrow 4\,H_2O + Ba^{2+} + 2\,NO_3^-$$

Net equation: H_3O^+ $+$ OH^- $\rightarrow$ $2\,H_2O$

5×10^{-4} mole $+$ 5×10^{-4} mol $\rightarrow$

For the molecular pictures show reactants and products in the indicated proportions. See Chapter 2 for a discussion of notations used. Hydrogen (or hydronium) ions will react with hydroxide ions in the ratio of 1 to 1. In the reaction described in this problem the hydrogen ion and hydroxide ion are being mixed in equal numbers of moles; therefore, they both will be all reacted in going from the reactants to the products side of your molecular picture. For your molecular picture show an equal number of hydrogen (hydronium), nitrate, and hydroxide ions and half that number of barium ions in the reacting solution. For the products show soluble barium and nitrate ions and the water molecules formed. The same number of each type of atom must be shown in the products as was present in the reactants.

4.41   Total Ionic Reaction: $Ba^{2+} + 2\,OH^- + 2\,H_3O^+ + 2\,Cl^- \rightarrow 4\,H_2O + Ba^{2+} + 2\,Cl^-$

Net Ionic:	H_3O^+	$+$	OH^-	$\rightarrow$	$2\,H_2O$
Start	5.00×10^{-3} mol		6.00×10^{-3} mol		
Change	-5.00×10^{-3} mol		-5.00×10^{-3} mol		
Finish	0		1.00×10^{-3} mol		

Final Volume 2.50×10^2 mL

Molarities: $H_3O^+ = 0$; $OH^- = \dfrac{1.00 \times 10^{-3} \text{ mol}}{0.250 \text{ L}} = 4.00 \times 10^{-3}$ M

$$Ba^{2+} = \frac{(1.50 \times 10^2 \text{ mL})(2.00 \times 10^{-2} \text{ M})}{(2.50 \times 10^2 \text{ mL})} = 1.20 \times 10^{-2} \text{ M}$$

$$Cl^- = \frac{(1.00 \times 10^2 \text{ mL})(5.00 \times 10^{-2} \text{ M})}{(2.50 \times 10^2 \text{ mL})} = 2.00 \times 10^{-2} \text{ M}$$

Note: Ba^{2+} and Cl^- are spectator ions and change concentration only by dilution.

4.43   mol KHP $= \dfrac{0.7996 \text{ g}}{204.2 \text{ g / mol}} = 3.916 \times 10^{-3}$ mol

$\therefore$ mol of NaOH added in titration $= 3.916 \times 10^{-3}$ mol

volume of NaOH added $= 43.75$ mL $- 0.15$ mL $= 43.60$ mL (or 43.60×10^{-3} L)

Molarity of NaOH solution $= \dfrac{3.916 \times 10^{-3} \text{ mol}}{43.60 \times 10^{-3} \text{ L}} = 0.08982$ M

4.45 a) $2 \, Al_{(s)} + 6 \, H_3O^+_{(aq)} \rightarrow 3 \, H_{2(g)} + 2 \, Al^{3+}_{(aq)} + 6 \, H_2O_{(l)}$

b) $3 \, Zn_{(s)} + 2 \, Au^{3+}_{(aq)} \rightarrow 2 \, Au_{(s)} + 3 \, Zn^{2+}_{(aq)}$

c) no reaction

d) $2 \, Na_{(s)} + 2 \, H_2O_{(l)} \rightarrow 2 \, Na^+_{(aq)} + H_{2(g)} + 2 \, OH^-_{(aq)}$

4.47 a) $2 \, Sr_{(s)} + O_{2(g)} \rightarrow 2 \, SrO_{(s)}$ b) $4 \, Cr_{(s)} + 3 \, O_{2(g)} \rightarrow 2 \, Cr_2O_{3(s)}$

c) $Sn_{(s)} + O_{2(g)} \rightarrow SnO_{2(s)}$

4.49 Counting atoms and molecules: $16 \, X + 6 \, Y \rightarrow 4 \, X + 6 \, YX_2$
Canceling unreacted atoms: $12 \, X + 6 \, Y \rightarrow 6 \, YX_2$
Reducing to smallest set of whole-number coefficients:

$2 \, X + Y \rightarrow YX_2$
Answer (d) is the best answer.

4.51

Reaction:	$Ca_3(PO_4)_2$	$+$	$2 \, H_2SO_4$	$\rightarrow$	$Ca(H_2PO_4)_2$	$+$	$2 \, CaSO_4$
Start	0.822 mol		1.539 mol		0		0
Change	-0.770 mol		-1.539 mol		+0.770 mol		+1.539 mol
Finish	0.052 mol		0		+0.770 mol		+1.539 mol

Grams of $Ca(H_2PO_4)_2$ = 0.770 mol x 234 g/mol = 180 g
Grams of $CaSO_4$ = 210 g
Grams of superphosphate = 210 + 180 = 390 g

4.53 $C_2H_5OH + 3 \, O_2 \rightarrow 2 \, CO_2 + 3 \, H_2O$

a) $\dfrac{4.6 \text{ g } C_2H_5OH}{46.07 \text{ g / mol}} \times \dfrac{3 \text{ mole } H_2O}{1 \text{ mol } C_2H_5OH} = 0.30 \text{ mol } H_2O$

b) 0.30 mole H_2O x 6.022×10^{23} molecules/mole = 1.8×10^{23} molecules H_2O

c) 0.30 mol H_2O x 18.02 g/mol = 5.4 g H_2O

4.55 a) The reaction is: $N_2 + 3 \, H_2 \rightarrow 2 \, NH_3$
Therefore, if the number of molecules shown is used there will be unreacted N_2
and H_2. The reaction becomes $6 \, N_2 + 16 \, H_2 \rightarrow 10 \, NH_3 + N_2 + H_2$
Your drawing should show 10 ammonia molecules, one nitrogen molecule, and one hydrogen.

b) Hydrogen (H_2) is the limiting reagent.

c)	N_2	$+$	$3 \, H_2 \rightarrow$	$2 \, NH_3$
Start:	6 mol		16 mol	
Change:	-5.333 mol		-16 mol	+10.667 mol
Finish:	0.667 mol		0	10.667 mol

Mass of NH_3 produced = (10.667 mol)(17.034 g/mol) = 181.7 g NH_3
Mass of excess N_2 = (0.667 mol)(28.02 g/mol) = 18.7 g N_2

4.57 Reaction: $B_{10}H_{18}$ + $12\,O_2$ $\rightarrow$

Start 3.96×10^5 mol 2.03×10^6 mol

Change -1.69×10^5 mol -2.03×10^6

Finish 2.27×10^5 mol   0 mol

O_2 will empty first. There will be 2.27×10^5 mol of $B_{10}H_{18}$ (or 2.87×10^5 kg) remaining after all the oxygen has been reacted.

4.59 1 mton earth x 95% eff x 15% ilmenite x 1,000 kg/mton = 142.5 kg ilmenite

Reaction: $FeTiO_3 \rightarrow Ti$

Start   142.5 kg

$$\frac{142.5 \text{ kg } FeTiO_3}{151.7 \text{ g / mol}} \times \frac{10^3 \text{ g}}{\text{kg}} \times \frac{1 \text{ mol Ti}}{1 \text{ mol } FeTiO_3} \times \frac{47.88 \text{ g Ti}}{\text{mol Ti}} \times \frac{10^{-3} \text{ kg}}{\text{g}} = 45.0 \text{ kg Ti}$$

4.61 $CH_{4(g)}$ + $2\,H_2O_{(g)}$ $\rightarrow$ $CO_{2(g)}$ + $4\,H_{2(g)}$

4.63 Reaction: As_2O_3 + $3\,C$ $\rightarrow$ $2\,As$ + $3\,CO$

Start   0.250 mol   0.600 mol

Change   -0.050 mol   -0.600   +0.400 mol   +0.600 mol

Finish   0.050 mol   0   0.400 mol   0.600 mol

Masses

at finish 9.9 g   0 g   30.0 g   16.8 g

4.65 $Ca(NO_3)_{2(aq)}$ + $(NH_4)_2SO_{4(aq)}$ $\rightarrow$ $CaSO_{4(s)}$ + $2\,NH_4NO_3$

precipitate = $CaSO_4$

net ionic equation = $Ca^{2+}_{(aq)}$ + $SO_4^{2-}_{(aq)}$ $\rightarrow$ $CaSO_{4(s)}$

Reaction: $Ca(NO_3)_2$ + $(NH_4)_2SO_4$ $\rightarrow$ $CaSO_4$ + $2\,NH_4NO_3$

Start   0.150 mol   0.225 mol   0   0

Change -0.150 mol   -0.150 mol   +0.150 mol   +0.300 mol

Finish   0   0.075 mol   0.150 mol   0.300 mol

These are in 175 mL of solution.

NH_4^+ as a spectator ion: $\dfrac{(75.0 \text{ mL})(3.00 \text{ mol / L})\left(\dfrac{2NH_4^+}{1 \text{ mol } (NH_4)SO_3}\right)}{175 \text{ mL}} = 2.57$ M

SO_4^{2-} = 0.075 mol/0.175 L = 0.43 M

NO_3^- as spectator ion: $\dfrac{(100.0 \text{ mL})(1.50 \text{ M})\,2 \text{ mol } NO_3^-/1 \text{ mol } Ca(NO_3)_2}{175 \text{ mL}} = 1.71$ M

$CaSO_4$ = 0.150 mol x 136.14 g/mol = 20.4 g

4.67 Reaction: $\quad$ CH_4O $\quad$ + $\quad$ C_4H_8 $\quad$ $\rightarrow$ $\quad$ $C_5H_{12}O$

Sold (10^9) $\qquad\qquad\qquad\qquad\qquad\qquad\qquad$ 3.4×10^2 g

Th. yield (10^9) $\qquad\qquad\qquad\qquad\qquad\qquad$ 3.96×10^2 g

Moles (10^9) $\qquad\qquad\qquad\qquad\qquad\qquad\qquad$ 4.49 mol

Reagents (10^9) $\quad$ 4.49 mol $\quad$ + $\quad$ 4.49 mol $\quad$ $\rightarrow$ $\quad$ 4.49 mol

a) Mass $\qquad\qquad\qquad\qquad\qquad\qquad$ 2.52×10^{11} g

b) $\dfrac{2.52 \times 10^{11} \text{ g } C_4H_8}{56.1 \text{ g / mol}}$ x $\dfrac{1 \text{ mol } C_5H_{12}O}{1 \text{ mol } C_4H_8}$ x 0.93 eff x $\dfrac{88.1 \text{ g}}{\text{mol}}$

$\qquad$ x $\dfrac{1 \text{ lb}}{454 \text{ g}}$ x $\dfrac{10^{-6} \text{ million pounds}}{\text{lb}}$ = 811 million pounds

(Note: could also use a ratio to solve this problem.)

c) (811 - 750) million lb x $\dfrac{10^6 \text{ pounds}}{\text{million lb}}$ x \$0.10 / lb x $\dfrac{10^{-6} \text{ million dollars}}{\text{dollar}}$

$\qquad$ = 6.1 million dollars

Note: This assumes that MTBE was sold at cost (no profit) the first year.

4.69 a) $Na^+{}_{(aq)}$, $Cl^-{}_{(aq)}$, $H_3O^+{}_{(aq)}$ and $NO_3^-{}_{(aq)}$, no reaction

b) $Ca^{2+}{}_{(aq)}$, $Cl^-{}_{(aq)}$, $Na^+{}_{(aq)}$ and $SO_4^{2-}{}_{(aq)}$
$\qquad$ $Ca^{2+}{}_{(aq)}$ + $SO_4^{2-}{}_{(aq)}$ $\rightarrow$ $CaSO_{4(s)}$

c) $K^+{}_{(aq)}$, $OH^-{}_{(aq)}$, $H_3O^+{}_{(aq)}$ and $Cl^-{}_{(aq)}$
$\qquad$ $H_3O^+{}_{(aq)}$ + $OH^-{}_{(aq)}$ $\rightarrow$ $2 H_2O_{(l)}$

d) $NH_{3(aq)}$, $H_3O^+{}_{(aq)}$ and $Cl^-{}_{(aq)}$
$\qquad$ $NH_{3(aq)}$ + $H_3O^+{}_{(aq)}$ $\rightarrow$ $NH_4^+{}_{(aq)}$ + $H_2O_{(l)}$

4.71 a) $Fe^{3+}{}_{(aq)}$ + $PO_4^{3-}{}_{(aq)}$ $\rightarrow$ $FePO_{4(s)}$
Fe^{3+} + 3 Cl^- +3 Na^+ + PO_4^{3-} $\rightarrow$ $FePO_{4(s)}$ + 3 Na^+ + 3 Cl^-

2.50 kg $FePO_4$ x $\dfrac{1000 \text{ g}}{1 \text{ kg}}$ x $\dfrac{1 \text{ mol } FePO_4}{150.77 \text{ g } FePO_4}$ = 16.6 mol $FePO_4$

16.6 mol $FePO_4$ x $\dfrac{1 \text{ mol } FeCl_3}{1 \text{ mol } FePO_4}$ x $\dfrac{162.2 \text{ g } FeCl_3}{1 \text{ mol } FeCl_3}$ = 2.69×10^3 g $FeCl_3$

$\qquad$ = 2.69 kg $FeCl_3$

16.6 mol $FePO_4$ x $\dfrac{1 \text{ mol } Na_3PO_4}{1 \text{ mol } FePO_4}$ x $\dfrac{163.94 \text{ g } Na_3PO_4}{1 \text{ mol } Na_3PO_4}$ = 2.72×10^3 g Na_3PO_4

$\qquad$ = 2.72 kg Na_3PO_4

(continued)

(4.71 continued)

  b) $Zn^{2+}_{(aq)} + 2\,OH^-_{(aq)} \rightarrow Zn(OH)_{2(s)}$

  $Zn^{2+} + 2\,NO_3^- + 2\,Na^+ + 2\,OH^- \rightarrow Zn(OH)_{2(s)} + 2\,Na^+ + 2\,NO_3^-$

  $2.50\ kg\ Zn(OH)_2\ \times \dfrac{1000\ g}{1\ kg} \times \dfrac{1\ mol\ Zn(OH)_2}{99.41\ g\ Zn(OH)_2} = 25.1\ mol\ Zn(OH)_2$

  $25.1\ mol\ Zn(OH)_2\ \times \dfrac{1\ mol\ Zn(NO_3)_2}{1\ mol\ Zn(OH)_2} \times \dfrac{189.41\ g\ Zn(NO_3)_2}{1\ mol\ Zn(NO_3)_2}$

  $= 4.75 \times 10^3\ g\ Zn(NO_3)_2 = 4.75\ kg\ Zn(NO_3)_2$

  $25.1\ mol\ Zn(OH)_2\ \times \dfrac{2\ mol\ NaOH}{1\ mol\ Zn(OH)_2} \times \dfrac{40.00\ g\ NaOH}{1\ mol\ NaOH} = 2.01 \times 10^3\ g\ NaOH$

  $= 2.01\ kg\ NaOH$

  c) $Ni^{2+}_{(aq)} + CO_3^{2-}_{(aq)} \rightarrow NiCO_{3(s)}$

  $Ni^{2+}_{(aq)} + 2\,Cl^-_{(aq)} + 2\,Na^+_{(aq)} + CO_3^{2-}_{(aq)} \rightarrow NiCO_{3(s)} + 2\,Na^+_{(aq)} + 2\,Cl^-_{(aq)}$

  $2.50\ kg\ NiCO_3\ \times \dfrac{1000\ g}{1\ kg} \times \dfrac{1\ mol\ NiCO_3}{118.70\ g\ NiCO_3} = 21.1\ mol\ NiCO_3$

  $21.1\ mol\ NiCO_3\ \times \dfrac{1\ mol\ NiCl_2}{1\ mol\ NiCO_3} \times \dfrac{129.59\ g\ NiCl_2}{1\ mol\ NiCl_2} = 2.73 \times 10^3\ g\ NiCl_2$

  $= 2.73\ kg\ NiCl_2$

  $21.1\ mol\ NiCO_3\ \times \dfrac{1\ mol\ Na_2CO_3}{1\ mol\ NiCO_3} \times \dfrac{105.99\ g\ Na_2CO_3}{1\ mol\ Na_2CO_3} = 2.24 \times 10^3\ g\ Na_2CO_3$

  $= 2.24\ kg\ Na_2CO_3$

4.73 $\dfrac{1.632\ mol\ NaOH}{1\ L} \times 0.05000\ L \times \dfrac{1}{1.000\ L} \times 0.04000\ L$

  $\times \dfrac{1\ mol\ H_2C_2O_4 \cdot 2H_2O}{2\ mol\ NaOH} \times \dfrac{126.1\ g\ H_2C_2O_4 \cdot 2H_2O}{mole} = 0.2058\ g\ H_2C_2O_4 \cdot 2H_2O$

  % purity = (0.2058 g/0.2500) x 100 = 82.3%

4.75 a) $Mg_{(s)} + 2\,HCl_{(aq)} \rightarrow MgCl_{2(aq)} + H_{2(g)}$

  Net ionic eq.: $Mg_{(s)} + 2\,H_3O^+_{(aq)} \rightarrow Mg^{2+}_{(aq)} + H_{2(g)} + 2\,H_2O_{(l)}$

  b) $KOH_{(aq)} + HCl_{(aq)} \rightarrow H_2O_{(l)} + KCl_{(aq)}$

  Net ionic eq.: $OH^-_{(aq)} + H_3O^+_{(aq)} \rightarrow 2\,H_2O_{(l)}$

  c) $BaCl_{2(aq)} + HCl_{(aq)} \rightarrow$ no reaction

4.77
$$\frac{1.25 \text{ kg S}}{32.07 \text{ g/mol}} \times \frac{\text{g}}{10^3 \text{kg}} \times \frac{1 \text{ mol SO}_2}{3 \text{ mol S}} \times \frac{6 \text{ mol H}_2\text{S}}{2 \text{ mol SO}_2} \times \frac{34.09 \text{ g H}_2\text{S}}{\text{mol}}$$

$$\times \frac{\text{kg}}{10^3 \text{ g}} = 1.33 \text{ kg H}_2\text{S}$$

4.79 a) $H_3PO_{4(aq)} + 3 KOH_{(aq)} \rightarrow 3 H_2O_{(l)} + K_3PO_{4(aq)}$

$H_3PO_{4(aq)} + 3 OH^-_{(aq)} \rightarrow 3 H_2O_{(l)} + PO_4^{3-}_{(aq)}$ acid/base

b) $2 Sr_{(s)} + O_{2(g)} \rightarrow 2 SrO_{(s)}$ redox

c) $2 C_4H_8O_{(l)} + 11 O_{2(g)} \rightarrow 8 CO_{2(g)} + 8 H_2O_{(g)}$ redox

d) $Mg_{(s)} + 2 HBr_{(aq)} \rightarrow MgBr_{2(aq)} + H_{2(g)}$

$Mg_{(s)} + 2 H_3O^+_{(aq)} \rightarrow Mg^{2+}_{(aq)} + H_{2(g)} + 2 H_2O_{(l)}$ redox

e) $Pb(NO_3)_{2(aq)} + (NH_4)_2S_{(aq)} \rightarrow PbS_{(s)} + 2 NH_4NO_{3(aq)}$

$Pb^{2+}_{(aq)} + S^-_{(aq)} \rightarrow PbS_{(s)}$ precipitation

f) $Ag_{(s)} + HCl_{(aq)} \rightarrow$ "nr"

g) $Ni_{(s)} + 2HCl_{(aq)} \rightarrow NiCl_{2(aq)} + H_{2(g)}$

$Ni_{(s)} + 2 H_3O^+_{(aq)} \rightarrow Ni^{2+}_{(aq)} + H_{2(g)} + 2 H_2O_{(l)}$ redox

h) $AgNO_{3(aq)} + KCH_3CO_{2(aq)} \rightarrow$ "nr"

4.81 $Fe + 2 HCl \rightarrow Fe^{2+} + 2 Cl^- + H_2$

$$\frac{5.8 \text{ g Fe}}{55.85 \text{ g/mol}} \times \frac{2 \text{ mol HCl}}{1 \text{ mol Fe}} \times \frac{1 \text{ L soln.}}{1.5 \text{ mol HCl}} = 0.14 \text{ L or } 140 \text{ mL}$$

4.83 $Al(OH)_3 + 3HCl \rightarrow AlCl_3 + 3H_2O$

$$\frac{0.175 \text{ mol HCl}}{\text{L}} \times 0.155 \text{ L} \times \frac{1 \text{ mol Al(OH)}_3}{3 \text{ mol HCl}} \times \frac{78.0 \text{ g Al(OH)}_3}{\text{mol}}$$

$$= 0.704 \text{ g Al(OH)}_3$$

4.85
$$\frac{375 \text{ kg C}_6\text{H}_{12}}{84.14 \text{ g/mol}} \times \frac{10^3 \text{ g}}{\text{kg}} \times \frac{2 \text{ mol C}_6\text{H}_{10}\text{O}_4}{2 \text{ mol C}_6\text{H}_{12}} \times \frac{146.16 \text{ g}}{\text{mol}}$$

$$\times \frac{10^{-3} \text{ kg}}{\text{g}} = 651 \text{ kg adipic acid}$$

4.87 Reaction: $CaCO_3$ + $2\,HCl$ → $CaCl_2$ + H_2O + $CO_{2(g)}$

Start	0.050 mol	0.050 mol	0	0
Change	-0.025 mol	-0.050 mol	+0.025 mol	+0.025
Finish	0.0250 mol	0	0.025 mol	0.025

$CaCO_3$: 2.5 g remain unchanged
CO_2: 1.1 g of gaseous CO_2 escaped
Solution: Ca^{2+} = 0.050 M, Cl^- = 0.10 M

4.89 1. Incomplete reaction   2. Non-quantitative isolation of the product
      3. Impure starting materials   4. The presence of a completing reaction

4.91 a) $CaO_{(s)}$ + $H_2O_{(l)}$ → $Ca^{2+}_{(aq)}$ + $2\,OH^-_{(aq)}$

b) $3\,HPO_4^{2-}$ + $5\,Ca^{2+}$ + $4\,OH^-$ → $Ca_5(PO_4)_3OH$ + $3\,H_2O$

c) 1.00×10^4 L x (0.0156 mol HPO_4^-/L) x (5 mol CaO/3 mol HPO_4^{2-})
   x (56.1 g CaO/mol) x (1 kg/10^3 g) = 14.6 kg lime

4.93 Zn + $2\,HCl$ → $ZnCl_2$ + $H_{2(g)}$

$$\frac{21.3\ mg\ H_2}{2.02\ g/mol} \times \frac{10^{-3}\ g}{mg} \times \frac{1\ mol\ Zn}{1\ mol\ HCl} \times \frac{65.39\ g\ Zn}{mol} = 0.690\ g\ Zn$$

% Zn = (0.690/5.73) x 100 = 12.0%   % Cu = [(5.73 - 0.690)/5.73] x 100 = 88.0%

4.95 a) $3\,Na_2CO_{3(aq)}$ + $2\,Fe(NO_3)_{3(aq)}$ → $6\,NaNO_{3(aq)}$ + $Fe_2(CO_3)_{3(s)}$

b) $HClO_{4(aq)}$ + $KOH_{(aq)}$ → $KClO_{4(aq)}$ + $H_2O_{(l)}$

c) $NaCl$ + $Ba(OH)_2$ → "nr"

4.97 Your drawing must be based upon the drawing given in the problem. The initial
      solution has $7\,HSO_4^-$ and $7\,H_3O^+$. The new drawings that show what the solution
      looks like after the reaction is complete will contain:

a) (after the addition of 6 OH^-) $7\,HSO_4^-$ + $12\,H_2O$ + H_3O^+
   $7\,HSO_4^-$ + $7\,H_3O^+$ + $6\,OH^-$ → $7\,HSO_4^-$ + H_3O^+ + $12\,H_2O$

b) (after the addition of 12 OH^-) $2\,HSO_4^-$ + $5\,SO_4^{2-}$ + $19\,H_2O$.
   $7\,HSO_4^-$ + $7\,H_3O^+$ + $12\,OH^-$ → $2\,HSO_4^-$ + $5\,SO_4^{2-}$ + $19\,H_2O$

c) (after the addition of 18 OH^-) $7\,SO_4^{2-}$ + $4\,OH^-$ + $21\,H_2O$.
   $7\,HSO_4^-$ + $7\,H_3O^+$ + $18\,OH^-$ → $7\,SO_4^{2-}$ + $4\,OH^-$ + $21\,H_2O$

4.99 $1.0 \times 6 \times 5 \times 1/44.0 \times 10^3 \times 2 \times 56.1 \times 10^{-3}$ = 76 kg KOH

4.101 $$\frac{5.96 \times 10^{-3} \text{ mol NaOH}}{1 \text{ L}} \times 0.00570 \text{ L} \times \frac{1 \text{ mol H}_2\text{SO}_4}{2 \text{ mol NaOH}} \times \frac{1 \text{ mol SO}_2}{1 \text{ mol H}_2\text{SO}_4}$$

$$\times \frac{64.1 \text{ g SO}_2}{\text{mol SO}_2} = 1.09 \times 10^{-3} \text{ g SO}_2$$

4.103 a) $2 \text{ C}_4\text{H}_{10} + 13 \text{ O}_2 \rightarrow 8 \text{ CO}_2 + 10 \text{ H}_2\text{O}$

  b) $2 \text{ C}_6\text{H}_6 + 15 \text{ O}_2 \rightarrow 12 \text{ CO}_2 + 6 \text{ H}_2\text{O}$

  c) $\text{C}_2\text{H}_6\text{O} + 3 \text{ O}_2 \rightarrow 2 \text{ CO}_2 + 3 \text{ H}_2\text{O}$

  d) $\text{C}_5\text{H}_{12} + 8 \text{ O}_2 \rightarrow 5 \text{ CO}_2 + 6 \text{ H}_2\text{O}$

  e) $2\text{C}_6\text{H}_{12}\text{O} + 17 \text{ O}_2 \rightarrow 12 \text{ CO}_2 + 12 \text{ H}_2\text{O}$

4.105 $\text{C}_9\text{H}_4\text{O}_3 + 11\text{O}_2 \rightarrow 9 \text{ CO}_2 + 7 \text{ H}_2\text{O}$

$$\frac{3.00 \text{ g C}_9\text{H}_{14}\text{O}_3}{170.2 \text{ g / mol}} \times \frac{11 \text{ mol O}_2}{1 \text{ mol C}_9\text{H}_{14}\text{O}_3} \times \frac{32.0 \text{ g O}_2}{\text{mol}} = 6.20 \text{ g O}_2 \text{ (consumed)}$$

$$\frac{3.00 \text{ g C}_9\text{H}_{14}\text{O}_3}{170.2 \text{ g / mol}} \times \frac{7 \text{ mol H}_2\text{O}}{1 \text{ mol C}_9\text{H}_{14}\text{O}_3} \times \frac{18.02 \text{ g H}_2\text{O}}{\text{mol}} = 2.22 \text{ g H}_2\text{O} \text{ (produced)}$$

CHAPTER 5: THE BEHAVIOR OF GASES

5.1

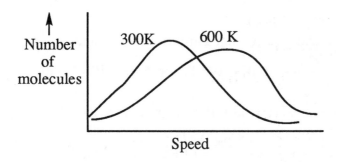

5.3

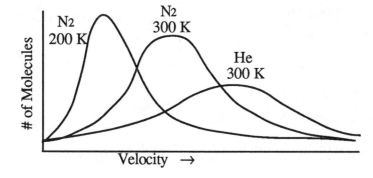

5.5 $k.E._{molar} = 3/2\ RT$
a) For He at 900 K, $k.E._{molar} = (3/2)\ (8.314\ J/mol\ K)\ (900\ K) = 1.12 \times 10^4\ J/mol$
b) For O_2 at 300 K, $k.E._{molar} = (3/2)\ (8.314\ J/mol\ K)\ (300\ K) = 3.74 \times 10^3\ J/mol$
c) For SF_6 at 900 K, $k.E._{molar} = (3/2)\ (8.314\ J/mol\ K)\ (900\ K) = 1.12 \times 10^4\ J/mol$

5.7   a) They have the same force (F) since F is directly proportional to temperature (T).
$\dfrac{F}{A} = P\alpha\dfrac{n}{V}T$. Pressure (P) is force per area (F/A) and is proportional to (n/V)
times T. As T varies so will both the k.E. and the F since k.E. is also directly
proportional to T.
b) The methane molecule will have the higher force at 700 K. $F \alpha v$ and $v \alpha T$
The molecule at the higher temperature will have the higher v and $\therefore$ greater F.
c) The F_2 will have the greater force because $F \alpha mv$, the mass of F_2 is greater than
that of H_2.

5.9   a) The pressure inside the container is 1 atm. Otherwise the piston will move. The collisions must be sufficient to keep the piston in position.
b) If the temperature in kelvins is doubled, the molecules will collide twice as many times per unit of time forcing the piston out until twice the volume is obtained and the number of collisions per unit time would then equal the original number and the pressure would be 1 atm again.
The redrawn picture would need to have the piston raised to a point that would yield twice the volume shown in the original figure in the textbook.

5.11   The level of the column of mercury would slowly drop until it reached the level of the mercury in the dish.

5.13   See Section 5.3 for conversion factors and a discussion of SI units.

a) $455 \text{ torr} \times \dfrac{1.01325 \times 10^5 \text{ Pa}}{760 \text{ torr}} = 6.07 \times 10^4 \text{ Pa}$ or $6.07 \times 10^4 \text{ N}/\text{m}^2$

b) $2.45 \text{ atm} \times \dfrac{1.01325 \times 10^5 \text{ Pa}}{1 \text{ atm}} = 2.48 \times 10^5 \text{ Pa}$ or $2.48 \times 10^5 \text{ N}/\text{m}^2$

c) $0.46 \text{ torr} \times \dfrac{1 \text{ atm}}{760 \text{ torr}} \times \dfrac{1.01325 \times 10^5 \text{ Pa}}{1 \text{ atm}} = 61 \text{ Pa}$ or $61 \text{ N}/\text{m}^2$

d) $1.33 \times 10^{-3} \text{ atm} \times \dfrac{1.01325 \times 10^5 \text{ Pa}}{1 \text{ atm}} = 135 \text{ Pa}$ or $135 \text{ N}/\text{m}^2$

5.15 $PV = nRT$

$$n = \frac{PV}{RT} = \frac{(5.00 \text{ atm})(20.0 \text{ L})}{(0.08206 \text{ L atm mol}^{-1} \text{ K}^{-1})(298 \text{ K})} = 4.09 \text{ mol}$$

$$V = \frac{nRT}{P} = \frac{(4.09 \text{ mol})(0.08206 \text{ L atm mol}^{-1} \text{ K}^{-1})(298 \text{ K})}{(1 \text{ atm})} = 100 \text{ L}$$

5.17   a) $\dfrac{P_i}{T_i} = \dfrac{P_f}{T_f}$ or $\dfrac{T_i}{P_i} = \dfrac{T_f}{P_f}$ b) $\dfrac{n_i}{V_i} = \dfrac{n_f}{V_f}$ or $\dfrac{V_i}{n_i} = \dfrac{V_f}{n_f}$
c) $P_i V_i = P_f V_f$

5.19 $V_i/T_i = V_f/T_f$ $V_f = (V_i/T_i) T_f = (V_i \times T_f)/T_i$
Given: $V_i = 0.255 \text{ L}$, $T_i = 273 + 25 = 298 \text{ K}$, $T_f = 273 - 10 = 263 \text{ K}$
$$V_f = \frac{(0.255 \text{ L})(263 \text{ K})}{298 \text{ K}} = 0.225 \text{ L}$$

5.21   Assume that the temperature of the ice bath is 0°C or 273 K.

V and n are constant, $\therefore P_i/T_i = P_f/T_f$ and $T_f = P_f(T_i/P_i)$.

$$T_f = \frac{(745 \text{ torr})(273 \text{ K})}{(345 \text{ torr})} = 590 \text{ K} \qquad T_f = 590 \text{ K} - 273 \text{ K} = 317°C$$

32

5.23 $PV = \dfrac{m}{MM}RT \therefore MM = \dfrac{mRT}{PV}$

$$MM = \dfrac{(2.55 \text{ g})(0.08206 \text{ L atm mol}^{-1} \text{ K}^{-1})(298 \text{ K})}{\left(\dfrac{262 \text{ torr}}{760 \text{ torr / atm}}\right)(1.50 \text{ L})} = 121 \text{ g / mol}$$

121 g/mol is the molecular mass of CCl_2F_2

5.25 $P_{NO_2} = X_{NO_2}P_{total} = \dfrac{0.78}{10^6}(758.4 \text{ torr}) = 5.9 \times 10^{-4} \text{ torr}$

5.27 For standard conditions $P_{total} = 760$ torr; $P_{gas} = X_{gas}P_{total}$.

$P_{N_2} = (0.7808)(760 \text{ torr}) = 593.4 \text{ torr}$

$P_{O_2} = (0.2095)(760 \text{ torr}) = 159.2 \text{ torr}$

$P_{Ar} = (9.34 \times 10^{-5})(760 \text{ torr}) = 7.10 \times 10^{-2} \text{ torr}$

$P_{CO_2} = (3.25 \times 10^{-4})(760 \text{ torr}) = 0.247 \text{ torr}$

5.29 $n_{CH_4} = \dfrac{1.57 \text{ g}}{16.04 \text{ g / mol}} = 9.79 \times 10^{-2} \text{ mol}$

$n_{C_2H_6} = \dfrac{0.41 \text{ g}}{30.068 \text{ g / mol}} = 1.36 \times 10^{-2} \text{ mol}$

$n_{C_3H_8} = \dfrac{0.020 \text{ g}}{44.094 \text{ g / mol}} = 4.54 \times 10^{-4} \text{ mol}$

$X_{CH_4} = 0.874 \qquad X_{C_2H_6} = 0.121 \qquad X_{C_3H_8} = 4.06 \times 10^{-3}$

$P_{CH_4} = (0.874)(2.35 \text{ atm}) = 2.05 \text{ atm}$

$P_{C_2H_6} = (0.121)(2.35 \text{ atm}) = 0.284 \text{ atm}$

$P_{C_3H_8} = (4.06 \times 10^{-3})(2.35 \text{ atm}) = 9.5 \times 10^{-3} \text{ atm}$

5.31 $C_6H_{12}O_{6(s)} + 6\,O_{2(g)} \rightarrow 6\,CO_{2(g)} + 6\,H_2O_{(l)}$

$MM_{glucose} = 180.16$

$n_{glucose} = \dfrac{4.65 \text{ g glucose}}{180.16 \text{ g / mol}} = 2.581 \times 10^{-2} \text{ mol glucose}$

$n_{CO_2} = (2.581 \times 10^{-2} \text{ mol glucose})\left(\dfrac{6 \text{ mol } CO_2}{1 \text{ mol glucose}}\right) = 0.1549 \text{ mol } CO_2$

$$V = \dfrac{(0.1549 \text{ mol } CO_2)(0.08206 \text{ L atm mol}^{-1} \text{ K}^{-1})(310 \text{ K})}{(1.00 \text{ atm})} = 3.94 \text{ L}$$

5.33 $V = 3.00 \times 10^3$ mL = 3.00 L

$$n_{Cl_2} = \frac{PV}{RT} = \frac{\left(\dfrac{1.25 \times 10^3 \text{ torr}}{760 \text{ torr / atm}}\right)(3.00 \text{ L})}{(0.08206 \text{ L atm mol}^{-1} \text{ K}^{-1})(300 \text{ K})} = 0.200 \text{ mol } Cl_2$$

$$n_{Na} = \frac{6.90 \text{ g}}{22.99 \text{ g / mol}} = 0.300 \text{ mol}$$

Reaction:	$2 \text{ Na}_{(s)}$	$+$	$Cl_{2(g)}$	$\rightarrow$	$2 \text{ NaCl}_{(s)}$
Initial amounts:	0.300 mol		0.200 mol		0 mol
Change in amounts:	-0.300 mol		-0.150 mol		+0.300 mol
Final amounts:	0 mol		0.050 mol		0.300 mol

Because Cl_2 is the only gas,

$$P_{total} = P_{Cl_2} = \frac{nRT}{V}$$

$$= \frac{(0.050 \text{ mol})(0.08206 \text{ L atm mol}^{-1}\text{K}^{-1})(320\text{K})}{(3.00 \text{ L})} = 0.438 \text{ atm}$$

5.35 Repeat Problem 5.34, then: $(24.3 \text{ kg } C_2H_4O_6)(0.65) = 16 \text{ kg}$

5.37 VP at 25°C = 23.765 torr

$$\text{relative humidity} = 78\% = \frac{P_{H_2O}}{VP_{H_2O}} \times 100 = \frac{P_{H_2O}}{23.756} \times 100$$

$$P_{H_2O} = \frac{(78)(23.756)}{100} = 18.5 \text{ torr}$$

This VP falls between 20°C and 25°C ($VP_{20°C} = 17.535$)

$$\left(\frac{18.5 - 17.535}{23.756 - 17.535}\right)(5°C) + 20°C = 20.8°C$$

5.39 At 35°C, $VP_{H2O} = 42.175$ torr; at 40°C, $VP_{H2O} = 55.324$ torr.
At 37°C, $VP_{H2O} = (2/5)(55.324 - 42.175)$ torr $+ 42.175$ torr $= 47.435$ torr

$$\frac{47.435 \text{ torr}}{55.324 \text{ torr}} \times 100\% = 85.74\% \approx 86\%$$

5.41 a) Particles must not suffer collisions while traveling their paths.
b) The materials will be used under the vacuum of space and, therefore, should be tested under such conditions.
c) Materials at high vacuum will not deposit on the metal surfaces but will, in fact, be released into the atmosphere.

5.43 $n_{Kr} = \dfrac{PV}{RT} = \dfrac{(10.0 \text{ atm})(0.600 \text{ L})}{(0.08206 \text{ L atm mol}^{-1} \text{ K}^{-1})(1273 \text{ K})} = 5.74 \times 10^{-2}$ mol

$m_{Kr} = n_{Kr}MM_{Kr} = (5.74 \times 10^{-2} \text{ mol})(83.80 \text{ g/mol}) = 4.81 \text{ g Kr}$
atoms of Kr $= (5.74 \times 10^{-2} \text{ mol})(6.022 \times 10^{23} \text{ atoms/mol}) = 3.46 \times 10^{22}$ atoms

5.45 Test whether 16 or 32 g of oxygen satisfied the ideal gas equation.

5.47 $X_{C_3H_6} = \dfrac{1}{1+4} = \dfrac{1}{5} = 0.200$

$X_{O_2} = \dfrac{4}{1+4} = \dfrac{4}{5} = 0.800$ (Assuming mole ratio to at least 3 sig. fig.)

$P_{C_3H_6} = (0.200)(1.00 \text{ atm}) = 0.200 \text{ atm}$

$P_{O_2} = (0.800)(1.00 \text{ atm}) = 0.800 \text{ atm}$

$n_{C_3H_6} = \dfrac{PV}{RT} = \dfrac{(0.200 \text{ atm})(2.00 \text{ L})}{(0.08206 \text{ L atm mol}^{-1} \text{ K}^{-1})(296 \text{ K})} = 1.647 \times 10^{-2} \text{ mol}$

$n_{O_2} = \dfrac{(0.800 \text{ atm})(2.00 \text{ L})}{(0.08206 \text{ L atm mol}^{-1} \text{ K}^{-1})(296 \text{ K})} = 6.587 \times 10^{-2} \text{ mol}$

$m_{C_3H_6} = (n_{C_3H_6})(MM_{C_3H_6}) = 1.647 \times 10^{-2} \text{ mol})(42.078 \text{ g/mol}) = 0.693 \text{ g C}_3\text{H}_6$

$m_{O_2} = (n_{O_2})(MM_{O_2}) = (6.587 \times 10^{-2} \text{ mol})(32.00 \text{ g/mol}) = 2.11 \text{ g O}_2$

5.49 $0°C = 273 \text{ K} \quad 22°C = 295 \text{ K}$

$(0.963 \text{ L})\left(\dfrac{273 \text{ K}}{295 \text{ K}}\right)\left(\dfrac{0.969 \text{ atm}}{1 \text{ atm}}\right) = 0.864 \text{ L}$

5.51 The decrease in pressure is due to the consumption of O_2 by the mouse.

∴ 760 torr - 720 torr = 40 torr of O_2 consumed.

$n_{O_2 \text{ cons.}} = \dfrac{PV}{RT} = \dfrac{\left(\dfrac{40 \text{ torr}}{760 \text{ torr / atm}}\right)(2.05 \text{ L})}{(0.08206 \text{ L atm mol}^{-1} \text{ K}^{-1})(300 \text{ K})} = 4.38 \times 10^{-3} \text{ mol}$

$m_{O_2} = n_{O_2}MM_{O_2} = (4.38 \times 10^{-3} \text{ mol})(32.0 \text{ g O}_2) = 0.14 \text{ g O}_2$

5.53 $n_{H_2} = \dfrac{PV}{RT} = \dfrac{(20.0 \text{ atm})(100 \text{ L})}{(0.0821 \text{ L atm mol}^{-1} \text{ K}^{-1})(600 \text{ K})} = 40.6 \text{ mol H}_2$

$n_{CO} = \dfrac{(10.0 \text{ atm})(100 \text{ L})}{(0.0821 \text{ L atm mol}^{-1} \text{ K}^{-1})(600 \text{ K})} = 20.3 \text{ mol CO}$

Reaction:	3 H$_2$	+	CO	→	CH$_4$	+	H$_2$O
Initial amounts:	40.6 mol		20.3 mol		0		0
Change in amounts:	-40.6 mol		-13.5 mol		13.5 mol		13.5 mol
Final amounts:	0 mol		6.8 mol		13.5 mol		13.5 mol

$m_{CH_4} = (n_{CH_4})(MM_{CH_4}) = (13.5 \text{ mol})(16.042 \text{ g/mol}) = 216.6 \text{ g CH}_4 = \text{theor. yield}$

$\text{percent yield} = \dfrac{150 \text{ g CH}_4}{216.6 \text{ g CH}_4} \times 100\% = 69.3\%$

5.55 (The MnO_2 is a catalyst in this reaction.)

a) $n_{O_2} = \dfrac{PV}{RT} = \dfrac{\left(\dfrac{759.2 \text{ torr}}{760 \text{ torr / atm}}\right)(0.02296 \text{ L})}{(0.08206 \text{ L atm mol}^{-1} \text{ K}^{-1})(298 \text{ K})} = 9.379 \times 10^{-4} \text{ mol } O_2$

b) $2KClO_3 \rightarrow 2KCl + 3O_2$

$9.379 \times 10^{-4} \text{ mol } O_2 \times \dfrac{2 \text{ mol } KClO_3}{3 \text{ mol } O_2} = 6.253 \times 10^{-4} \text{ mol } KClO_3$

c) $(6.253 \times 10^{-4} \text{ mol } KClO_3)(122.55 \text{ g/mol}) = 0.07663 \text{ g } KClO_3$

$\text{mass \%} = \dfrac{0.07663 \text{ g } KClO_3}{0.1054 \text{ g mixture}} \times 100\% = 72.70\%$

5.57 a) for dry air

$m_{N_2} = (28.02 \text{ g/mol})(0.7808 \text{ mol}) \qquad = 21.88 \text{ g } N_2$

$m_{O_2} = (32.00 \text{ g/mol})(0.2095 \text{ mol}) \qquad = 6.704 \text{ g } O_2$

$m_{Ar} = (39.948 \text{ g/mol})(9.34 \times 10^{-3} \text{ mol}) = 0.373 \text{ g Ar}$

$m_{CO_2} = (44.01 \text{ g/mol})(3.25 \times 10^{-4} \text{ mol}) = 0.0143 \text{ g } CO_2$

$\text{Total mass} = 28.97 \text{ g}$

$V = \dfrac{nRT}{P} = \dfrac{(1)(0.0821 \text{ L atm mol}^{-1} \text{ K}^{-1})(300 \text{ K})}{1 \text{ atm}} = 24.6 \text{ L}$

density of dry air = 28.97 g/24.6 L = 1.18 g/L

b) VP_{H_2O} at 300 K = (31.824 - 23.756) torr (2/5) + 23.756 torr = 26.983 torr

or (26.983 torr) ÷ (760 torr/atm) = 0.0355 atm

$P_{\text{dry air}}$ = (760 torr - 26.983 torr) = 733 torr = (733 torr)/(760 torr/atm) = 0.964 atm

$n_{\text{dry air}} = \dfrac{PV}{RT} = \dfrac{(0.964 \text{ atm})(24.6 \text{ L})}{(0.0821 \text{ L atm mol}^{-1} \text{ K}^{-1})(300 \text{ K})} = 0.963 \text{ mol}$

$n_{H_2O} = \dfrac{(0.0355 \text{ atm})(24.6 \text{ L})}{(0.0821 \text{ L atm mol}^{-1} \text{ K}^{-1})(300 \text{ K})} = 0.0355 \text{ mol}$

mass of 0.963 mol of dry air $\approx$ (0.963 mol)(28.97 g/mol) = 27.90 g

mass of 0.0355 mol of H_2O = (0.0355 mol)(18.016 g/mol) = 0.64 g

mass of wet air = 27.90 g + 0.64 g = 28.54 g

density of wet air = 28.54 g/24.6 L = 1.16 g/L

5.59 $\text{Relative humidty} = \dfrac{P_{H_2O}}{VP_{H_2O}} \times 100 \qquad P_{H_2O} = \dfrac{(\text{rel. hum.})(VP_{H_2O})}{100}$

a) $P_{H_2O} = \dfrac{(80)(42.175 \text{ torr})}{100} = 33.74 \text{ torr}$

(continued)

36

(5.59 continued)

$$\left(\frac{33.74 - 31.824}{42.175 - 31.824}\right)(5) + 30°C = 30.9°C \approx 31°C$$

b) $P_{H_2O} = \dfrac{(50)(12.788 \text{ torr})}{100} = 6.394 \text{ torr}$

$$\left(\frac{6.394 - 4.579}{6.543 - 4.579}\right)(5) + 0°C = 4.62°C \approx 4.6°C$$

c) $P_{H_2O} = \dfrac{(30)(23.756 \text{ torr})}{100} = 7.1268 \text{ torr}$

$$\left(\frac{7.1268 - 6.543}{9.209 - 6.543}\right)(5) + 5°C = 6.09°C \approx 6.1°C$$

5.61 $n_{unk} = \dfrac{PV}{RT} = \dfrac{\left(\dfrac{435 \text{ torr}}{760 \text{ torr / atm}}\right)(0.150 \text{ L})}{(0.0821 \text{ L atm mol}^{-1} \text{ K}^{-1})(423 \text{ K})} = 2.472 \times 10^{-3} \text{ mol}$

$MM_{unkn} = \dfrac{m}{n} = \dfrac{0.250 \text{ g}}{2.472 \times 10^{-3} \text{ mol}} = 101 \text{ g / mol}$

H: 14.94 14.94/1.008 = 14.82; 14.82 ÷ 0.9879 = 15.00

C: 71.22 71.22/12.01 = 5.93; 5.93 ÷ 0.9879 =   6.00

N: 13.84 13.84/14.01 = 0.9879; 0.9879 ÷ 0.9879 = 1.00

$C_6H_{15}N$

$MM_{C_6H_{15}N}$ = 101 g/mol ∴ molecular formula = $C_6H_{15}N$

5.63 $He_A : He_B : He_C = 6 : 12 : 9$ Assume each chamber contains only He.

a) B has the highest pressure because it contains the greatest amount of gas (He).

b) The pressure of A would be 0.5 atm (1/2 of 1.0 atm) because it contains half as much He (6 : 12).

c) The new pressure would be (1.0 atm)(27/6) = 4.5 atm because the amount of He is increased 4.5 (27/6) times.

d) If the valves were opened, the 27 portions of He (6 + 12 + 9) would divide equally between the chambers, 27/3 = 9. The contents and, therefore, the pressure would be 3/4 (9/12) of the original pressure. The pressure would be (0.50 atm) x (9/12) or 0.38 atm.

5.65 $n_{air} = \dfrac{PV}{RT} = \dfrac{(170 \text{ atm})(12.5 \text{ L})}{(0.0821 \text{ L atm mol}^{-1} \text{ K}^{-1})(300 \text{ K})} = 86.28 \text{ mol air}$

$n_{O2} = (X_{O2})(n_{air}) = (0.20)(86.28 \text{ mol}) = 17.3 \text{ mol } O_2$

$m_{O2} = (32.00 \text{ g/mol})(17.3 \text{ mol } O_2) = 554 \text{ g } O_2$

$(554 \text{ g } O_2) \div (14.0 \text{ g } O_2/\text{min}) = 39.6 \text{ min}$

39.6 min - 6.0 min = 33.6 min of diving

5.67 $n_{Ar} = \dfrac{1.00 \text{ g}}{39.95 \text{ g / mol}} = 0.0250 \text{ mol}$

$n_{Ne} = \dfrac{0.050 \text{ g}}{20.18 \text{ g / mol}} = 0.0025 \text{ mol}$

$X_{Ar} = \dfrac{0.0250 \text{ mol}}{0.0250 \text{ mol} + 0.0025 \text{ mol}} = 0.9091$

$X_{Ne} = \dfrac{0.0025 \text{ mol}}{0.0250 \text{ mol} + 0.0025 \text{ mol}} = 0.0909$

$P_{total} = \dfrac{nRT}{V}$

$= \dfrac{(0.0250 \text{ mol} + 0.0025 \text{ mol})(0.08206 \text{ L atm mol}^{-1} \text{ K}^{-1})(275 \text{ K})}{5.00 \text{ L}} = 0.124 \text{ atm}$

$P_{Ar} = (X_{Ar})(P_{total}) = (0.9091)(0.124 \text{ atm}) = 0.113 \text{ atm}$

$P_{Ne} = (X_{Ne})(P_{total}) = (0.0909)(0.124 \text{ atm}) = 0.0113 \text{ atm}$

5.69 The sulfur burns when the coal is burned: $S + O_2 \rightarrow SO_2$

The SO_2 then either dissolves in rain: $SO_2 + H_2O \rightarrow H_2SO_3$ (sulfurous acid)

or is oxidized by O_2 and uv light: $2SO_2 + O_2 \rightarrow 2SO_3$

The SO_3 then dissolves in rain: $SO_3 + H_2O \rightarrow H_2SO_4$ (sulfuric acid)

5.71 a) ~400 m/s

b) $k.E. = 1/2 mv^2 = \left(\dfrac{1}{2}\right)\left(\dfrac{17.03 \times 10^{-3} \text{ kg / mol}}{6.022 \times 10^{23} \text{ molecules / mol}}\right)(400 \text{ m / s})^2$

$= 2.26 \times 10^{-21} \text{ kg m}^2 / \text{s}^2 \text{ molecule}$

$= 2.26 \times 10^{-21} \text{ J/molecule}$

5.73 $n = \dfrac{PV}{RT} = \dfrac{m}{MM}$ $m = \dfrac{MM \, PV}{RT}$ $m_{He} + 300 \times 10^3 \text{ g} = m_{air}$

$MM_{He} = 4.0026$ g/mol; $MM_{air} = 28.97$ g/mol (See Prob. 5.57)

$\dfrac{MM_{He} \, PV}{RT} + 300 \times 10^3 \text{ g} = \dfrac{MM_{air} \, PV}{RT}$

$\dfrac{(4.0026 \text{ g / mol})(1.0 \text{ atm})V}{(0.0821 \text{ L atm mol}^{-1} \text{ K}^{-1})(298 \text{ K})} + 300 \times 10^{-3} \text{ g}$

$= \dfrac{(28.97 \text{ g / mol})(1.0 \text{ atm})V}{(0.0821 \text{ L atm mol}^{-1} \text{ K}^{-1})(298 \text{ K})}$

(continued)

(5.73 continued)

$$300 \times 10^3 \text{ g} = (28.97 - 4.0026 \text{ g/mol})\left(\frac{(1.0 \text{ atm})}{(0.0821 \text{ L atm mol}^{-1} \text{ K}^{-1})(298 \text{ K})}\right)V$$

$$300 \times 10^3 \text{ g} = (24.97 \text{ g/mol})(0.04087 \frac{\text{mol}}{\text{L}})V$$

$$V = \frac{300 \times 10^3 \text{ g}}{(24.97 \text{ g/mol})(0.04087 \text{ mol/L})} = 2.9 \times 10^5 \text{ L}$$

5.75   a) $MM_{TNT} = 227.14$ g/mol

$$n_{TNT} = \frac{1.0 \times 10^3 \text{ g}}{227.14 \text{ g/mol}} = 4.40 \text{ mol}$$

$2C_7H_5(NO_2)_{3(s)} \rightarrow 12CO_{(g)} + 2C_{(s)} + 5H_{2(g)} + 3N_{2(g)}$
4.40 mol   26.4 mol   4.4 mol   11.0 mol   6.60 mol
moles of gas = 26.4 + 11.0 mol + 6.60 mol = 44.0 mol

b) $V = \frac{nRT}{P} = \frac{(44.0 \text{ mol})(0.0821 \text{ L atm mol}^{-1} \text{ K}^{-1})(298 \text{ K})}{1.0 \text{ atm}} = 1076 \text{ L} = 1.1 \times 10^3 \text{ L}$

c) $X_{CO} = \frac{26.4 \text{ mol}}{44.0 \text{ mol}} = 0.60$ $X_{H_2} = \frac{11.0 \text{ mol}}{44.0 \text{ mol}} = 0.25$

$X_{N_2} = \frac{6.60 \text{ mol}}{44.0 \text{ mol}} = 0.15$

$P_{CO} = (0.60)(1.0 \text{ atm}) = 0.60 \text{ atm}$
$P_{H_2} = (0.25)(1.0 \text{ atm}) = 0.25 \text{ atm}$
$P_{N_2} = (0.15)(1.0 \text{ atm}) = 0.15 \text{ atm}$

5.77 $P_{SF_6} = \left(\frac{1}{10^9}\right)(1 \text{ atm}) = 10^{-9} \text{ atm}$

$$n = \frac{PV}{RT} = \frac{(10^{-9} \text{ atm})(10^{-3} \text{ L})}{(0.0821 \text{ L atm mol}^{-1} \text{ K}^{-1})(294 \text{ K})} = 4.14 \times 10^{-14} \text{ mol}$$

molecules of $SF_6 = (4.14 \times 10^{-14} \text{ mol})(6.022 \times 10^{23} \text{ molecules/mol})$
$= 2.49 \times 10^{10}$ molecules

5.79 a)

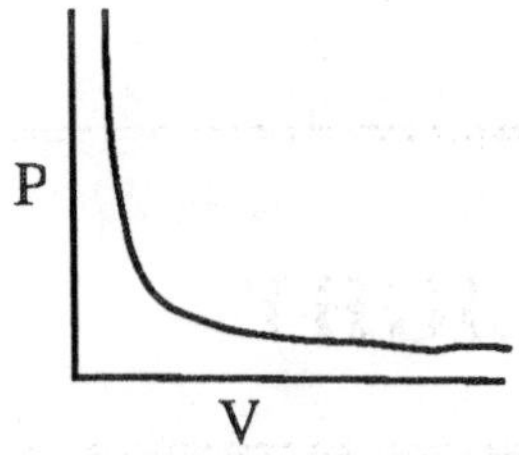

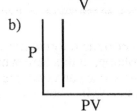

$V = nRT / P$

The product of P x V is equal to the constant nRT.

b)

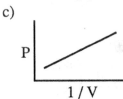

$PV = nRT$

$PV = \text{constant}$

c)

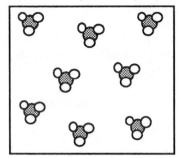

$P \times (1/nRT) = 1/V$

5.81 The figure in the textbook shows: $N_2 \quad + \quad 3H_2 \quad \rightarrow \quad 2NH_3$
$\qquad\qquad\qquad\qquad\qquad\qquad$ 4 mol $\quad$ 12 mol $\qquad$ 8 mol
Your drawing of the system after reaction must show 8 representations of NH_3.

5.83 $n_{CO_2} = \dfrac{15.00 \text{ g}}{44.01 \text{ g} / \text{mol}} = 0.3408 \text{ mol}$

$P = \dfrac{nRT}{V} = \dfrac{(0.3408 \text{ mol})(0.0821 \text{ L atm mol}^{-1} \text{ K}^{-1})(273 \text{ K})}{(0.750 \text{ L})} = 10.2 \text{ atm}$

5.85 See Problem 5.84 for the formula that can be used in this problem.

$\text{Density of SF}_6 = \dfrac{P \text{ MM}_{SF_6}}{RT} = \dfrac{\left(\dfrac{755 \text{ torr}}{760 \text{ torr} / \text{atm}}\right)(146.05 \text{ g} / \text{mol})}{(0.0821 \text{ L atm mol}^{-1} \text{ K}^{-1})(300 \text{ K})} = 5.89 \text{ g} / \text{L}$

5.87 in auto: $\qquad\qquad N_2 + O_2 \rightarrow 2 \text{ NO}$

in atmosphere: $\qquad 2 \text{ NO} + O_2 \rightarrow 2 \text{ NO}_2$

40

CHAPTER 6: ATOMS AND LIGHT

6.1 Knowing the grams per volume (density) and the grams per mole (atomic mass), one can calculate the volume per mole of atoms (molar volume, in this case where they are all treated as single atom units). From molar volume one can conclude which element has the largest atoms since each molar volume will contain the same number of atoms.

Molar Volume = MM ÷ density

Molar Volume of Al = 26.98 g/mol ÷ 2.70 g/cm^3 = 9.99 cm^3/mol

Molar Volume of Hg = 200.6 g/mol ÷ 13.55 g/cm^3 = 14.80 cm^3/mol

Molar Volume of Pb = 207.2 g/mol ÷ 11.34 g/cm^3 = 18.27 cm^3/mol

Each mole has the same number of atoms, $\therefore$ Pb has the largest atoms;

Al has the smallest atoms.

6.3 Everything is made of atoms and everything has mass and volume.

6.5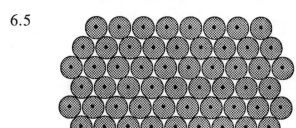
This picture is only 2-dimensional. The thinnest layer would be only one atom thick.
The nuclei would also be smaller and exactly in the center of each atom.

6.7 $E_{\text{mole of photons}} = E_{\text{photon}} \times N_A$

a) $E_{\text{mole of photons}} = (4.05 \times 10^{-19} \text{ J})(6.022 \times 10^{23} \text{ mol}^{-1})(1 \text{ kJ} / 10^3 \text{ J})$

$$= 2.44 \times 10^2 \text{ kJ} / \text{mol}$$

b) $E_{\text{mole of photons}} = (2.99 \times 10^{-19} \text{ J})(6.022 \times 10^{23} \text{ mol}^{-1})(1 \text{ kJ} / 10^3 \text{ J})$

$$= 1.80 \times 10^2 \text{ kJ} / \text{mol}$$

c) $E_{\text{mole of photons}} = (7.79 \times 10^{-18} \text{ J})(6.022 \times 10^{23} \text{ mol}^{-1})(1 \text{ kJ} / 10^3 \text{ J})$

$$= 4.69 \times 10^3 \text{ kJ} / \text{mol}$$

d) $E_{\text{mole of photons}} = (1.59 \times 10^{-19} \text{ J})(6.022 \times 10^{23} \text{ mol}^{-1})(1 \text{ kJ} / 10^3 \text{ J})$

$$= 95.7 \text{ kJ} / \text{mol}$$

6.9   Given: E_{photon} = 745 kJ/mole. From the value of E_{photon} per photon one can calculate its wavelength.

$$E_{photon} = \frac{(745 \text{ kJ} / \text{mol})(10^3 \text{ J} / \text{kJ})}{(6.022 \times 10^{23} \text{ mol}^{-1})} = 1.237 \times 10^{-18} \text{ J}$$

$$\lambda = \frac{hc}{E} = \frac{(6.626 \times 10^{-34} \text{ J} \cdot \text{s})(2.998 \times 10^8 \text{ m} \cdot \text{s}^{-1})}{(1.237 \times 10^{-18} \text{ J})}$$

$$= 1.61 \times 10^{-7} \text{ m or 161 nm}$$

6.11   a) Given: $c = \nu\lambda$ and $\lambda = \frac{c}{\nu}$

$$\lambda = \frac{(2.998 \times 10^8 \text{ m} \cdot \text{s}^{-1})}{(1.30 \times 10^{15} \text{ s}^{-1})} = 2.31 \times 10^{-7} \text{ m or 231 nm}$$

b) $h\nu = h\nu_0 + \text{k.E.}_{electron}$

$$h\nu_0 = h\nu - \text{k.E.}_{electron} = (6.626 \times 10^{-34} \text{ J} \cdot \text{s})(1.30 \times 10^{15} \text{ s}^{-1}) - 5.2 \times 10^{-19} \text{ J}$$

$$= 8.6 \times 10^{-19} \text{ J} - 5.2 \times 10^{-19} \text{ J} = 3.4 \times 10^{-19} \text{ J}$$

c) $3.4 \times 10^{-19} \text{ J} = \frac{hc}{\lambda}$ $\lambda = \frac{hc}{3.4 \times 10^{-19} \text{ J}}$

$$\lambda = \frac{(6.626 \times 10^{-34} \text{ J} \cdot \text{s})(2.998 \times 10^8 \text{ m} \cdot \text{s}^{-1})}{(3.4 \times 10^{-19} \text{ J})} = 5.8 \times 10^{-7} \text{ m or 580 nm}$$

6.13   For this answer the drawing should look like Figure 6-8 except the answer needs to have a binding energy and a kinetic energy of cesium and a separate indication for the binding energy and kinetic energy of chromium.

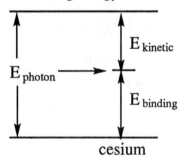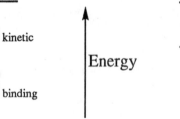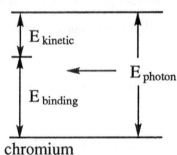

6.15   To elevate an electron to the state from which it can emit 404 nm radiation involves absorption of 254 nm radiation then 436 nm radiation.

$$E = \frac{hc}{\lambda} N_A \qquad \Delta E = \Delta E_1 + \Delta E_2$$

$$\Delta E = hcN_A\left(\frac{1}{\lambda_1} + \frac{1}{\lambda_2}\right) = (6.626 \times 10^{-34} \text{ J} \cdot \text{s})(2.998 \times 10^8 \text{ m} \cdot \text{s}^{-1}) \text{ x}$$

$$(6.022 \times 10^{23} \text{ mol}^{-1})\left(\frac{1 \text{ kJ}}{10^3 \text{ J}}\right)\left(\frac{1}{\lambda_1} + \frac{1}{\lambda_2}\right) = 1.196 \times 10^{-4} \frac{\text{kJ} \cdot \text{m}}{\text{mol}}\left(\frac{1}{\lambda_1} + \frac{1}{\lambda_2}\right)$$

$$\Delta E = 1.196 \times 10^{-4} \frac{\text{kJ} \cdot \text{m}}{\text{mol}}\left(\frac{1}{436 \text{ nm}(10^{-9}\text{m}/\text{nm})} + \frac{1}{254 \text{ nm}(10^{-9}\text{m}/\text{nm})}\right)$$

$$= 745 \text{ kJ}/\text{mol}$$

6.17   The energy needed to reach the $n = 8$ state is the difference between the energy for $n = 1$ and the energy for $n = 8$. A similar calculation is needed for the $n = 1$ to the $n = 9$ transition. Equation 6-5 $\left(E_n = -\dfrac{2.18 \times 10^{-18} \text{ J}}{n^2}\right)$ is used to calculate these energies.

For $n = 8$: $\Delta E = E_8 - E_1 = \left(-\dfrac{2.18 \times 10^{-18} \text{ J}}{8^2}\right) - \left(-\dfrac{2.18 \times 10^{-18} \text{ J}}{1^2}\right)$

$= (2.18 \times 10^{-18} \text{ J})(1/1^2 - 1/8^2) = (2.18 \times 10^{-18} \text{ J})(0.984375) = 2.146 \times 10^{-18} \text{ J}$

$$\lambda = \frac{hc}{E} = \frac{\left(6.626 \times 10^{-34} \text{ J} \cdot \text{s}\right)\left(2.998 \times 10^8 \text{ m} \cdot \text{s}^{-1}\right)}{\left(2.146 \times 10^{-18} \text{ J}\right)} = 9.26 \times 10^{-8} \text{ m}$$

$$= 92.6 \times 10^{-9} \text{ m} = 92.6 \text{ nm} = 0.926 \times 10^{-7} \text{ m}$$

$$\nu = \frac{E}{h} = \frac{2.146 \times 10^{-18} \text{ J}}{6.626 \times 10^{-34} \text{ J} \cdot \text{s}} = 3.239 \times 10^{15} \text{ s}^{-1} \qquad \text{ultraviolet photons (See Table 6-1)}$$

For $n = 9$: $\Delta E = E_9 - E_1 = (2.18 \times 10^{-18} \text{ J})(1/1^2 - 1/9^2)$

$$= (2.18 \times 10^{-18} \text{ J})(0.987654) = 2.153 \times 10^{-18} \text{ J}$$

$$\lambda = \frac{hc}{E} = \frac{\left(6.626 \times 10^{-34} \text{ J} \cdot \text{s}\right)\left(2.998 \times 10^8 \text{ m} \cdot \text{s}^{-1}\right)}{\left(2.153 \times 10^{-18} \text{ J}\right)} = 9.23 \times 10^{-8} \text{ m}$$

$$= 92.3 \times 10^{-9} \text{ m} = 92.3 \text{ nm} = 0.923 \times 10^{-7} \text{ m}$$

$$\nu = \frac{E}{h} = \frac{2.153 \times 10^{-18} \text{ J}}{6.626 \times 10^{-34} \text{ J} \cdot \text{s}} = 3.25 \times 10^{15} \text{ s}^{-1} \qquad \text{ultraviolet photons (See Table 6-1)}$$

43

6.19

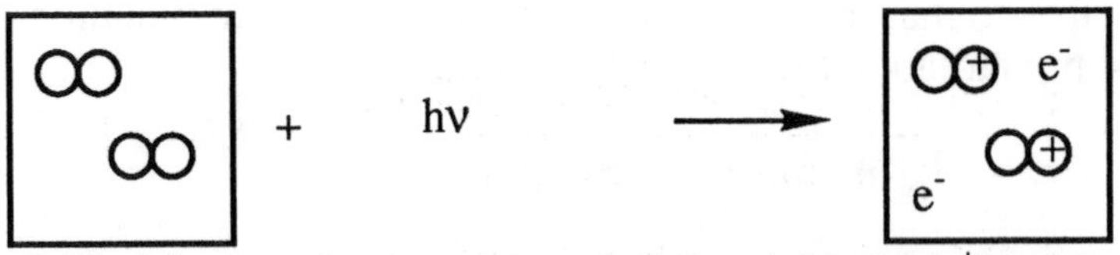

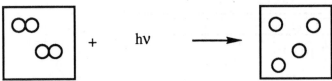

2 N_2 (g) plus short wavelength light yields 2 N_2^+ and 2 e^-

2 O_2 (g) plus short wavelength light yields 4 O

6.21

Spectral region	Gas in atmosphere that absorbs
a) <200 nm	O_2, N_2
b) 200 - 240 nm	O_3, O_2
c) 240 - 310 nm	O_3
d) 310 - 700 nm	O_3
e) 700 - 2000 nm	none

6.23   -30°C = 243 K   This is not sufficient information to determine the altitude.
Possibliities are: ~7 km, ~38 km, ~68 km, ~107 km.

6.25   Mass of one e^- = 9.109 x 10^{-31} kg

Mass of one mole of e^- = (9.109 x 10^{-31} kg/e^-)(6.022 x 10^{23} e^-/mol)

= 5.485 kg/mol

6.27   From Table 6-2: λ = h/mv   J•s = (kg m^2/s^2)•s = kg m^2/s

$$\lambda = \frac{(6.626 \times 10^{-34} \text{ kg m}^2 / \text{s})}{(9.109 \times 10^{-31} \text{ kg})(4.8 \times 10^5 \text{ m} / \text{s})} = 1.5 \times 10^{-9} \text{ m} = 1.5 \text{ nm}$$

6.29   **H**: Hamiltonian, expression for the total energy
E: Total energy of the system and has only certain definite values (quantized)

ψ: The wave function has wave properties and is generally a function of all spatial
variables (x, y and z). Its square, ψ^2, describes the distribution of an electron in
space.

6.31   The emission transitions can occur in steps back to the ground state while the
absorptions occur from the ground state to the excited state in one step.

6.33 $E = hc / \lambda$

For 488 nm: $E = \dfrac{(6.626 \times 10^{-34} \text{ J} \cdot \text{s})(2.998 \times 10^8 \text{ m / s})}{(488 \text{ nm})(10^{-9} \text{ m / nm})} = 4.07 \times 10^{-19} \text{ J}$

For 514 nm: $E = \dfrac{(6.626 \times 10^{-34} \text{ J} \cdot \text{s})(2.998 \times 10^8 \text{ m / s})}{(514 \text{ nm})(10^{-9} \text{ m / nm})} = 3.86 \times 10^{-19} \text{ J}$

a)
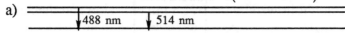

b) After the above emission, the Ar$^+$ ion is in an energy level 2.76×10^{-18} J above the ground state. The radiation emitted on returning to the ground state:

$$\nu = E/h = \dfrac{(2.76 \times 10^{-18} \text{ J})}{(6.626 \times 10^{-34} \text{ J} \cdot \text{s})} = 4.17 \times 10^{15} \text{ s}^{-1}$$

$$\lambda = \dfrac{c}{\nu} = \dfrac{2.998 \times 10^8 \text{ m / s}}{4.17 \times 10^{15} \text{ s}^{-1}} = 7.19 \times 10^{-8} \text{ m} = 71.9 \times 10^{-9} \text{ m} = 71.9 \text{ nm}$$

6.35 E from wavelength =

$$E = \dfrac{hc}{\lambda} = \dfrac{(6.626 \times 10^{-34} \text{ J} \cdot \text{s})(2.998 \times 10^8 \text{ m / s})}{(\lambda (\text{nm}))(10^{-9} \text{ m / nm})} = \dfrac{1.9865 \times 10^{-16} \text{ J}}{\lambda (\text{nm})}$$

422.7 nm: E = 4.700×10^{-19} J   272.2 nm: E = 7.298×10^{-19} J

239.9 nm: E = 8.281×10^{-19} J   671.8 nm: E = 2.957×10^{-19} J

504.2 nm: E = 3.940×10^{-19} J

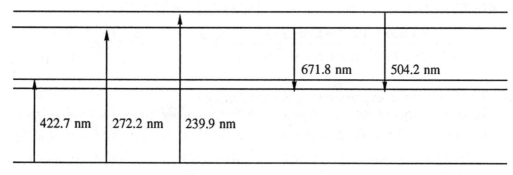

a) $E_{239.9 \text{ nm}} - E_{504.2 \text{ nm}} = 8.281 \times 10^{-19} \text{ J} - 3.940 \times 10^{-19} \text{ J} = 4.341 \times 10^{-19} \text{ J}$

$$\lambda = \dfrac{hc}{E} = \dfrac{(6.626 \times 10^{-34} \text{ J} \cdot \text{s})(2.998 \times 10^8 \text{ m / s})}{(4.341 \times 10^{-19} \text{ J})} = 4.576 \times 10^{-7} \text{ m}$$

$$= 457.6 \times 10^{-9} \text{ m} = 457.6 \text{ nm}$$

(continued)

(6.35 continued)

b) $E_{272.2\ nm} - E_{422.7\ nm} = 7.298 \times 10^{-19}\ J - 4.700 \times 10^{-19}\ J = 2.598 \times 10^{-19}\ J$

$$\lambda = \frac{hc}{E} = \frac{(6.626 \times 10^{-34}\ J \cdot s)(2.998 \times 10^8\ m/s)}{(2.598 \times 10^{-19}\ J)} = 7.646 \times 10^{-7}\ m$$

$$= 764.6 \times 10^{-9}\ m = 764.6\ nm$$

c) The 272.2 nm absorption followed by the 671.8 nm emission and the 239.9 nm absorption followed by the 504.2 nm emission both lead to an energy level of 4.34×10^{-19} J above the lowest state.

6.37 Equation 6-5: $E_n = -\dfrac{2.18 \times 10^{-18}\ J}{n^2}$

a) for n = 4:

$$E_4 = -\frac{2.18 \times 10^{-18}\ J}{4^2} = -1.36 \times 10^{-19}\ J$$

b) for n = 2:

$$E_2 = -\frac{2.18 \times 10^{-18}\ J}{2^2} = -5.45 \times 10^{-19}\ J$$

$E_{emitted\ photon} = E_2 - E_4 = (-5.45 \times 10^{-19}\ J) - (-1.36 \times 10^{-19}\ J) = -4.09 \times 10^{-19}\ J$

$$\lambda = \frac{hc}{E} = \frac{(6.626 \times 10^{-34}\ J \cdot s)(2.998 \times 10^8\ m/s)}{(4.09 \times 10^{-19}\ J)} = 4.86 \times 10^{-7}\ m$$

$$= 486 \times 10^{-9}\ m = 486\ nm$$

c) $E_{emitted\ photon} = E_{threshold} + E_{of\ ejected\ electron}$

$E_{of\ ejected\ electron} = E_{emitted\ photon} - E_{threshold} = 4.09 \times 10^{-19}\ J - 3.2 \times 10^{-19}\ J$

$$= 0.9 \times 10^{-19}\ J = 9 \times 10^{-20}\ J$$

d) $\lambda = \dfrac{h}{mv} = \dfrac{h}{m}\left(\dfrac{m}{2E}\right)^{\frac{1}{2}} = \left(\dfrac{h^2}{2mE}\right)^{\frac{1}{2}} = \left[\dfrac{(6.626 \times 10^{-34}\ J \cdot s)^2}{(2)(9.109 \times 10^{-31}\ kg))(9 \times 10^{-20}\ J)}\right]^{\frac{1}{2}}$

$$= \left[2.678 \times 10^{-18}\ \frac{kg\ m^2/s^2 \cdot s^2}{kg}\right]^{\frac{1}{2}} = \sqrt{2.7 \times 10^{-18}\ m^2}$$

$$= 1.6 \times 10^{-9}\ m = 1.6\ nm$$

6.39 $E_{\text{photon at 337.1 nm}} = \dfrac{hc}{\lambda}$

$$= \frac{(6.626 \times 10^{-34} \text{ J} \bullet \text{s})(2.998 \times 10^8 \text{ m/s})}{(337.1 \text{ nm})(10^{-9} \text{ m/nm})} = 5.893 \times 10^{-19} \text{ J}$$

$\dfrac{E_{\text{laser}}}{E_{\text{photon}}} = \text{number of photons} = \dfrac{10 \text{ mJ}}{5.893 \times 10^{-19} \text{ J/photons}}$

$$= \frac{10 \times 10^{-3} \text{ J}}{5.893 \times 10^{-19} \text{ J/photons}} = 1.7 \times 10^{16} \text{ photons}$$

6.41 Adapting equations for electrons from Table 6-2:

a) $\nu = \dfrac{h}{m\lambda} = \dfrac{(6.626 \times 10^{-34} \text{ kg m}^2/\text{s})}{(1.6749 \times 10^{-27} \text{ kg})(75 \text{ pm})(10^{-12} \text{ m/pm})} = 5.3 \times 10^3 \text{ m/s}$

b) $\lambda = \dfrac{h}{m\nu} = \dfrac{(6.626 \times 10^{-34} \text{ kg m}^2/\text{s})}{(1.6749 \times 10^{-27} \text{ kg})\left(\dfrac{1.25}{100}\right)(2.998 \times 10^8 \text{ m/s})} = 1.06 \times 10^{-13} \text{ m}$

$$= 0.106 \times 10^{-12} \text{ m} = 0.106 \text{ pm}$$

6.43 The H_2O in clouds absorbs the heat being radiated from the earth and keeps it from being lost.

6.45 $E_{\text{per electron}} = (216.4 \text{ kJ/mol})(10^3 \text{ J/kJ}) \div (6.022 \times 10^{23} \text{ electrons/mol})$

$$= 3.593 \times 10^{-19} \text{ J/electron}$$

$\lambda = \dfrac{hc}{E} = \dfrac{(6.626 \times 10^{-34} \text{ J} \bullet \text{s})(2.998 \times 10^8 \text{ m/s})}{(3.593 \times 10^{-19} \text{ J})} = 5.529 \times 10^{-7} \text{ m}$

$$= 552.9 \times 10^{-9} \text{ m} = 552.9 \text{ nm}$$

6.47 As we move upward through the Earth's atmosphere, the pressure continually decreases because there are less and less molecules present.

The temperature of the troposphere decreases as we go upward because the portion near the earth is heated by the earth and has enough molecules present to hold that heat. As we go higher and the troposphere "thins", there are fewer molecules to hold heat.

The stratosphere becomes warmer as we move upward because O_3 strongly absorbs ultraviolet light. Although the concentration of O_3 in the stratosphere decreases on moving upward, it absorbs uv light so efficiently that the uv light is almost completely absorbed by the O_3 in the upper part of the stratosphere and little uv light reaches the more concentrated O_3 in the lower part.

(continued)

(6.47 continued)

The mesosphere's temperature decreases on moving upward. Sunlight is not absorbed in the mesosphere and the temperature decreases as the density of molecules decreases on moving upward.

The thermosphere's temperature increases on moving upward. Molecules of N_2 and O_2 absorb high frequency solar light. As the solar light passes through the thermosphere, its high energy photons are progressively removed by this absorption. Thus, the intensity of high-energy light decreases and less energy is deposited on moving downward in the thermosphere.

6.49   a) The binding (threshold) energy of the electrons =
$$6.00 \times 10^{-19} \text{ J} - 2.70 \times 10^{-19} \text{ J} = 3.30 \times 10^{-19} \text{ J}$$

b) $E = hc/\lambda$ $\lambda = hc/E$

$$\lambda = \frac{(6.626 \times 10^{-34} \text{ J} \cdot \text{s})(2.998 \times 10^8 \text{ m/s})}{(6.00 \times 10^{-19} \text{ J})} = 3.31 \times 10^{-7} \text{ m}$$

$$= 331 \times 10^{-9} \text{ m} = 331 \text{ nm}$$

c) Given: $\lambda_e = \dfrac{h}{mv}$, $v = \left(\dfrac{2E}{m}\right)^{\frac{1}{2}}$, and $\lambda_e = \left(\dfrac{h}{m}\right)\left(\dfrac{m}{2E}\right)^{\frac{1}{2}}$

$$\lambda_e = \frac{(6.626 \times 10^{-34} \text{ kg m}^2/\text{s})}{(9.109 \times 10^{-31} \text{ kg})}\left[\frac{9.109 \times 10^{-31} \text{ kg}}{(2)(2.70 \times 10^{-19} \text{ kg} \cdot \text{m}^2/\text{s}^2)}\right]^{\frac{1}{2}}$$

$$= 9.45 \times 10^{-10} \text{ m} = 945 \times 10^{-12} \text{ m} = 945 \text{ pm}$$

6.51   5% velocity of light = $(0.05)(3.0 \times 10^8 \text{ m/s}) = 1.5 \times 10^7 \text{ m/s}$
$m_e = 9.109 \times 10^{-31} \text{ kg}$ $m_p = 1.673 \times 10^{-27} \text{ kg}$

$\lambda = h/mv$

$$\lambda_e = \left(\frac{(6.626 \times 10^{-34} \text{ kg} \cdot \text{m}^2 \cdot \text{s}/\text{s}^2)}{(9.109 \times 10^{-31} \text{ kg})(1.5 \times 10^7 \text{ m/s})}\right) = 4.8 \times 10^{-11} \text{ m}$$

$$= 48 \times 10^{-12} \text{ m} = 48 \text{ pm}$$

$$\lambda_p = \left(\frac{(6.626 \times 10^{-34} \text{ kg} \cdot \text{m}^2 \cdot \text{s}/\text{s}^2)}{(1.673 \times 10^{-27} \text{ kg})(1.5 \times 10^7 \text{ m/s})}\right) = 2.6 \times 10^{-14} \text{ m}$$

$$= 26 \times 10^{-15} \text{ m} \ (= 26 \text{ fm})$$

6.53 From Equation 6-5: $E_n = 2.18 \times 10^{-18}\,\text{J}\left(\dfrac{1}{n_f^2} - \dfrac{1}{n_i^2}\right)$

$$\lambda = \frac{hc}{E} = \frac{(6.626 \times 10^{-34}\ \text{J} \cdot \text{s})(2.998 \times 10^8\ \text{m/s})}{E\ (\text{in J})}$$

$$\lambda = (1.9865 \times 10^{-25}/\,E)$$

For the Paschen Series $n_{final} = 3$ and $n_{initial}$ is equal to integers 4 or greater.

$n = 4$: $E_n = 2.18 \times 10^{-18}\,\text{J}\left(\dfrac{1}{3^2} - \dfrac{1}{4^2}\right) = 1.06 \times 10^{-19}$ J

$\lambda = 1.9 \times 10^{-6}$ m $= 1.9$ μm

$n = 5$: $E_n = 2.18 \times 10^{-18}\,\text{J}\left(\dfrac{1}{3^2} - \dfrac{1}{5^2}\right) = 1.55 \times 10^{-19}$ J

$\lambda = 1.3 \times 10^{-6}$ m $= 1.3$ μm

$n = 6$: $E_n = 2.18 \times 10^{-18}\,\text{J}\left(\dfrac{1}{3^2} - \dfrac{1}{6^2}\right) = 1.81 \times 10^{-19}$ J

$\lambda = 1.1 \times 10^{-6}$ m $= 1.1$ μm

$n = 7$: $E_n = 2.18 \times 10^{-18}\,\text{J}\left(\dfrac{1}{3^2} - \dfrac{1}{7^2}\right) = 1.98 \times 10^{-19}$ J

$\lambda = 1.0 \times 10^{-6}$ m $= 1.0$ μm

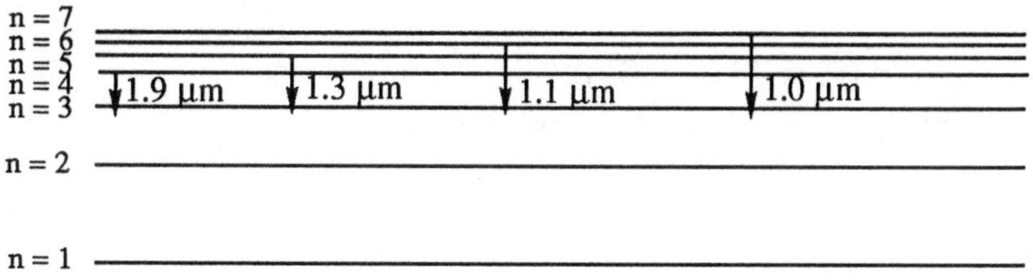

6.55 27.3 MHz $= 27.3 \times 10^6$ Hz $= 27.3 \times 10^6$ s^{-1}

$$\lambda = \frac{c}{\nu} = \frac{(2.998 \times 10^8\ \text{m/s})}{(27.3 \times 10^6\ \text{s}^{-1})} = 11.0\ \text{m}$$

$E = h\nu = (6.626 \times 10^{-34}\ \text{J} \cdot \text{s})(27.3 \times 10^6\ \text{s}^{-1}) = 1.81 \times 10^{-26}$ J

6.57   frequencies: $v = \dfrac{c}{\lambda} = \dfrac{2.998 \times 10^8 \text{ m / s}}{\lambda}$

$v_{487 \text{ nm}} = \dfrac{2.998 \times 10^8 \text{ m / s}}{(487 \text{ nm})(10^{-9} \text{ m / nm})} = 6.156 \times 10^{14} \text{ s}^{-1} = 6.16 \times 10^{14} \text{ s}^{-1}$

$v_{514 \text{ nm}} = 5.833 \times 10^{14} \text{ s}^{-1} = 5.83 \times 10^{14} \text{ s}^{-1}$

$v_{543 \text{ nm}} = 5.521 \times 10^{14} \text{ s}^{-1} = 5.52 \times 10^{14} \text{ s}^{-1}$

$v_{553 \text{ nm}} = 5.421 \times 10^{14} \text{ s}^{-1} = 5.42 \times 10^{14} \text{ s}^{-1}$

$v_{578 \text{ nm}} = 5.187 \times 10^{14} \text{ s}^{-1} = 5.19 \times 10^{14} \text{ s}^{-1}$

Energies per mole:

$E = h v N_A = (6.626 \times 10^{-34} \text{ J} \cdot \text{s}) v (6.022 \times 10^{23} \text{ mol}^{-1})(\text{kJ} / 10^3 \text{ J})$

$$= 3.99 \times 10^{-13} \text{ kJ} \cdot \text{s / mol} \times v$$

$E_{487 \text{ nm}} = (3.99 \times 10^{-13} \text{ kJ} \cdot \text{s / mol})(6.156 \times 10^{14} \text{ s}^{-1}) = 246 \text{ kJ / mol}$

$E_{514 \text{ nm}} = (3.99 \times 10^{-13} \text{ kJ} \cdot \text{s / mol})(5.833 \times 10^{14} \text{ s}^{-1}) = 233 \text{ kJ / mol}$

$E_{543 \text{ nm}} = (3.99 \times 10^{-13} \text{ kJ} \cdot \text{s / mol})(5.521 \times 10^{14} \text{ s}^{-1}) = 220 \text{ kJ / mol}$

$E_{553 \text{ nm}} = (3.99 \times 10^{-13} \text{ kJ} \cdot \text{s / mol})(5.421 \times 10^{14} \text{ s}^{-1}) = 216 \text{ kJ / mol}$

$E_{578 \text{ nm}} = (3.99 \times 10^{-13} \text{ kJ} \cdot \text{s / mol})(5.187 \times 10^{14} \text{ s}^{-1}) = 207 \text{ kJ / mol}$

CHAPTER 7: ATOMIC STRUCTURE AND PERIODICITY

7.1   For 6p; $n = 6$, $l = 1$

n	l	m_l	m_s
6	1	1	+1/2
6	1	1	-1/2
6	1	0	+1/2
6	1	0	-1/2
6	1	-1	+1/2
6	1	-1	-1/2

7.3   For $n = 3$

$l = 0$	$m_l = 0$	$m_s = -1/2, +1/2$
$l = 1$	$m_l = -1, 0, +1$	$m_s = -1/2, +1/2$
$l = 2$	$m_l = -2, -1, 0, +1, +2$	$m_s = -1/2, +1/2$

7.5   a) nonexistent, m_s can be only -1/2 or +1/2

b) describes actual orbital

c) nonexistent, largest possible value of l is $(n–1)$, 2 in this case

d) describes actual orbital

7.7   a) See Figure 7-8 (a) for a similar drawing.

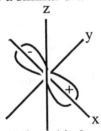

A 2p orbital
(could also have been
the $2p_y$ or $2p_z$ orbital)

b) See Figure 7-5 (d) for a similar drawing.

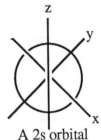

A 2s orbital

(continued)

(7.7 continued)

c) See Figure 7-5 (d) for a similar drawing except 3s will be larger than 2s.

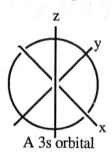

A 3s orbital

d) See Figure 7-9 for a similar drawing.

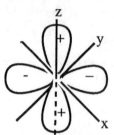

A $3d_{xy}$ orbital (could also have been one of the other d orbitals.)

7.9   1s Figure C;   2s Figure B;   2p Figure D;   3p Figure A

7.11

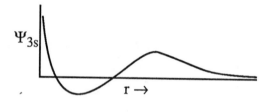

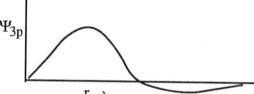

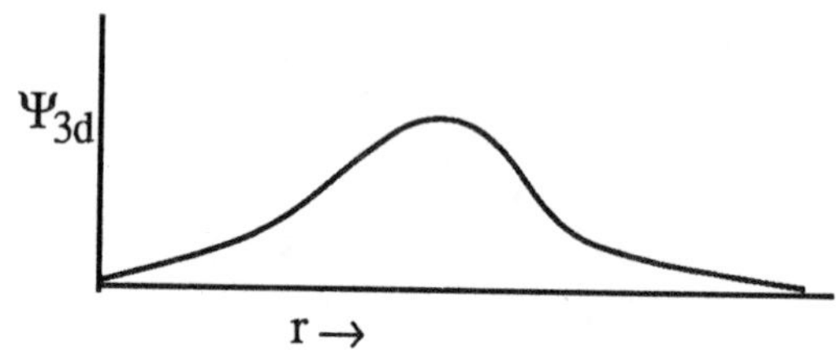

7.13   a) He 1s is more stable than the 2s orbital because orbital stability decreases with increasing n value.

b) Kr 5p is more stable than Kr 6p because orbital stability decreases with increasing n value

c) He^+ 2s is more stable because the electron in the 2s orbital of a He atom has a 1s electron screening the electron in the 2s orbital. The Z_{eff} is approximately +1 in the He atom and +2 in the He^+.

d) Ar 4s is more stable because the higher the l quantum number, the more that orbital is screened by electrons in smaller more stable orbitals.

52

7.15 In a multi-electron atom (such as helium) each electron affects the properties of all other electrons. These electron-electron interactions make the orbital energies different and unique for each element. The lower the l for an orbital, the more stable the orbital because of the difference in screening by the inner electron. The hydrogen atom has no inner electron, no screening, and the energy of orbitals depends only on the size of the orbitals (therefore, on the value of n).

7.17 For Be: $Z = 4$ $1s^2 2s^2$

n	l	m_l	m_s
1	0	0	+1/2
1	0	0	-1/2
2	0	0	+1/2
2	0	0	-1/2

For O: $Z = 8$ $1s^2 2s^2 2p^4$

n	l	m_l	m_s
1	0	0	+1/2
1	0	0	-1/2
2	0	0	+1/2
2	0	0	-1/2
2	1	+1	+1/2
2	1	+1	-1/2
2	1	0	+1/2
2	1	-1	+1/2

For P: $Z = 15$ $1s^2 2s^2 2p^6 3s^2 3p^3$

n	l	m_l	m_s
1	0	0	+1/2
1	0	0	-1/2
2	0	0	+1/2
2	0	0	-1/2
2	1	+1	+1/2
2	1	+1	-1/2
2	1	0	+1/2
2	1	0	-1/2
2	1	-1	+1/2
2	1	-1	-1/2
3	0	0	+1/2
3	0	0	-1/2
3	1	+1	+1/2
3	1	0	+1/2
3	1	-1	+1/2

For Ne: $Z = 10$ $1s^2 2s^2 2p^6$

n	l	m_l	m_s
1	0	0	+1/2
1	0	0	-1/2
2	0	0	+1/2
2	0	0	-1/2
2	1	+1	+1/2
2	1	+1	-1/2
2	1	0	+1/2
2	1	0	-1/2
2	1	-1	+1/2
2	1	-1	-1/2

7.19 $Z = 4$ for Be $1s^2 2s^2$

a) $1s^3 2s^1$ Pauli-forbidden

b) $1s^1 2s^3$ Pauli-forbidden

c) $1s^1 2p^3$ excited state

d) $1s^2 2s^1 2p^1$ excited state

e) $1s^2 2s^2$ ground state

f) $1s^2 1p^2$ nonexistent orbital

g) $1s^2 2s^1 2d^1$ nonexistent orbital

7.21   Mo   [Kr] $5s^14d^5$ Tc   [Kr] $5s^24d^5$

5s ↑   4d ↑↓ ↑ ↑ ↑ ↑   5s ↑↓   4d ↑ ↑ ↑ ↑ ↑

7.23   O is paramagnetic   and   P is paramagnetic

2s ↑↓   2p ↑↓ ↑ ↑   3s ↑↓   3p ↑ ↑ ↑

[He]   [Ne]

7.25   N has the ground state of $(1s^22s^22p^3)$. It has 7 excited states in which no electron has $n > 2$; they are: $1s^22s^12p^4$, $1s^22s^02p^5$, $1s^12s^22p^4$, $1s^12s^12p^5$, $1s^12s^02p^6$, $1s^02s^12p^6$ and $1s^02s^22p^5$

7.27   Your drawing would include the following features: (a) the elements that are one electron short of having a filled set of p orbitals are all the elements of column VII (F, Cl, Br, I and At), (b) the elements for which $n = 3$ is filling are: Na, Mg, Al, Si, P, S, Cl, Ar, Sc, Ti, V, Cr, Mn, Fe, Co, Ni, Cu and Zn, (c) the elements with half filled d orbitals are: C r, Mn, Mo, Tc, Re and Uns, and (d) the first element that contains a 5s electron is Rb.

7.29   Using Figure 7-20, one should conclude that since $Z = 111$ the configuration might be: $1s^22s^22p^63s^23p^64s^23d^{10}4p^65s^24d^{10}5p^66s^24f^{14}5d^{10}6p^67s^25f^{14}6d^9$. One might also have predicted it to be $7s^15f^{14}6d^{10}$ since the three elements above it have the s^1d^{10} configuration.

7.31   Ar > Cl > K > Cs

7.33   A negative electron affinity means that the neutral atom is more stable than the anion.

Be   2s ↑↓   2p — — —   Forming the anion Be⁻ requires adding an electron to a p orbital which is of higher energy than the s orbital (screening less for p orbitals).

N   2s ↑↓   2p ↑ ↑ ↑   Forming the anion N⁻ requires pairing two electrons in one of the p orbitals. This requires energy; therefore, the atom is more stable than the anion.

Mg   3s ↑↓   3p — — —   Forming the anion Mg⁻ requires adding an electron to a p orbital which is of higher energy than the s orbital (screening less for p orbitals).

(continued)

(7.33 continued)

Ar   4s ——

3s ↑↓   3p ↑↓ ↑↓ ↑↓

Forming the anion Ar⁻ requires adding an electron to an s orbital with n one greater than the valence orbitals; therefore, the energy of the electron in the ion is greater than the energy of the electrons in the atom.

Zn   4p —— —— ——

3d ↑↓ ↑↓ ↑↓ ↑↓ ↑↓

4s ↑↓

Forming the anion Zn⁻ requires adding an electron to a 4p orbital which is of higher energy than the 4s or 3d orbitals; therefore, the anion is more unstable than the neutral atom.

7.35 $Y^{3+} < Sr^{2+} < Rb^+ < Kr < Br^- < Se^{2-} < As^{3-}$

7.37 $K_{(g)} + 1/2 I_{2(g)} \rightarrow K^+I^-_{(g)}$

$$E_{coulomb} = \frac{(1.389 \times 10^5 \text{ kJ pm mol}^{-1})(q^+)(q^-)}{d}$$

$$= \frac{(1.389 \times 10^5 \text{ kJ pm mol}^{-1})(+1)(-1)}{(133 + 220) \text{ pm}} = -393 \text{ kJ / mol}$$

$K_{(g)} \rightarrow K^+_{(g)} + e^-$ $\Delta E = IE = 418.8$ kJ/mol

$I_{(g)} + e^- \rightarrow I^-_{(g)}$ $\Delta E = -EA = -295.3$ kJ/mol

$\Delta E = IE + (-EA) + E_{coulomb} = 418.8$ kJ/mol $- 295.3$ kJ/mol $- 393$ kJ/mol $= -270$ kJ/mol

7.39 From the list given, the stable cations and anions would be: Ca^{2+}, Cu^+ or Cu^{2+}, Cs^+, Cl^- and Cr^{3+} or Cr^{2+}.

7.41 As n increases, the electrons in the p orbitals are held more and more loosely and, therefore, can be easily removed. This is especially true if these p orbitals contain 1 or 2 electrons.

7.43 From column VI the nonmetals are O, S and Se; the only metalloid is Te; Po is a metal. See Problem 7.42.

7.45 There are 6 sets of quantum numbers for any set of np electrons. Therefore, there would also be 6 possible sets of quantum numbers for any 4p electron.

7.47 For 4d

n	l	m_l	m_s
4	2	-2	±1/2
4	2	-1	±1/2
4	2	0	±1/2
4	2	+1	±1/2
4	2	+2	±1/2

7.49 After 7p the next two orbitals to fill would be 8s and 5g.

7.51 From the following electron configurations one can see that S^+ will have the most unpaired electrons of these three since it has one electron in each of the 3p orbitals. S has 2 unpaired electrons (and 2 paired), while S^- has 1 unpaired electron (and 4 paired electrons).

S $1s^22s^22p^63s^23p^4$ S^+ $1s^22s^22p^63s^23p^3$
S^- $1s^22s^22p^63s^23p^5$

7.53 Rivers and most lakes have some dissolved salts in their waters. These waters eventually flow into the oceans. Some of the oceans' waters evaporate leaving behind the salts. This evaporated water forms precipitation which supplies water for the rivers and lakes. This water dissolves more salts as it flows to the oceans. This process continues, thus increasing the concentrations of the dissolved salts in the ocean. This same process can make a lake 'salty' if the lake has no outflow.

7.55

Quantum Number	Associated Property	Restrictions
n	energy of electron	positive integers
l	shape of atomic orbitals	positive integers less than n
m_l	direction orientation of orbitals	all integers from $-l$ to $+l$
m_s	spin orientation of electrons	either +1/2 or -1/2

7.57 F^{2+} $1s^22s^22p^3$ Ca^{2+} $1s^22s^22p^63s^23p^6$
      Fe^{2+} $1s^22s^22p^63s^23p^63d^6$ As^{2+} $1s^22s^22p^63s^23p^64s^23d^{10}4p^1$

7.59 Below are drawings of the contour diagrams for the 1s orbital and the $2p_z$ orbital drawn to about the same scale.

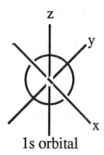
1s orbital

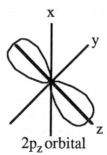
$2p_z$ orbital

7.61   The electron configuration for P is: $1s^22s^22p^63s^23p^3$. Its total spin = 3(1/2) = 3/2 or 3p $\underline{\uparrow}$ $\underline{\uparrow}$ $\underline{\uparrow}$.

The electron configuration for Br⁻ is: $1s^22s^22p^63s^23p^64s^23d^{10}4p^6$. Its total spin = 0 or 4p $\underline{\uparrow\downarrow}$ $\underline{\uparrow\downarrow}$ $\underline{\uparrow\downarrow}$.

The electron configuration for Cu⁺ is: $1s^22s^22p^63s^23p^64s^03d^{10}$. Its total spin = 0 or 3d $\underline{\uparrow\downarrow}$ $\underline{\uparrow\downarrow}$ $\underline{\uparrow\downarrow}$ $\underline{\uparrow\downarrow}$ $\underline{\uparrow\downarrow}$. (See Problem 7.60 above)

The electron configuration for Gd is: [Xe] $6s^24f^75d^1$. Its total spin = 8(1/2) = 4 or 4f $\underline{\uparrow}$ $\underline{\uparrow}$ $\underline{\uparrow}$ $\underline{\uparrow}$ $\underline{\uparrow}$ $\underline{\uparrow}$ $\underline{\uparrow}$ 5d $\underline{\uparrow}$ $\underline{\ }$ $\underline{\ }$ $\underline{\ }$ $\underline{\ }$.

The electron configuration for Sr is: [Kr] $5s^2$ Its total spin = 0 or 5s $\underline{\uparrow\downarrow}$.

7.63   Ca > Br > Cl⁻ > K⁺   Valence electrons of Ca and Br occupy orbitals with $n = 4$ which means their radii are greater than the radii for Cl⁻ and K⁺ with occupied orbitals with $n = 3$. Br is smaller than Ca because Br has more protons (Z = 35) than Ca (Z = 20) which attract the electrons more strongly and make the radius smaller. Cl⁻ and K⁺ have the same electron configuration ($1s^22s^22p^63s^23p^6$). Again, the extra protons of K⁺ (Z = 19) vs. Cl⁻ (Z = 17) pull the electrons closer to the nucleus making K⁺ smaller.

7.65   Francium should most closely resemble Cesium. It is an alkali metal. It should form a plus one ion (Fr⁺). It should be a soft, silvery to slightly golden, corrosive metal with relatively low melting and boiling points. It should be a good electrical and thermal conductor and show good photoelectric properties. Its first ionization energy should be small but the second ionization energy should be very large. Its compounds should virtually all be ionic and exhibit the +1 oxidation state. Unlike the other alkali metals, it is radioactive.

7.67   Four electrons is the maximum number of valence p electrons in the metal polonium (Po).

7.69   Make a plot of Figure 7.7 superimposed on Figure 7.6. Your plot should be similar to the answer to Section Exercise 7.3.1. Using your plot (drawing), describe the reduced screening arising from the electron density lobes of the 3s and 3p orbitals that lie close to the nucleus.

7.71   5d electron $l = 2$ $m_l = +2, +1, 0, -1, -2$
       4f electron $l = 3$ $m_l = +3, +2, +1, 0, -1, -2, -3$

7.73   Ce³⁺ [Xe] $4f^1$ Ce [Xe] $4f^15d^16s^2$
       La²⁺ [Xe] $5d^1$ La [Xe] $5d^16s^2$
       Ba⁺ [Xe] $6s^1$ Ba [Xe] $6s^2$
       Ground state configurations are different. Orbitals are most effectively screened by electrons occupying orbitals with smaller values of n. Therefore, 5d and 4f electrons are lower energy than 6s electrons.

7.75   Removing an electron from Be $(1s^22s^2, Z = 4)$ is more difficult than removing an electron from Li $(1s^22s^1, Z = 3)$ because the Z_{eff} is greater for Be. It is more difficult to remove an electron from $Li^+(1s^2)$ than from Be^+ $(1s^22s^1)$ because for Li^+ the electron is being removed from the much more stable 1s orbital $(n = 1)$.

7.77 $E = \dfrac{hc}{\lambda}$ $E_{molar} = \left(\dfrac{hc}{\lambda}\right)N_A$

$$E = \frac{(6.626 \times 10^{-34}\ J\ s)(2.998 \times 10^8\ m/s)}{(589\ nm)\left(10^{-9}\ m/nm\right)}(6.022 \times 10^{23}\ mol^{-1})$$

$$= 203.1\ kJ/mol$$

Ionization energy of excited atom = 496.5 kJ/mol - 203.1 kJ/mol = 293.4 kJ/mol

7.79   a) There would be 9 g orbitals when $l = 4$. This can be calculated by:
number of orbitals $= (2l + 1) = 2(4) + 1 = 9$

b) The possible values of m_l for the g orbitals are: +4, +3, +2, +1, 0, -1, -2, -3, -4.

c) The lowest principal quantum number for which g orbitals could exist is $n = 5$.

d) The 8s orbital is the orbital closest to the lowest energy g orbital and may turn out to be nearly degenerate with that lowest g orbital.

e) For the requested drawing extend Figure 7-17 by adding the following orbitals as one goes to higher and higher energy levels: one 5s orbital, three 5p orbitals, one 6s orbital, seven 4f's, five 5d's, three 6p's, one 7s, seven 5f's, five 6d's, three 7p's, one 8s, and nine 5g's.
The atomic number of the first element which could include a g electron would be 121 if 5g fills after 8s. If 1 (or 2) electrons enter the 6f and/or the 7d before the 5g fills (analogous to lanthanides and actinides), the atomic number could be 124.

7.81   Nonexistent wave function: Reason for nonexistence:

4g   For g, $l = 4$, but $n = 4$ and
largest value of $l = n - 1 = 3$

2d   For d, $l = 2$, but $n = 2$ and
largest value of $l = n - 1 = 1$

7.83   a)

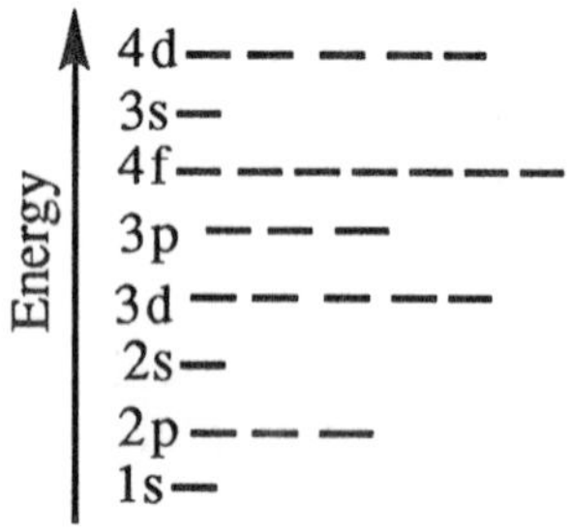

b) In this system the ground-state configuration for an atom containing 27 electrons would be: $1s^22p^62s^23d^{10}3p^64f^1$.

(continued)

58

(7.83 continued)
    c)

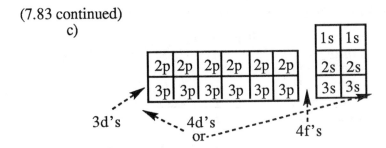

7.85   a) Element 18 ($n = 3$) should have a larger ionization energy than element 30 ($n = 4$) because the electron is held more loosely as n increases and the orbitals become larger.

b) Element 15 and cation 17^{2+} have the same electronic configuration:
$1s^3 2s^3 2p^9$
This electronic configuration is analogous to a noble gas electronic configuration. Since 17^{2+} has 2 more protons than 15, it would be expected that the radius of 17^{2+} would be smaller than 15. Comparison of the measurement of the radius of an ion with the radius of the atom of a noble gas presents problems because the measurement of an ion is made on a salt containing that ion while it is not possible to measure the radius of the noble gas atom under similar conditions.

c) The electronic configuration of element 47 is: $1s^3 2s^3 2p^9 3s^3 3p^9 4s^3 3d^{15} 4p^2$.
The electronic configuration of element 48 is: $1s^3 2s^3 2p^9 3s^3 3p^9 4s^3 3d^{15} 4p^3$.
The electron affinity should be larger (more exothermic) for element 47 than for element 48 because adding an electron to 47 results in 3 unpaired electrons in the 4p orbitals while adding an electron to 48 requires pairing of 2 electrons in a 4 p orbital.

7.87   a) Elements with 3, 15 and 27 electrons would be the first three noble gases.

b) Element 4:

n	l	m_l	m_s
1	0	0	+1/2
1	0	0	-1/2
1	0	0	0
2	0	0	+1/2

Element 14:

n	l	m_l	m_s
1	0	0	+1/2
1	0	0	-1/2
1	0	0	0
2	0	0	+1/2
2	0	0	-1/2
2	0	0	0
2	1	-1	+1/2
2	1	0	+1/2
2	1	+1	+1/2
2	1	-1	-1/2
2	1	0	-1/2
2	1	+1	-1/2
2	1	-1	0
2	1	0	0

Element 16:

n	l	m_l	m_s
1	0	0	+1/2
1	0	0	-1/2
1	0	0	0
2	0	0	+1/2
2	0	0	-1/2
2	0	0	0
2	1	-1	+1/2
2	1	0	+1/2
2	1	+1	+1/2
2	1	-1	-1/2
2	1	0	-1/2
2	1	+1	-1/2
2	1	-1	0
2	1	0	0
2	1	+1	0
3	0	0	+1/2

Element 18:

n	l	m_l	m_s
1	0	0	+1/2
1	0	0	-1/2
1	0	0	0
2	0	0	+1/2
2	0	0	-1/2
2	0	0	0
2	1	-1	+1/2
2	1	0	+1/2
2	1	+1	+1/2
2	1	-1	-1/2
2	1	0	-1/2
2	1	+1	-1/2
2	1	-1	0
2	1	0	0
2	1	+1	0
3	0	0	+1/2
3	0	0	-1/2
3	0	0	0

c) The electrons of Morspin would probably more effectively screen a 4s electron because there are 3/2 as many electrons (in presumably the same space) as in our universe.

CHAPTER 8: FUNDAMENTALS OF CHEMICAL BONDING

8.1   The electron configuration of Be is: $1s^22s^2$. The 2s electrons will be involved in bond formation.

8.3   The electron configuration of Na is: $1s^22s^22p^63s^1$. The bond is formed by the overlap of the 3s orbitals of the 2 Na atoms.

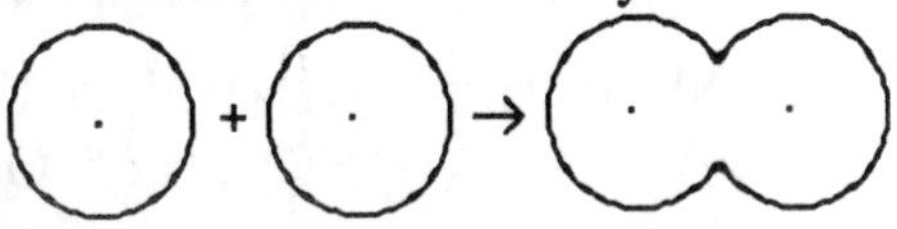

8.5   The electron configuration of Br is: $1s^22s^22p^63s^23p^64s^23d^{10}4p^5$. The bond is formed by the overlap of two 4p orbitals (one from each atom) pointing along the bond axis (line joining the nuclei).

8.7   From the list given one would conclude, based upon the electronegativity differences of the two atoms involved, that F_2 is non-polar, HF and NaH are polar, and NaF and CaO are ionic.

8.9   From smallest electronegativity difference to largest electronegativity difference:
Si—H < C—H < N—H < O—H < F—H

8.11   The following molecules are arranged in order of increasing bond polarity:
PH_3 < H_2S < NH_3 < H_2O

8.13
Compound	Valence Electrons
HBr	H : 1 Br : 7 total = 8
KBr	K : 1 Br : 7 K^+ : 1 - 1 = 0 Br^- : 7 + 1 = 8
NH_4Br	NH_4^+ : N : 5 H : 1 total: 5 + 4(1) -1 = 8 Br : 7 Br^- : 7 + 1 = 8
CBr_4	C : 4 Br : 7 total: 4 + 4(7) = 32
Br_2	Br : 7 total: (2)(7) = 14

8.15 Compound   Valence Electrons

Na$_2$SO$_4$ Na:1 each Na$^+$:1 - 1 = 0   SO$_4^{2-}$: S:6 O:6
                   total SO$_4^{2-}$: 6 + 4(6) + 2 = 32

SO$_3$ S : 6 O : 6   total: 6 + 3(6) = 24

AlCl$_3$ Al : 3 Cl : 7 total: 3 + 3(7) = 24

PCl$_3$ P : 5 Cl : 7 total: 5 + 3(7) = 26

PCl$_5$ P : 5 Cl : 7 total: 5 + 5(7) = 40

NH$_4$PCl$_6$ NH$_4^+$: N : 5 H:1 total 5 + 4(1) - 1 = 8
                  PCl$_6^-$: P : 5 Cl : 7 total: 5 + 6(7) + 1 = 48

8.17 Compound   Valence Electrons

NH$_3$ N : 5 H : 1   total: 5 + 3(1) = 8

NH$_4$NO$_3$ NH$_4^+$: N : 5 H : 1 total: 5 + 4(1) - 1 = 8
                  NO$_3^-$: N : 5 O : 6 total: 5 + 3(6) + 1 = 24

HNO$_3$ H : 1 N : 5 O : 6 total: 1 + 5 + 3(6) = 24

KNO$_2$ K$^+$: K : 1   total: 1 - 1 = 0
                  NO$_2^-$: N : 5 O : 6 total = 5 + 2(6) + 1 = 18

NO$_2$ N : 5   O : 6   total: 5 + 2(6) = 17

8.19 Compound   Polyatomic Ions Present

LiOH   hydroxide anion, OH$^-$

KH$_2$PO$_4$ dihydrogenphosphate anion, H$_2$PO$_4^-$

NaBF$_4$ fluoroborate anion (or tetrafluoroborate anion), BF$_4^-$

LiIO$_4$ periodate anion, IO$_4^-$

NaCN   cyanide anion, CN$^-$

(NH$_4$)$_2$Cr$_2$O$_7$ ammonium cation, NH$_4^+$ dichromate anion, Cr$_2$O$_7^{2-}$

8.21 Compound   Valence Electrons

NH$_4$(HSO$_4$)   NH$_4^+$: N : 5 H : 1   total: 5 + 4(1) - 1 = 8
                   HSO$_4^-$: H : 1 S : 6 O : 6 total: 1 + 6 + 4(6) + 1 = 32

NaClO$_3$ ClO$_3^-$: Cl : 7 O : 6   total 7 + 3(6) + 1 = 26

LiNO$_3$ NO$_3^-$: N : 5 O : 6   total: 5 + 3(6) + 1 = 24

(NH$_4$)$_2$CO$_3$ NH$_4^+$: N : 5 H : 1   total: 5 + 4(1) - 1 = 8
                   CO$_3^{2-}$: C : 4 O : 6   total 4 + 3(6) + 2 = 24

KPF$_6$ PF$_6^-$: P : 5 F : 7   total: 5 + 6(7) + 1 = 48

KMnO$_4$ MnO$_4^-$: Mn : 7 O : 6   total: 7 + 4(6) + 1 = 32

62

8.23

Compound	Bonding Framework
N_2O	N—N—O
O_3	O—O—O
CCl_4	$\begin{array}{c} Cl \\ | \\ Cl-C-Cl \\ | \\ Cl \end{array}$
H_2S	H—S—H
PH_3	$\begin{array}{c} H-P-H \\ | \\ H \end{array}$

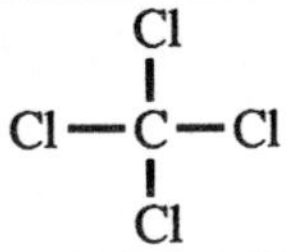

8.25   Compound   Provisional Lewis Structures and Formal Charges

N_2O

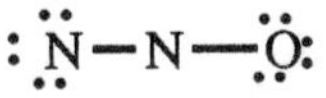

:N̈—N—Ö:

16 e⁻ - 2(2) - 2(6) = 0 $FC_{N—} = 5 - 7 = -2$

$FC_{—N—} = 5 - 2 = +3$ $FC_O = 6 - 7 = -1$

O_3

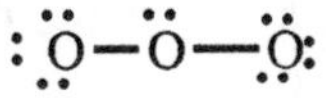

:Ö—Ö—Ö: 18 e⁻ - 2(2) - 2(6) - 2 = 0

$FC_{O—} = 6 - 7 = -1$ $FC_{—O—} = 6 - 4 = +2$

CCl_4

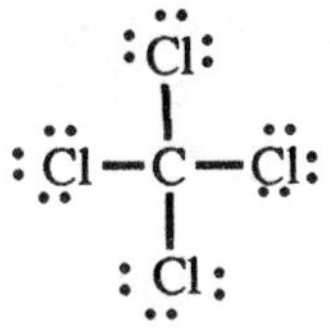

$\begin{array}{c} :\ddot{C}l: \\ | \\ :\ddot{C}l-C-\ddot{C}l: \\ | \\ :\ddot{C}l: \end{array}$

32 e⁻ - 4(2) - 4(6) = 0

$FC_C = 4 - 4 = 0$ $FC_{Cl} = 7 - 7 = 0$

H_2S

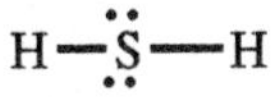

H—S̈—H   8 e⁻ - 2(2) - 4 = 0

$FC_S = 6 - 6 = 0$ $FC_H = 1 - 1 = 0$

PH_3

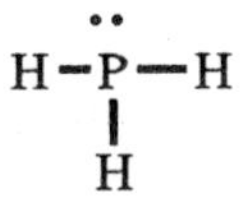

$\begin{array}{c} \overset{..}{H-P-H} \\ | \\ H \end{array}$ 8 e⁻ - 3(2) - 2 = 0

$FC_P = 5 - 5 = 0$ $FC_H = 1 - 1 = 0$

8.27   (See Problem 8.25)

Compound   Adjusted Lewis Structures

N_2O

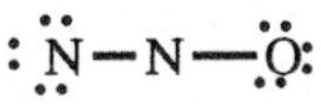

:N̈—N—Ö:

:N̈=N=Ö: ⟷ :N≡N—Ö:

$FC_{N=} = 5 - 6 = -1$ $FC_{N≡} = 5 - 5 = 0$

$FC_{=N=} = 5 - 4 = +1$ $FC_{≡N—} = 5 - 4 = +1$

$FC_{O=} = 6 - 6 = 0$ $FC_{—O} = 6 - 7 = -1$

As also described in Problem 8.26, N is limited to 4 available orbitals and, therefore, the formal charge cannot be minimized further.

(continued)

(8.27 continued)

O_3

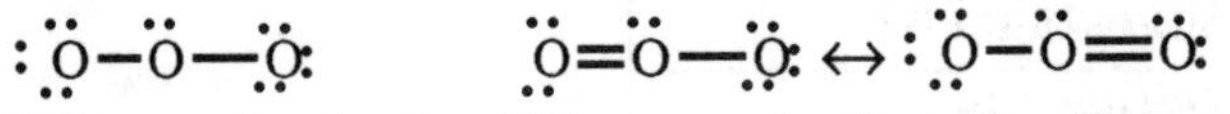

$$:\ddot{O}-\ddot{O}-\ddot{O}: \qquad \ddot{O}=\ddot{O}-\ddot{O}: \leftrightarrow :\ddot{O}-\ddot{O}=\ddot{O}:$$

$FC_{O=} = 6 - 6 = 0 \quad FC_{=O-} = 6 - 5 = +1 \quad FC_{O-} = 6 - 7 = -1$

Note: O also is limited to 4 available orbitals and the formal charges cannot be minimized further.

CCl_4 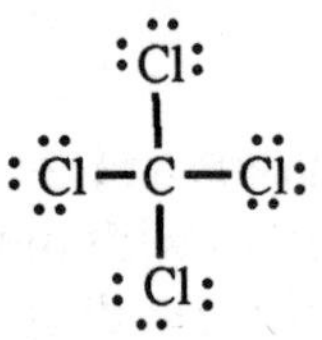   no change

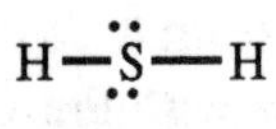

$$:\ddot{C}l-\overset{\displaystyle :\ddot{C}l:}{\underset{\displaystyle :\ddot{C}l:}{C}}-\ddot{C}l:$$

H_2S $H-\ddot{S}-H$ no change

PH_3 $H-\overset{\displaystyle \cdots}{\underset{\displaystyle H}{P}}-H$ no change

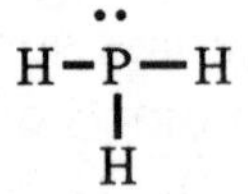

8.29   The geometry of CH_2Cl_2 is a tetrahedron with the C atom at the center of the tetrahedron and the H and Cl atoms each at a corner of the tetrahedron. This is not a "perfect" tetrahedron because the C—H bond and C—Cl bonds are not exactly the same in length or polarity. The C—H bonds result from the overlap of an sp^3 orbital on the C and the 1s orbital of the H. The C—Cl bonds result from the overlap of an sp^3 orbital on the C and a 3p orbital of the Cl.

8.31   The steps followed in the solution below are like those steps followed in Sample Problem 8-10 in the text.

For C_6H_{12} 1. covalent   2. e^- H : 1   C : 6   total $e^- = 6(4) + 12(1) = 36\ e^-$

3. 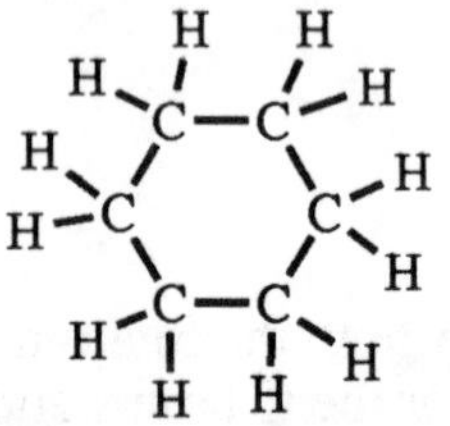 $36e^- - 18(2) = 0$

4. and 5. These steps do not change the structure given in step 3.

6. $FC_C = 4 - 4 = 0$ $FC_H = 1 - 1 = 0$

7. Based upon the FC's in step 6, there is no need to change the structure worked out in steps 3, 4, and 5.

"Ball-and-Stick" Model. See Figures 8-14 through 8-17 for examples of ball-and-stick models

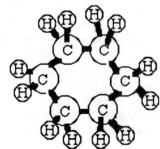

8.33   <u>Formulas</u>   <u>Use</u>

C_8H_{18} automobile fuel

C_3H_8 cooking gas

$C_{30}H_{62}$ asphalt

$C_{18}H_{38}$ lubricant

$C_{15}H_{32}$ jet fuel

8.35 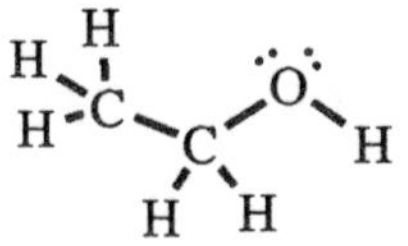   The hybridization of the two C atoms and the O atom are sp^3. For the end C, 3 sp^3 orbitals overlap with the 1s orbitals of H atoms to form C—H bonds and the fourth sp^3 orbital overlaps with an sp^3 orbital of the other C atom to form a C—C bond.

The other C atom uses 2 of its sp^3 orbitals to overlap with the 1s orbitals of H atoms to form C—H bonds and the last sp^3 orbital overlaps with an sp^3 from the O atom to form the C—O bond. Two of the other 3 sp^3 orbitals of the O atom hold lone pairs of electrons and the last sp^3 orbital overlaps the 1s orbital of H to form the O—H bond. The geometry of the orbitals is tetrahedral around the 2 C atoms and the O atom. The molecular geometry is tetrahedral around the C atoms but bent around the O atom.

The ball-and-stick model requested in this problem should reflect the similarities of methane and ethyl alcohol by showing, as shown in the structural formulas below, that ethyl alcohol's structure can be compared to methane with two hydrogens replaced: one with a methyl group, the other with a O–H group. See Figures 8-14 through 8-17 in the textbook for examples of how to draw ball-and-stick models.

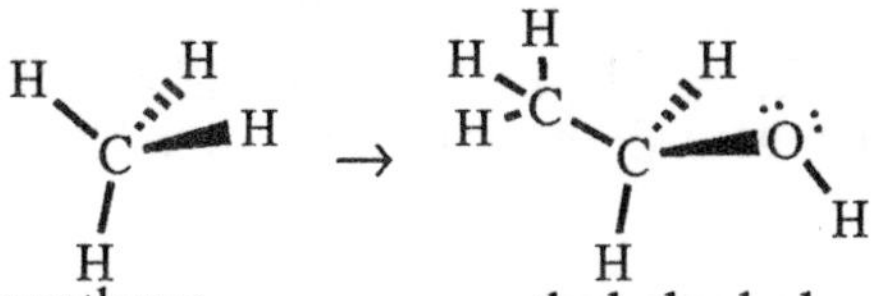

methane   ethyl alcohol

8.37 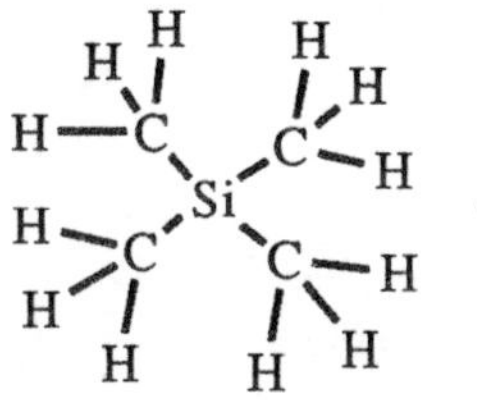   The hybridizations are sp^3 for both the Si atom and the C atoms (using 2s and 2p's for the C atoms and 3s and 3p's for the Si atom). The orbital and molecular geometries about the Si and C atoms are all tetrahedral.

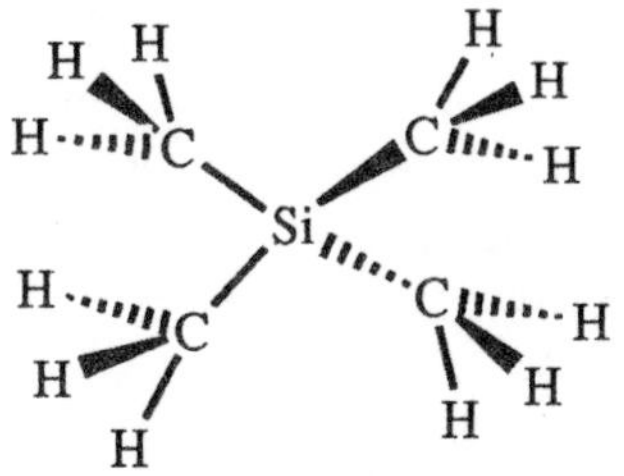

  The ball-and-stick model requested in this problem should reflect features included in the structural formula shown below. See Figures 8-14 through 8-17 in the textbook for examples of how to draw ball-and-stick models.

65

8.39 XeOF$_4$ $e^- = 8 + 6 + 4(7) = 42$

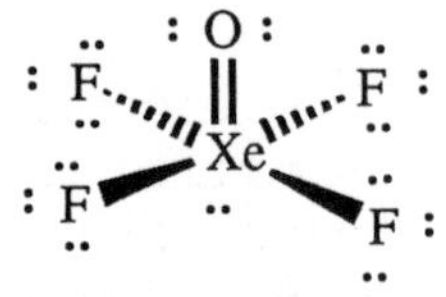

$FC_{Xe} = 8 - 8 = 0$ $FC_F = 7 - 7 = 0$ $FC_O = 6 - 6 = 0$
Steric No. = 6
Orbital Arrangement: octahedron
Molecular Geometry: square pyramid
Hybridization: sp^{3}d^2 (for Xe); no hybridization for other atoms

(Second bond for Xe=O uses overlap of 2p on O atom and 4d on the Xe atom.)

8.41

Compound:	GeF$_4$	SeF$_4$	XeF$_4$
Lewis Structure:	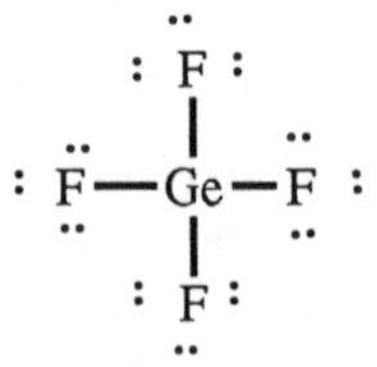	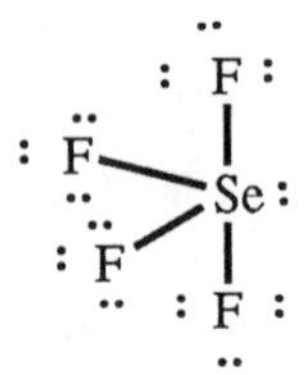	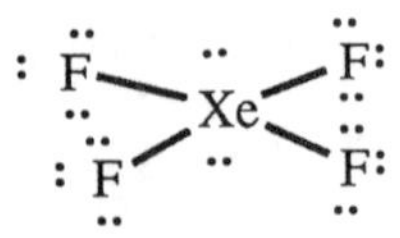
Orbital Orientation:	tetrahedron	trigonal bipyramid	octahedron
Molecular Shape:	tetrahedron	seesaw	square plane
Hybridization:	sp^3	sp^{3}d	sp^{3}d^2

8.43

Compound:	GeF$_4$	SeF$_4$	XeF$_4$
Shape:	tetrahedron	seesaw	square plane
Bond angles	109.5°	<90°* <120°*	90°

* Lone pair decreases bond angles slightly from idealized angles.

8.45

Compound	Lewis Structure	Dipole Moment
a) CF$_4$	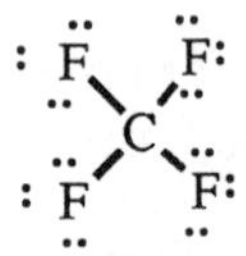	no dipole moment
b) H$_2$S	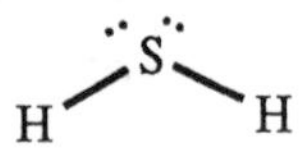	has a dipole moment
c) XeF$_2$	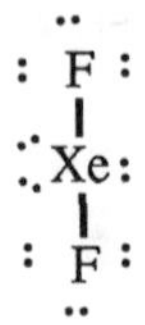	no dipole moment
d) NF$_3$		has a dipole moment

66

8.47   Compound   Lewis Structures   3-D Ball-and-Stick
       a) Cl_2O

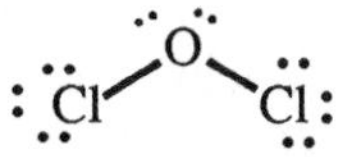

$$:\overset{..}{\underset{..}{Cl}}\,\overset{\overset{..}{O}}{}\,\overset{..}{\underset{..}{Cl}}:$$

       b) C_6H_6

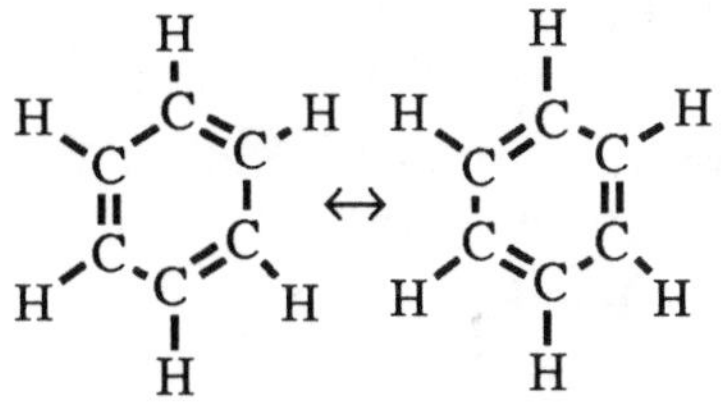

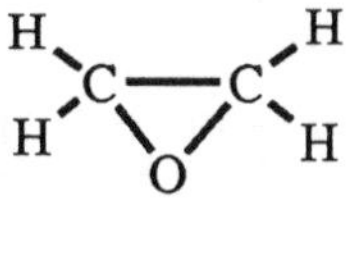

For the three-dimensional ball-and-stick structures one needs to draw 3-D drawings like those in Figures 8-14 through 8-17 in the textbook.

$\leftrightarrow$

       c) C_2H_4O

8.49

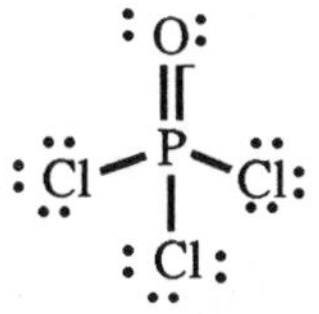

total $e^- = 5 + 6 + 3(7) = 32$
$FC_P = 5 - 5 = 0$
$FC_O = 6 - 6 = 0$
$FC_{Cl} = 7 - 7 = 0$

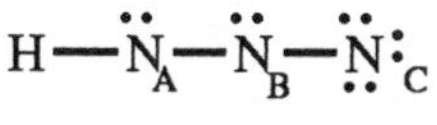

$$H-\overset{..}{\underset{A}{N}}-\overset{..}{\underset{B}{N}}-\overset{..}{\underset{C}{N}}:$$

total $e^- = 1 + 3(5) = 16$
$16 - 3(2) - 6 - 2(2) = 0$
$FC_{N_C} = 5 - 7 = -2 \quad FC_{N_{A,B}} = 5 - 4 = +1$

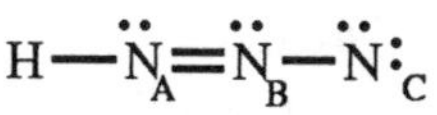

$$H-\overset{..}{\underset{A}{N}}=\overset{..}{\underset{B}{N}}-\overset{..}{\underset{C}{N}}:$$

$FC_{N_{A,B}} = 5 - 5 = 0 \quad FC_{N_C} = 5 - 5 = 0$

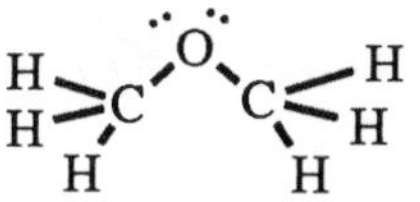

total $e^- = 6(1) + 6 + 2(4) = 20\ e^-$
$FC_O = 6 - 6 = 0 \qquad FC_C = 4 - 4 = 0$
$FC_H = 1 - 1 = 0$

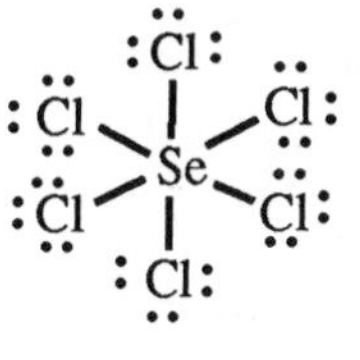

total $e^- = 6(7) + 6 = 48\ e^-$
$FC_{Se} = 6 - 6 = 0$
$FC_{Cl} = 7 - 7 = 0$

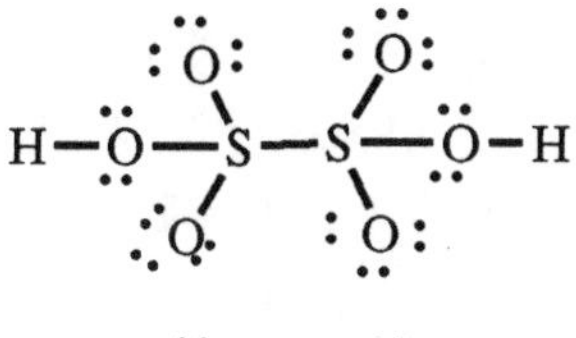

total $e^- = 2(1) + 6(6) + 2(6) = 50\ e^-$
$50e^- - 9(2) - 4(6) - 2(4) = 0$
$FC_S = 6 - 4 = +2 \qquad FC_{-O-} = 6 - 6 = 0$
$FC_{O-} = 6 - 7 = -1$

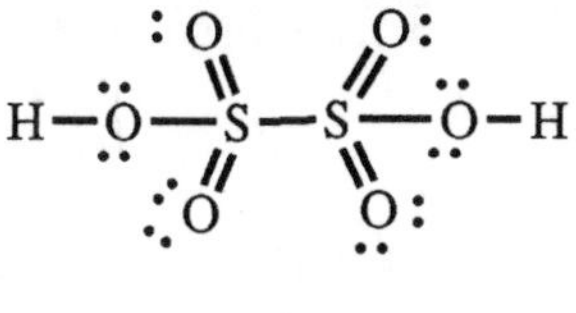

$FC_S = 6 - 6 = 0$
$FC_O = 6 - 6 = 0$

67

8.51 P may form hybrid orbitals from 3s, 3p and 3d atomic orbitals. N can form hybrid orbitals only from 2s and 2p atomic orbitals. Therefore, N is limited to 4 hybrid orbitals and to a total of 4 bonds and/or lone pairs. NF_5 would require 5 hybrid orbitals and; therefore, would not be expected to exist. PF_5 is possible because P can form more than 4 hybrid orbitals.

8.53 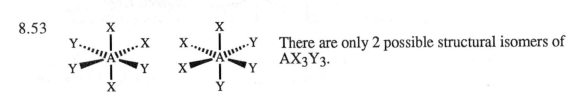 There are only 2 possible structural isomers of AX_3Y_3.

8.55

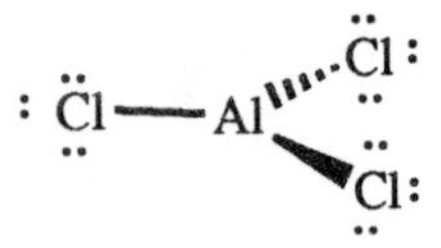

: C̈l—Al⟨ Cl̈: / Cl̈:   Bonds are formed by overlap of each of the sp^2 hybrid orbitals of the Al with a 3p orbital of Cl. The molecule is planar and the bond angles are 120°.

8.57

Molecular Geometry	Formula	Lewis Structure	
a) bent	TeF_2	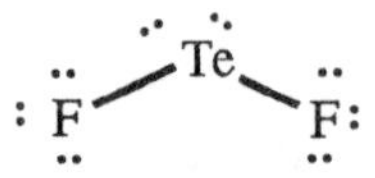 :F̈—Te—F̈:	total available e⁻ from atoms = 6 + 2(7) = 20 e⁻ e⁻ needed = 2(2) + 2(6) + 4 = 20 e⁻
b) T-shape	TeF_3^-	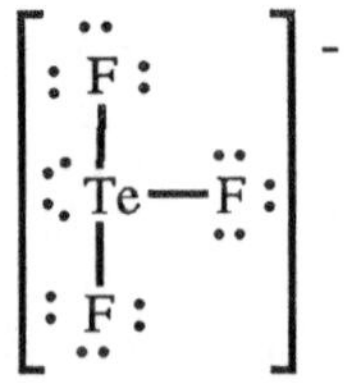 [:F̈: / :Te—F̈: / :F̈:]⁻	total available e⁻ = 6 + 3(7) + 1 = 28 e⁻ e⁻ needed = 3(2) + 3(6) + 4 = 28 e⁻
c) square pyramid	TeF_5^-	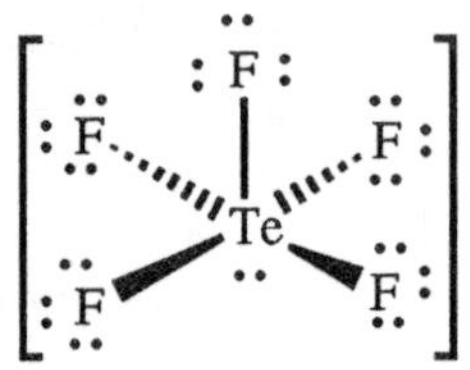 [:F̈: / :F̈—Te—F̈: / :F̈—F̈:]⁻	total available e⁻ = 6 + 5(7) + 1 = 42e⁻ e⁻ needed = 5(2) + 5(6) + 2 = 42 e⁻
d) trigonal bipyramid	TeF_5^+	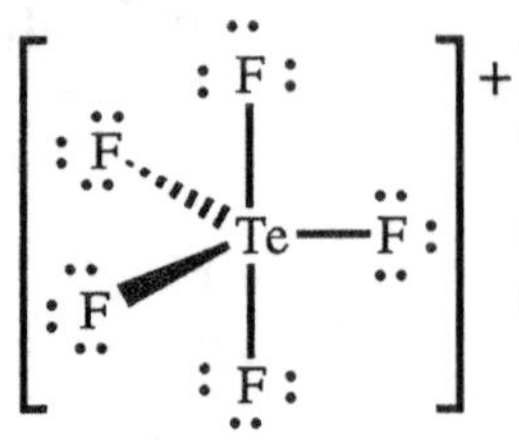 [:F̈: / :F̈—Te—F̈: / :F̈—F̈:]⁺	total available e⁻ = 6 + 5(7) - 1 = 40e⁻ e⁻ needed = 5(2) + 5(6) = 40 e⁻

(continued)

68

(8.57 continued)

e) octahedron TeF_6 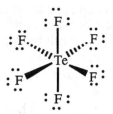   total available e⁻ from
atoms = 6 + 6(7) = 48 e⁻

e⁻ needed
= 6(2) + 6(6) = 48 e⁻

f) seesaw TeF_4 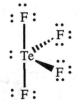   total available e⁻ from
atoms = 6 + 4(7) = 34 e⁻

e⁻ needed
= 4(2) + 4(6) + 2 = 34 e⁻

8.59   Examples of sulfur-containing compounds with various steric numbers:

Steric Number	Example	Lewis Structure
2	none	—
3	SO_2	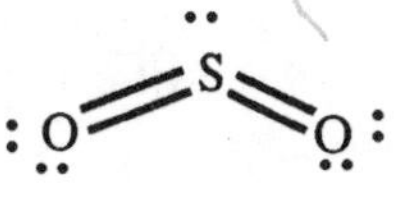
4	SO_4^{2-}	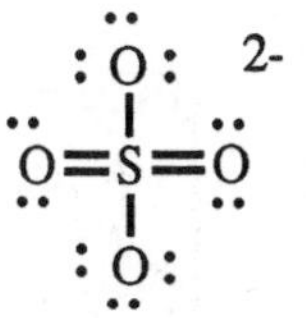
5	SF_4	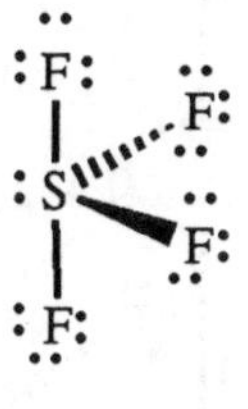
6	SF_6	

69

8.61   Compound   Lewis Structures

a) NO_2

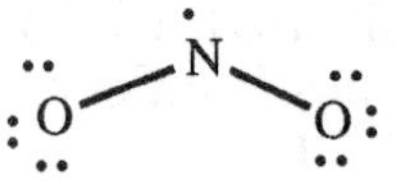

total e⁻ = 5 + 2(6) = 17 e⁻

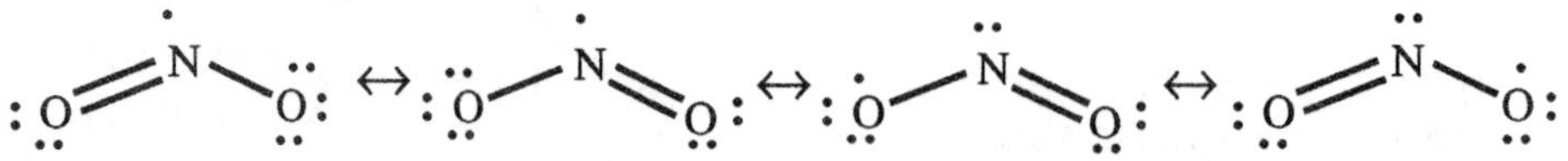

b) CH_3NCO

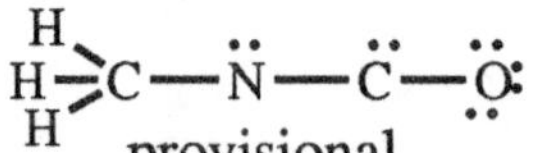

provisional

total e⁻ = 3(1) + 2(4) + 5 + 6 = 22 e⁻
22 e⁻ -6(2) - 6 - 2(2) = 0
FC_{CH_3} = 4 - 4 = 0 FC_N = 5 - 4 = +1
FC_{-C-} = 4 - 4 = 0 FC_O = 6 - 7 = -1

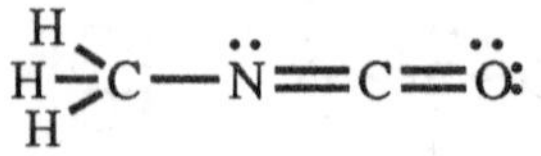

FC_N = 5 - 5 = 0 $FC_{=C=}$ = 4 - 4 = 0
FC_O = 6 - 6 = 0

c) ClO_2

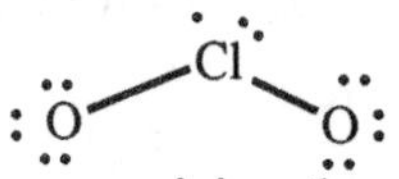

provisional

total e⁻ = 7 + 2(6) = 19 e⁻
19 e⁻ -2(2) - 2(6) - 2 - 1 = 0
FC_{Cl} = 7 - 5 = +2 FC_O = 6 - 7 = -1

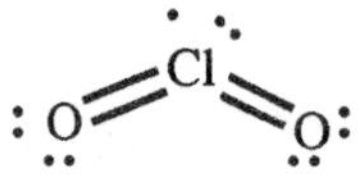

FC_{Cl} = 7 - 7 = 0 FC_O = 6 - 6 = 0

d) N_2F_4

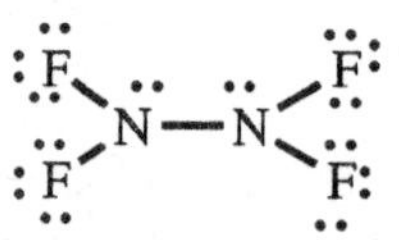

FC_F = 7 - 7 = 0 FC_N = 5 - 5 = 0

8.63

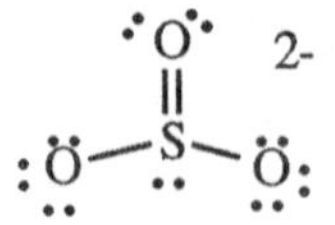

FC_S = 6 - 6 = 0
FC_{O-} = 6 - 7 = -1
$FC_{O=}$ = 6 - 6 = 0
Net Charge = 0 + 0 + 2(-1) = -2

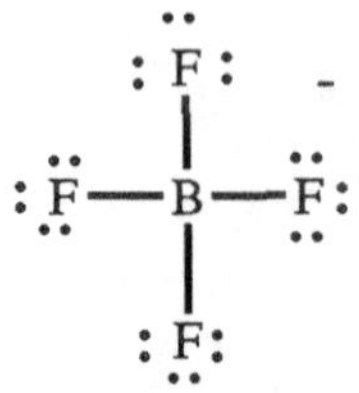

FC_B = 3 - 4 = -1
FC_F = 7 - 7 = 0

Net Charge = -1 + 4 (0) = -1

FC_{Cl} = 7 - 7 = 0
FC_O = 6 - 7 = -1

Net Charge = 1(0) + 1(-1) = -1

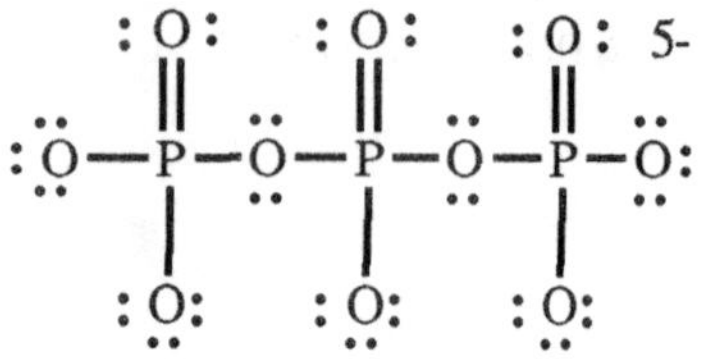

FC_P = 5 - 5 = 0 $FC_{O=}$ = 6 - 6 = 0
FC_{-O-} = 6 - 6 = 0 FC_{-O} = 6 - 7 = -1

Net Charge = 3(0) + 3(0) + 2(0) + 5(-1) = -5

70

8.65   For your ball-and-stick drawings refer to Figures 8-14 through 8-17 for examples. Then draw ball-and-stick models of orthosilicate (SiO_4^{4-}) and metasilicate ($Si_2O_6^{4-}$) that also reflect their 3-dimensional properties as reflected in Figures 8-17 and 8-18, respectively.

8.67   quartz   — silica
       mica   — silicate sheets
       zircon   — orthosilicate
       asbestos   — silicate fibers
       jade   — metasilicate

8.69

	Polyatomic Ion	Formula	Lewis Structure	

a) bromate BrO_3^-

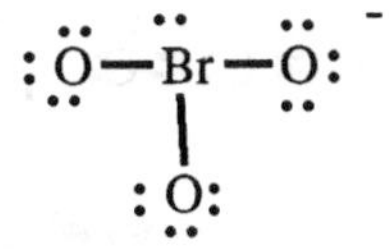

$:\!\ddot{O}\!-\!Br\!-\!\ddot{O}\!:$ $^-$
with $:\ddot{O}:$ below

$FC_{Br} = 7 - 5 = +2$
$FC_O = 6 - 7 = -1$
provisional

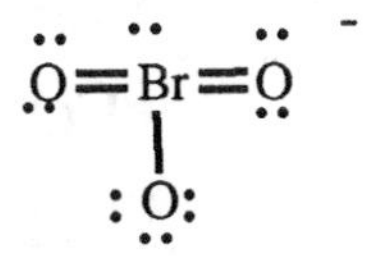

$\ddot{O}\!=\!Br\!=\!\ddot{O}$ $^-$
with $:\ddot{O}:$ below

$FC_{Br} = 7 - 7 = 0$
$FC_{O=} = 6 - 6 = 0$
$FC_{O-} = 6 - 7 = -1$
(2 other resonance structures)

b) cyanide CN^-

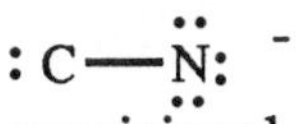

$:C\!-\!\ddot{N}:$ $^-$
provisional

$FC_C = 4 - 3 = +1$
$FC_N = 5 - 7 = -2$

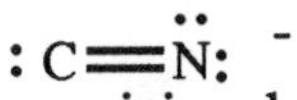

$:C\!=\!\ddot{N}:$ $^-$
provisional

$FC_C = 4 - 4 = 0$
$FC_N = 5 - 6 = -1$

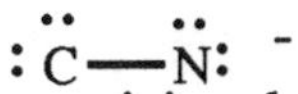

$:\ddot{C}\!-\!\ddot{N}:$ $^-$
provisional

$FC_C = 4 - 5 = -1$
$FC_N = 5 - 5 = 0$

$:C\!\equiv\!N:$ $^-$

$FC_C = 4 - 5 = -1$
$FC_N = 5 - 5 = 0$

c) nitrate NO_3^-

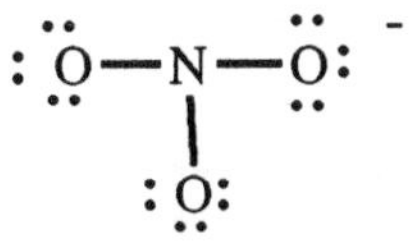

$:\ddot{O}\!-\!N\!-\!\ddot{O}:$ $^-$
with $:\ddot{O}:$ below

$FC_N = 5 - 3 = +2$
$FC_O = 6 - 7 = -1$
provisional

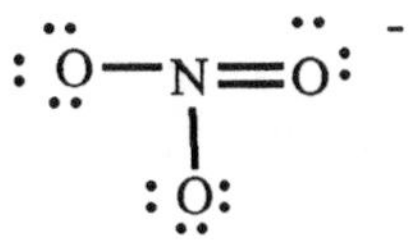

$:\ddot{O}\!-\!N\!=\!\ddot{O}:$ $^-$
with $:\ddot{O}:$ below

$FC_N = 5 - 4 = +1$
$FC_{O=} = 6 - 6 = 0$
$FC_{O-} = 6 - 7 = -1$
(N limited to 4 valence orbitals)
(2 other resonance structures)

(continued)

71

(8.69 continued)

d) nitrite NO_2^-

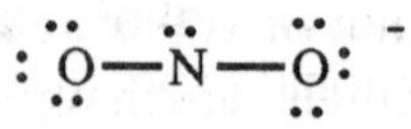

$$:\ddot{O}-\dot{N}-\ddot{O}:^-$$

$FC_N = 5 - 4 = +1$
$FC_O = 6 - 7 = -1$
provisional

$$:\ddot{O}-\dot{N}=\ddot{O}:^-$$

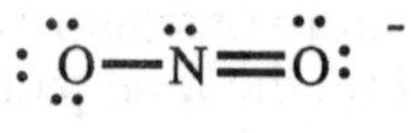

$FC_N = 5 - 5 = 0$
$FC_{O=} = 6 - 6 = 0$
$FC_{O-} = 6 - 7 = -1$
(1 other resonance structure)

e) phosphate PO_4^{3-}

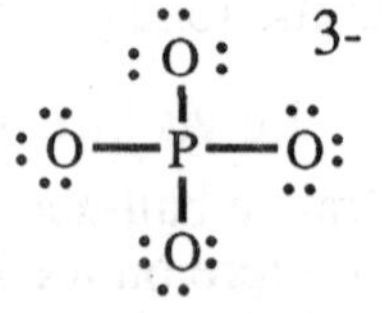

$$:\ddot{O}-\underset{\underset{:\ddot{O}:}{|}}{\overset{\overset{:\ddot{O}:}{|}}{P}}-\ddot{O}:^{3-}$$

$FC_P = 5 - 4 = +1$
$FC_O = 6 - 7 = -1$
provisional

$$:\ddot{O}-\underset{\underset{:\ddot{O}:}{|}}{\overset{\overset{:O:}{\|}}{P}}-\ddot{O}:^{3-}$$

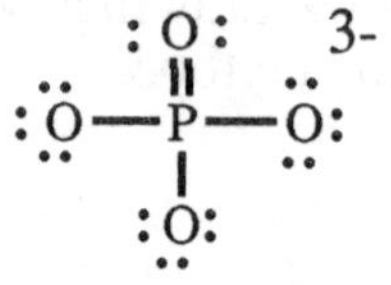

$FC_P = 5 - 5 = 0$
$FC_{O=} = 6 - 6 = 0$
$FC_{O-} = 6 - 7 = -1$
(3 other resonance structures)

f) hydrogen carbonate HCO_3^-

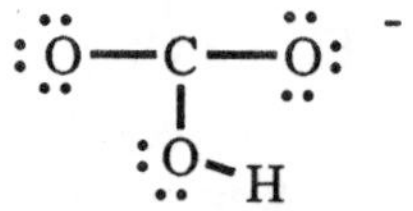

$$:\ddot{O}-\underset{\underset{:\ddot{O}_{\diagdown H}}{|}}{C}-\ddot{O}:^-$$

$FC_C = 4 - 3 = +1$
$FC_O = 6 - 7 = -1$
$FC_{-O-} = 6 - 6 = 0$
provisional

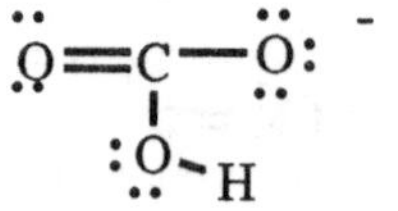

$$\ddot{O}=\underset{\underset{:\ddot{O}_{\diagdown H}}{|}}{C}-\ddot{O}:^-$$

$FC_C = 4 - 4 = 0$
$FC_{O-} = 6 - 7 = -1$
$FC_{-O-} = 6 - 6 = 0$
$FC_{O=} = 6 - 6 = 0$
(1 other resonance structure)

8.71 $SiCl_4$ tetrahedron

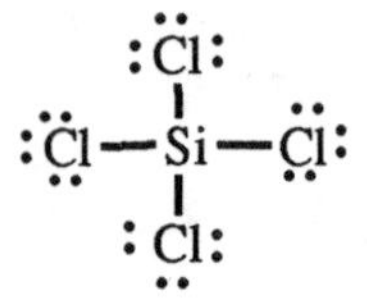

$$:\ddot{Cl}-\underset{\underset{:\ddot{Cl}:}{|}}{\overset{\overset{:\ddot{Cl}:}{|}}{Si}}-\ddot{Cl}:$$

SeF_4 seesaw

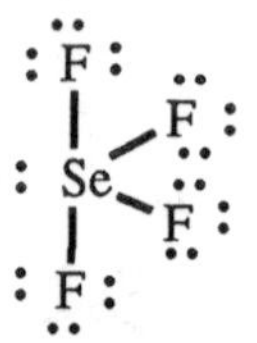

CI_4 tetrahedron

$$:\ddot{I}-\underset{\underset{:I:}{|}}{\overset{\overset{:I:}{|}}{C}}-\ddot{I}:$$

$CdCl_4^{2-}$ tetrahedron

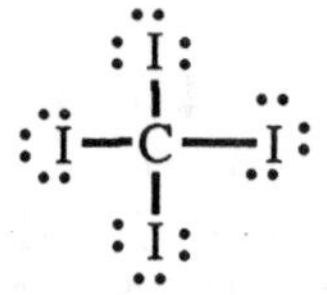

$$:\ddot{Cl}-\underset{\underset{:Cl:}{|}}{\overset{\overset{:\ddot{Cl}:}{}}{Cd}}-\ddot{Cl}:^{2-}$$

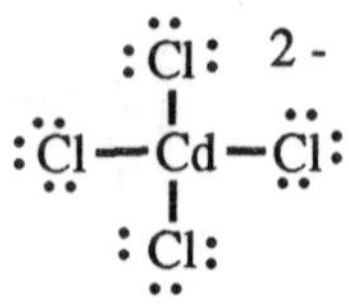

XeF_4 square plane

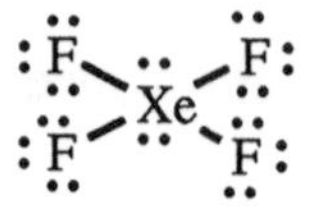

$BeCl_4^{2-}$ tetrahedron

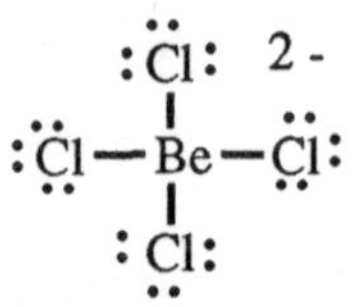

$$:\ddot{Cl}-\underset{\underset{:Cl:}{|}}{\overset{\overset{:\ddot{Cl}:}{}}{Be}}-\ddot{Cl}:^{2-}$$

8.73 The hybridization of the central atom (N or P) of each of these molecules is sp^3 resulting in an orbital orientation of 109.5°. Each molecule has a lone pair of electrons in one of the sp^3 orbitals which repels the bonded electrons and atoms forcing them closer together and decreasing their bond angles. The sp^3 hybrid orbitals of N (from 2s and 2p atom orbitals) are smaller than the sp^3 of P (from 3s and 3p atom orbitals). Therefore, the lone pair of electrons in NH_3 is held more closely to the N atom and occupies less space. The repulsion of the bonding electrons is less which results in a smaller decrease in bond angles for NH_3 than for PH_3. The large Cl atoms (with 3 lone pairs) repel each other more strongly than H atoms and, therefore, counteract the lone pair repulsion and result in less decrease of the bond angles.

8.75 H_3CNH_2

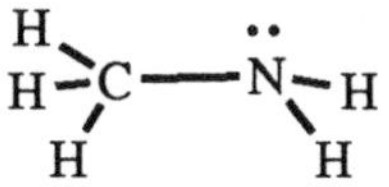

Draw a ball-and-stick model of the Lewis structure on the left. For examples of ball-and-stick drawings see Figures 8-14 through 8-17 in the textbook.

8.77

a) H_2O $FC_O = 6 - 6 = 0$

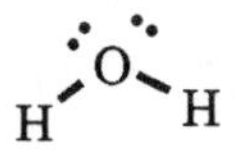

b) C_2H_2 $H-C\equiv C-H$ $FC_C = 4 - 4 = 0$

c) HCN $H-C-\ddot{N}:$ $FC_C = 4 - 2 = +2$
                           $FC_N = 5 - 7 = -2$
                           provisional

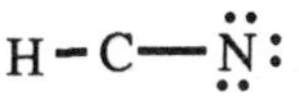

                  $H-C\equiv N:$ $FC_C = 4 - 4 = 0$
                           $FC_N = 5 - 5 = 0$

d) CH_2O $FC_C = 4 - 3 = +1$
                           $FC_O = 6 - 7 = -1$
                           provisional

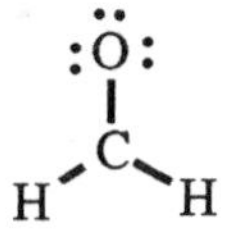

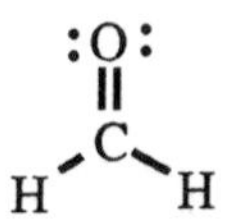

                      $FC_C = 4 - 4 = 0$
                      $FC_O = 6 - 6 = 0$

e) H_2S $FC_S = 6 - 6 = 0$

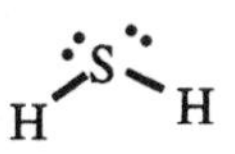

f) H_3CCN $FC_{H_3C-} = 4 - 4 = 0$
                           $FC_{-C-} = 4 - 2 = +2$
                           $FC_N = 5 - 7 = -2$
                           provisional

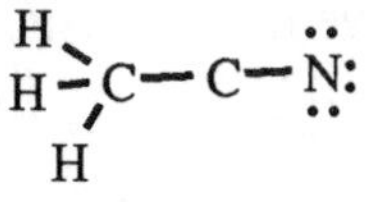

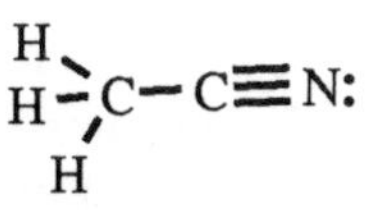

                      $FC_{H_3C-} = 4 - 4 = 0$
                      $FC_{-C\equiv} = 4 - 4 = 0$
                      $FC_N = 5 - 5 = 0$

g) NH_3 $H-\ddot{N}-H$ $FC_N = 5 - 5 = 0$

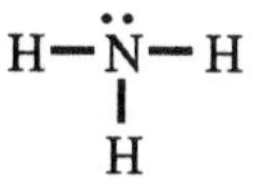

73

8.79 SF₂ ... $FC_S = 6 - 6 = 0$
$FC_F = 7 - 7 = 0$

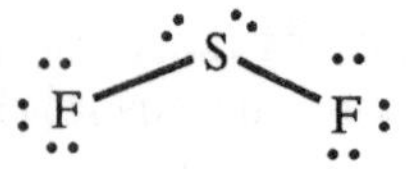

SSF₂ ... $FC_{S\_} = 6 - 7 = -1$
$FC_{\_S} = 6 - 5 = +1$
$FC_F = 7 - 7 = 0$
provisional

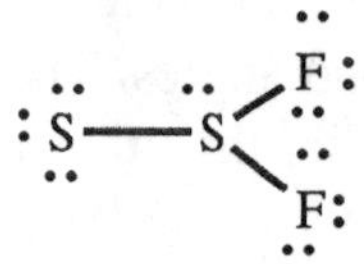

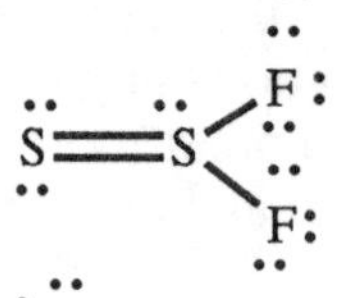

... $FC_{S=} = 6 - 6 = 0$
$FC_{\_S=} = 6 - 6 = 0$
$FC_F = 7 - 7 = 0$

FSSF ... $FC_S = 6 - 6 = 0$
$FC_F = 7 - 7 = 0$

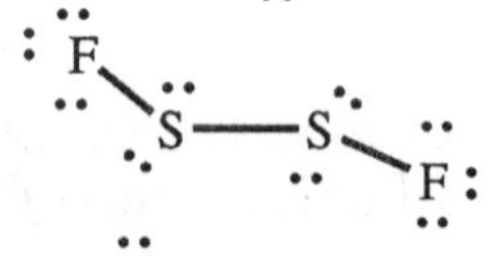

F₃SSF ... $FC_F = 7 - 7 = 0$
$FC_{\_S\_} = 6 - 6 = 0$
$FC_{\equiv S\_} = 6 - 6 = 0$

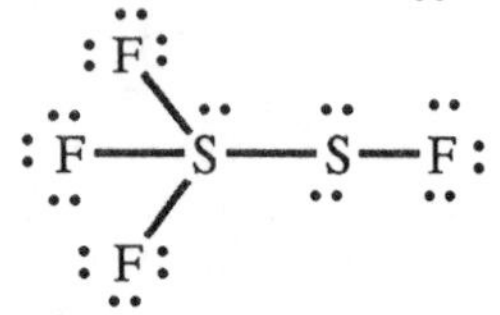

SF₄ ... $FC_S = 6 - 6 = 0$
$FC_F = 7 - 7 = 0$

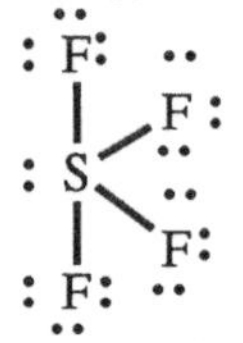

F₅SSF₅ ... $FC_F = 7 - 7 = 0$
$FC_S = 6 - 6 = 0$

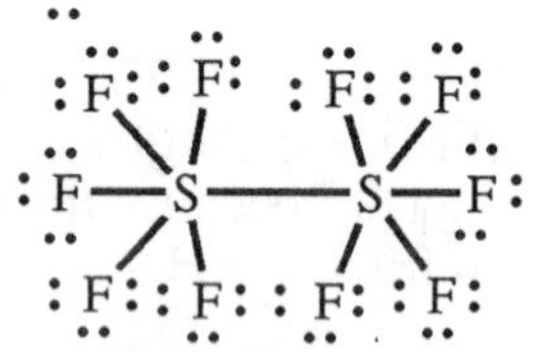

SF₆ ... $FC_F = 7 - 7 = 0$
$FC_S = 6 - 6 = 0$

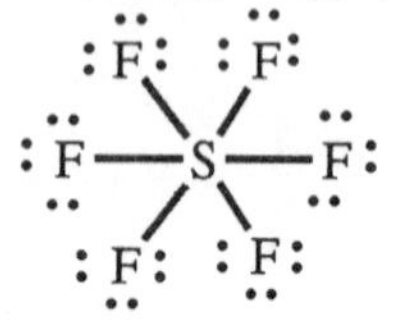

8.81 Silica has the empirical formula of SiO_2 and consists of a continuous network of Si—O bonds. Each Si is bonded to 4 O atoms. Each O atom is bonded to 2 Si atoms. Silica has no ionic units. Orthosilicates have discrete SiO_4^{4-} ionic units. Each Si is bonded to 4 outer oxygen atoms. Metasilicates have the empirical formula of SiO_3^{2-} and consist of linear chains or ring systems of Si—O—Si linkages with negative charges. Each Si atom is bonded to 2 outer O atoms and 2 inner O atoms. (Also see Problem 8.65.)

8.83 The hybridization of the central atom O of water molecules is sp^3 resulting in an orbital orientation of 109.5°. Each molecule has a lone pair of electrons in one of the sp^3 orbitals which repels the bonded electrons and atoms forcing them closer together and decreasing their bond angles. The larger atoms (orbitals) of sulfur allow it to bond to two hydrogens in H_2S without hybridizing. The repulsion due to like charges of the bonding hydrogens results in a small increase in bond angles for H_2S, from the 90° of p orbitals to the observed 92.2°.

CHAPTER 9: CHEMICAL BONDING: MULTIPLE BONDS

9.1 The outer C atoms have sp^3 hybrid orbitals and the center C atom has sp^2 hybrid orbitals. The C—C bonds are formed by overlap of an outer C sp^3 orbital with a center C sp^2 orbital. The C—H bonds are formed by overlap of outer C sp^3 orbitals with H 1s orbitals. The center C atom forms a σ bond with the O atom by end-to-end overlap of an sp^2 orbital of the C with a p orbital of the O. A second C—O bond, a π bond, is formed by the side-by-side overlap of the unused p orbital of the C and a p orbital of the O. Two lone pairs of electrons on the O atom occupy the 2s orbital and the other 2p orbital. The sketches of all the bonding orbitals should include the above information and the structural formula:

See Figures 8-1 through 8-5 for examples of sketches of the various bonding orbitals.

9.3 (See Sample Problem 9-1.) The hybridization at both the C and N atoms is sp^2 with "ideal" angles of 120°. The H—C—H angle is less than 120° because the electrons of the π bond occupy space and force the H—C—H angle to be smaller. These π bonds also tend to force the C=N—H to open to greater than 120°, but this is counteracted by the lone pair of electrons on the N atom. The actual angle depends on which of these effects is greater.

9.5 H—N < N≡N < C≡N < N—N < Cl—N
H—N is the shortest because the bond is polar and the small 1s orbital of H ($n = 1$) is used in forming the bond. The N≡N and C≡N with $n = 2$ for both N and C are longer than H—N but shorter than N—N and Cl—N because they possess triple bonds. Because the effective nuclear charge of N is greater than C, N≡N is shorter than C≡N. The length of N—N is shorter than Cl—N because $n = 2$ for N but $n = 3$ for Cl meaning Cl uses larger orbitals for bonding.

9.7

Table 9-2	N—N	< Cl—N	< H—N	< C≡N	< N≡N
energy (kJ/mol):	165	315	390	890	942.7

Cl—N is stronger than N—N because of greater electronegativity difference. H—N is stronger than Cl—N because $n = 1$ for H versus $n = 3$ for Cl. C≡N is stronger than H—N because more electrons are shared between C≡N than for H—N. N≡N is stronger than C≡N because the effective nuclear charge is greater for N than for C.

9.9 N_2 + $2 O_2$ →   (Lewis structure of N_2O_4)

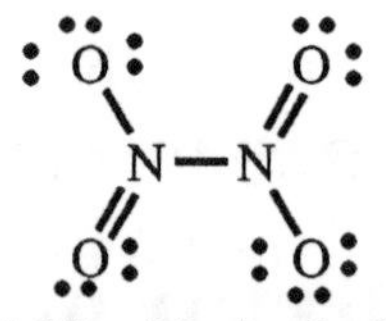

1 N≡N triple bond   2 O=O double bond

1 N—N single bond
2 N—O single bonds
2 N=O double bonds

$\Delta E_{reaction} = \Sigma BE_{(bonds\ broken)} - \Sigma BE_{(bonds\ formed)}$

$= [(2\ mol\ N≡N)(942.7\ kJ/mol\ N≡N) + (2\ mol\ O=O)(493.6\ kJ/mol\ O=O)]$
$- [(1\ mol\ N—N)(165\ kJ/mol\ N—N) + (2\ mol\ N—O)(200\ kJ/mol\ N—O) +$
$(2\ mol\ N=O)(607\ kJ/mol\ N=O)] = 2873\ kJ - 1779\ kJ = 1094\ kJ$
This energy change is per mole of N_2O_4.

9.11 $2\ CH_3—CH=CH_2 + 2\ NH_3 + 3\ O_2 → 2\ H_2C=CH—C≡N + 6\ H_2O$

12 C—H single bonds   6 C—H single bonds
2 C—C single bonds   2 C=C double bonds
2 C=C double bonds   2 C—C single bonds
6 N—H single bonds   2 C≡N triple bonds
3 O=O double bonds   12 H—O single bonds

$\Delta E = [(12\ mol\ C—H)(415\ kJ/mol\ C—H) + (2\ mol\ C—C)(345\ kJ/mol\ C—C)$
$+ (2\ mol\ C=C)(615\ kJ/mol\ C=C) + (6\ mol\ N—H)(390\ kJ/mol\ N—H)$
$+ (3\ mol\ O=O)(493.6\ kJ/mol\ O=O)] - [(6\ mol\ C—H)(415\ kJ/mol\ C—H)$
$+ (2\ mol\ C=C)(615\ kJ/mol\ C=C) + (2\ mol\ C—C)(345\ kJ/mol\ C—C)$
$+ (2\ mol\ C≡N)(890\ kJ/mol\ C≡N) + (12\ mol\ H—O)(460\ kJ/mol\ H—O)]$
$= 10721\ kJ - 11710\ kJ = -989\ kJ$

9.13   (MO diagram: σ^* empty, σ filled)

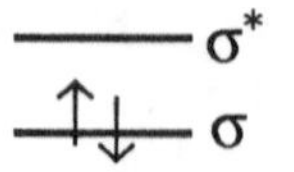

orbitals generated by 3s atomic orbitals

Na_2 has 2 valence electrons. The Bond Order is $(2 - 0)/2 = 1$ since the 2 valence electrons occupy a bonding orbital and Na_2 is stable.

9.15   (MO diagram: σ^* empty, π^* filled, π filled, σ filled)

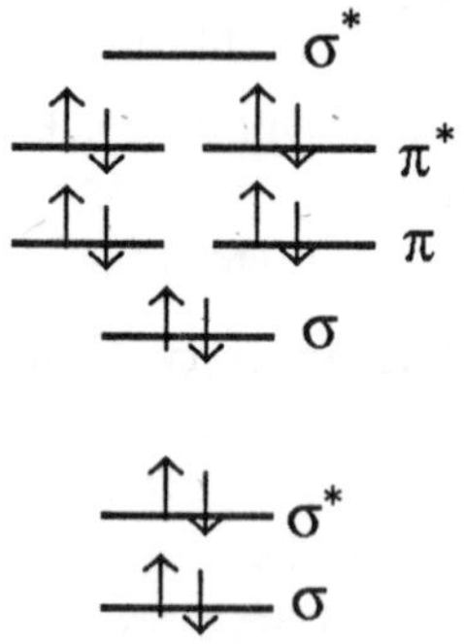

orbitals generated by 3p of Cl and 2p of O

(Lewis structure of ClO^-)

(MO diagram: σ^* filled, σ filled)

orbitals generated by 3s of Cl and 2s of O

For ClO^-

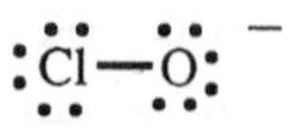

Bond Order = $(8 - 6)/2 = 1$

76

9.17 NO: $(\sigma_s)^2 \ (\sigma_s^*)^2 \ (\sigma_p)^2 \ (\pi_x)^2 \ (\pi_y)^2 \ (\pi_x^*)^1$ B.O. = (8 - 3)/2 = 2.5

       NO$^+$: $(\sigma_s)^2 \ (\sigma_s^*)^2 \ (\sigma_p)^2 \ (\pi_x)^2 \ (\pi_y)^2$ B.O. = (8 - 2)/2 = 3

       NO$^-$: $(\sigma_s)^2 \ (\sigma_s^*)^2 \ (\sigma_p)^2 \ (\pi_x)^2 \ (\pi_y)^2 \ (\pi_x^*)^1 \ (\pi_y^*)^1$ B.O. = (8 - 4)/2 = 2

NO has one unpaired electron and NO$^-$ has two unpaired electrons and both, therefore, show magnetism. All of the electrons of NO$^+$ are paired and, therefore, it is not magnetic.

9.19 :Ö–C̈–C—C̈–Ö: valence electrons = 3 x 4 (from C) + 2 x 6 (from O) = 24 e$^-$

                       FC$_O$ = 6 - 7 = -1   FC$_{C:}$ = 4 - 4 = 0   FC$_C$ = 4 - 2 = +2

:Ö=c=c=c=Ö: FC$_O$ = 6 - 6 = 0   FC$_C$ = 4 - 4 = 0

sp hybrid orbitals are used by the C atoms along with $2p_z$ atomic orbitals of O atoms to form a totally linear molecule. The $2p_x$ and $2p_y$ atomic orbitals of the C and O atoms form 2 sets of delocalized π orbitals. Make the sketches that this question calls for similar to the σ bonding shown in Figure 9-24 (except it needs to include 5 atoms) and, as separate sketches, the π bonding shown in Figure 9-18 for the π bonding in first the y plane, then the z plane (2 of the π bonds will be in each plane).

9.21 HN$_3$ H–N̈–N̈–N̈: valence electrons

           FC$_{-N-}$ = 5 - 4 = +1   = 1 x 1 (from H) + 3 x 5 (from N) = 16 e$^-$

           FC$_{-N}$ = 5 - 7 = -2   16 e$^-$ - 6 e$^-$ - 6 e$^-$ - 4 e$^-$ = 0 e$^-$

           H–N̈=N=N̈:

           FC$_{-N=}$ = 5 - 5 = 0   A triple bond between the end nitrogen atom and the

           FC$_{=N=}$ = 5 - 4 = +1   central nitrogen atom is not possible because the

           FC$_{N=}$ = 5 - 6 = -1   central N is limited to 4 orbitals.

           H–N̈=N=N̈: ↔ H–N̈–N≡N:
                 1   2   3

N$_1$: has sp^2 hybridization and H—N=N has a bond angle of ~ 120°.
N$_2$: has sp hybridization and N=N=N has a bond angle of 180°.
There is a localized π orbital set for N$_2$=N$_3$ and a delocalized π orbital set for all three of the 3 N atoms.

9.23

All three C atoms use sp^2 hybridization to form 3 σ bonds (with H and C and one with O). The geometry of the molecule at each C atom is triangular with bond angles ~ 120°. There is a set of delocalized π orbitals extending over the 3 C atoms and the O atom as each C atom has one unused 2p orbital and the O atom has 1 unused 2p orbital.

9.25 a) All O's in the OH groups are sp^3 hybridized. The O in the ring is also sp^3 hybridized. The doubly bonded O is not hybridized. The 3 carbon atoms that contain π bonds are sp^2 hybridized. The other 3 C atoms are sp^3 hybridized.

b) In the formula of vitamin C there are 4 delocalized electrons, corresponding to the two π bonds shown in the Lewis structure of the molecule. Delocalization extends over the three sp^2 hybridized C atoms and the unhybridized O atom adjacent to them.

9.27 SO_2

18 valence electrons - 4 e^- (in σ bonds) - 12 e^- (on O) - 2 e^- (on S) = 0 $FC_O = 6 - 7 = -1$ $FC_S = 6 - 4 = +2$

The steric number of S is 3; the molecule is bent; S uses sp^2 hybridization to form σ bonds with the O atoms by overlap with 2p orbitals of O and to hold a lone pair of electrons. The O—S—O bond angle should be about 120°. The unused 3p orbital on S and the 2p orbitals on the O atoms which are perpendicular to the plane of the molecule overlap to form delocalized π orbitals. There are 4 valence electrons in the 2 σ bonds and 2 valence electrons as a lone pair localized in the sp^2 hybrid orbital on the S. Four electrons are in the 2s orbitals on the outer O atoms and 4 more occupy the O 2p orbitals that lie in the plane of the molecule. This leaves 4 electrons to be placed in the delocalized π orbitals. The antibonding π orbital is empty.

SO_3

24 valence electrons - 6 e^- (in σ bonds) - 18 e^- (on O) = 0

$FC_O = 6 - 7 = -1$ $FC_S = 6 - 3 = +3$

$FC_O = 0$ $FC_S = 0$

The steric number of S is 3; the molecule is triangular; S uses sp^2 hybridization to form σ bonds with the O atoms by overlap with 2p orbitals of O atoms. The O—S—O bond angles should be 120°. The unused 3p orbital on S and the 2p orbitals on the O atoms which are perpendicular to the plane of the molecule overlap to form delocalized π orbitals. There are 6 valence electrons in the 3 σ bonds. There are 6 electrons in the 2s orbitals on the outer O atoms and 6 more occupy the O 2p orbitals that lie in the plane of the molecules. This leaves 6 electrons to be placed in the delocalized π orbitals. The antibonding π orbital is empty.

9.29

2- provisional

34 valence e^- - 10 e^- (in σ bonds) - 24 e^- (on O) = 0

$FC_O = 6 - 7 = -1$ $FC_C = 4 - 3 = +1$

All 6 atoms contribute p orbitals to the delocalized π system.

9.31

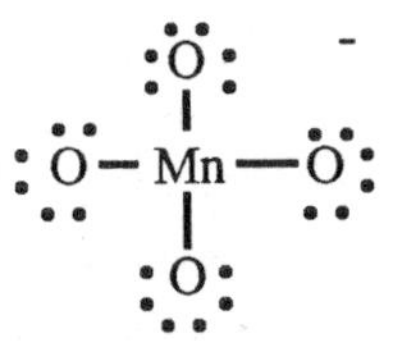

:O:⁻

:O—Mn—O:

:O:

32 valence e⁻ - 8 e⁻ (in σ bonds) - 24 e⁻ (on O) = 0

$FC_{Mn} = 7 - 4 = +3$ $FC_O = 6 - 7 = -1$

provisional

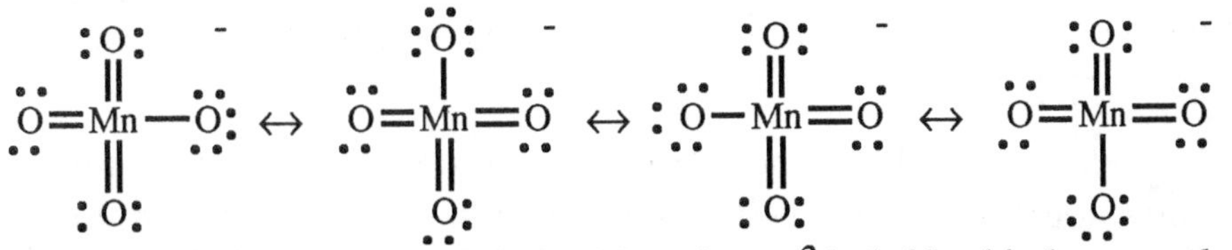

O=Mn—O: ↔ :O—Mn=O ↔ :O—Mn=O ↔ O=Mn=O

The MnO_4^- is tetrahedral with the Mn using sp^3 hybrid orbitals to overlap 2p orbitals of O to form σ bonds and 3d orbitals to overlap other 2p orbitals of O to form π bonds.

9.33   Silicon doped with antimony would be an n-type semiconductor, because Sb has more valence electrons than Si. Its band gap diagram would look like that on the top left portion of Figure 9-36.

9.35   The double bonded compounds have a π bond with electron density above and below the plane of the molecule. In order to rotate about the internuclear axis to convert between the *cis* and *trans* forms of the 1, 2-dichloroethylene, the π bond must be broken and reformed. This would require approximately 279 kJ/mol (615 kJ/mol - 345 kJ/mol, difference in strengths of C=C and C—C). The 1, 2-dichloroethane molecule has only a σ C—C bond with the electron density symmetrical around the internuclear axis. No bond breakage is necessary for rotation about this C—C bond.

9.37   There are 60 sp^2-hybridized carbon atoms in buckminsterfullerene, leaving one p orbital per atom available to form π orbitals. Thus there are 60 π orbitals, 30 of which are bonding and 30 of which are antibonding. The Lewis structure contains 30 π bonds and has the resonance structures, indicating that all 30 of these orbitals are delocalized.

9.39   a) Ge doped with S is an n-type semiconductor.
b) As doped with Si is a p-type semiconductor
c) Si doped with In is a p-type semiconductor.

9.41

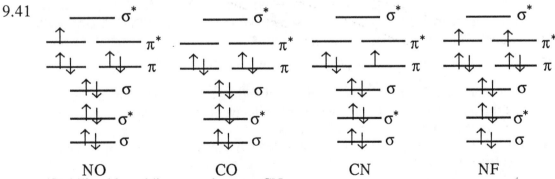

Stabilized by adding one electron: CN
Stabilized by removing one electron: NO, NF

9.43

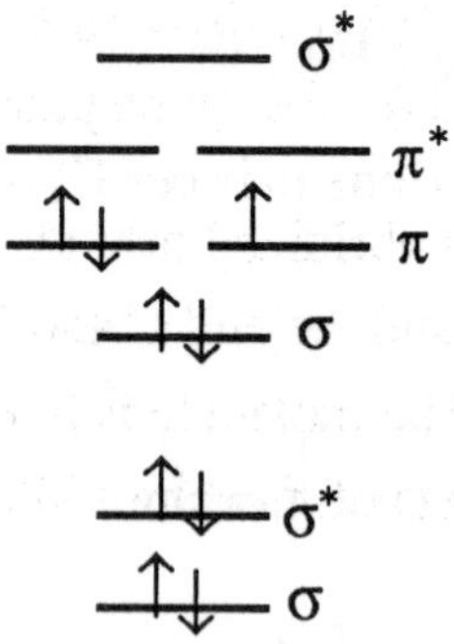

The bonding in CN involves the net effective bonding of one σ bond and two and a half π bonds. These bonds will look like the central σ and π bonds shown in Figure 9-8.

CN

9.45

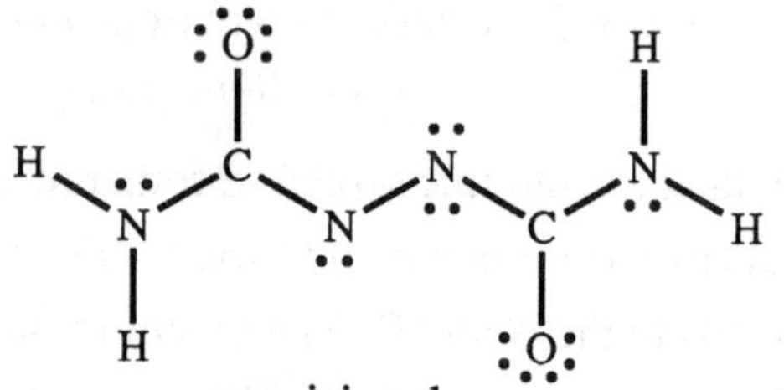

provisional

valence e⁻
= 4 (1 e⁻ per H) + 4 (5 e⁻ per N)
  + 2 (4 e⁻ per C) + 2(6 e⁻ per O) = 44 e⁻

44 e⁻ - 22 (2 per σ bond) - 12 (6 per O)
  - 8 (2 per N) - 2 (on an inner N) = 0

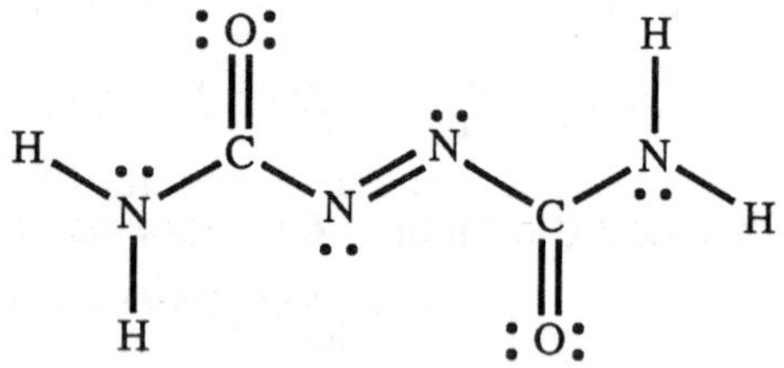

Both C atoms are triangular (bond angles ~ 120°) and the hybridization is sp². The two outer N atoms are tetrahedral and hybridization of sp³. The two inner N's are bent with sp² hybridization.

9.47 ethanol

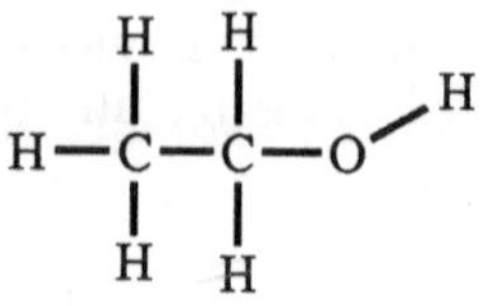

5 C—H bonds	5(415 kJ/mol) = 2075 kJ/mol
1 H—O bond	1(460 kJ/mol) = 460 kJ/mol
1 C—C bond	1(345 kJ/mol) = 345 kJ/mol
1 C—O bond	1(360 kJ/mol) = 360 kJ/mol
	= 3240 kJ/mol

dimethyl ether

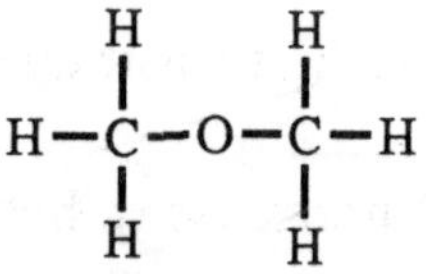

6 C—H bonds	6(415 kJ/mol) = 2490 kJ/mol
2 C—O bonds	2(360 kJ/mol) = 720 kJ/mol
	= 3210 kJ/mol

The ethanol is 30 kJ/mol more stable than the dimethyl ether.

9.49 The energy gap between bonding and antibonding orbitals would be associated with the threshold energy. No, they would not all have the same kinetic energy. The bands are made up of a series of orbitals spread over a continuous range of energies. Therefore, the electrons that filled the band have a range of energies, and ejected electrons would also have a range of energies.

9.51 a) CCCO $\overset{..}{\underset{..}{C}}-\overset{..}{\underset{..}{C}}-\overset{..}{\underset{..}{C}}-\overset{..}{\underset{..}{O}}:$ valence $e^- = 3(4 \text{ per C}) + 1(6 \text{ per O}) = 18\ e^-$

FC +1   0   0   -1 $18\ e^- - 6\ e^-$ (for 3 σ orbitals) $- 6\ e^-$ (lone pairs

on O) $- 6\ e^-$ (lone pair per C) $= 0$

$\overset{.}{\underset{.}{C}}=C=C=\overset{..}{\underset{..}{O}}$ The 2 middle C's use sp hybridized orbitals to form

σ bonds with each other and the end C and O.

The end C and O use p atomic orbitals to form σ bonds. The molecule is linear. The 2 sets of p orbitals perpendicular to the σ bonds form 2 sets of π orbitals which extend over the whole molecule.

b) HNCO $H-\overset{..}{\underset{..}{N}}-C-\overset{..}{\underset{..}{O}}:$ valence $e^- = 1(1 \text{ per H}) + 1(5 \text{ per N}) + 1(4 \text{ per C}) +$

FC 0   -1   +2   -1 $1(6 \text{ per O}) = 16\ e^-$

$16\ e^- - 6\ e^-$ (for 3 σ orbitals) $- 6\ e^-$ (lone pairs on O)

$- 4\ e^-$ (lone pair per N) $= 0$

$H-\overset{..}{N}=C=\overset{..}{\underset{..}{O}}:$ The H uses a 1s atomic orbital to form a σ bond

to N. The N uses sp^2 hybridized orbitals to form σ bonds to H and C. The C atom uses sp hybridized orbitals to form σ bonds to the N and O. The O atom uses a p atomic orbital to form a σ bond with C. One π system extends over the N, C and O atoms and the other only over the C and O atoms.

c) OCS $:\overset{..}{\underset{..}{O}}-C-\overset{..}{\underset{..}{S}}:$ valence $e^- = 1(6 \text{ per O}) + 1(4 \text{ per C}) + 1(6 \text{ per S})$

FC -1   +2   -1 $= 16\ e^-$

$16\ e^- - 4\ e^-$ (for 2 σ orbitals) $- 6\ e^-$ (lone pairs on O)

$- 6\ e^-$ (lone pairs on S) $= 0$

$:\overset{..}{O}=C=\overset{..}{\underset{..}{S}}:$ The bonding is like CO_2. The carbon atom uses sp

hybridized orbitals to form σ bonds by overlapping p_z orbitals on O and S atoms. The molecule is linear. Two sets of π orbitals (π_x set and π_y set) are each delocalized over all three atoms.

d) HCCCCH $H-\overset{..}{C}-\overset{..}{C}-\overset{..}{C}-\overset{..}{C}-H$

FC 0   0   0   0   0   0   valence $e^- = 2(1 \text{ per H}) + 4(4 \text{ per C}) = 18\ e^-$

$18\ e^- - 10\ e^-$ (for 5 σ orbitals) $- 8\ e^-$ (lone pair on each C) $= 0$

$H-C\equiv C-C\equiv C-H$

FC 0   0   0   0   0   0   The C atoms use sp hybrid orbitals to

form σ bonds with other C atoms or with 1s orbitals of H. The unused p orbitals on the C atoms form 2 sets of π orbitals extending over the carbon atoms.

9.53   The bond energy for the H—Cl bond was measured for the compound HCl, the only example of the H—Cl bond. The energy for the H—C bond is an average of measured energies for this bond in many different compounds containing this bond. Also, measurements on HCl do not involve any other bonds while most compounds containing H—C bonds have other kinds of bonds present which must be accounted for in the measurement.

81

9.55 a) The bonding in CO is: $:C\equiv O:$. The lone pair on the carbon will occupy an sp hybrid orbital when it bonds to a metal. The drawing of the bonding of the sp orbital of CO and the d_{z^2} orbital of a metal would look like other σ bonds (such as the one shown in Figure 9-1 in the textbook) except the shapes of the overlapping orbitals would reflect the shapes of the sp and d_{z^2} orbitals.

b) The π bonding between the π^* orbital of CO and the d_{xz} orbital of the metal would look like other π bonds such as the one shown in Figure 9-31 in the textbook.

9.57 The F_2 bond is a single bond; the O_2 bond is a double bond; and the N_2 bond is a triple bond. Each succeeding extra pair of electrons supplies a stronger attraction for the nuclei resulting in a stronger and shorter bond. Looking at the molecular orbital diagrams for these three molecules (See answer for Problem 9.14; a, b and c), the bond orders are: $F_2 = 1$, $O_2 = 2$ and $N_2 = 3$. As the bond order increases, the bond strength increases and the bond length decreases.

9.59 F—F < Br—Br < Br—Cl < Cl—Cl < Cl—F
     154.8   190.0   216   239.7   249 kJ/mol

The F—F bond is the weakest because the small size and high effective nuclear charge of the fluorine atoms cause decreased overlap of the bonding orbitals and increased repulsion between the non-bonding electrons on the F atoms. The high principal quantum number ($n = 4$) resulting in more diffuse valence orbitals and a larger spread of electron density makes the Br—Br bond weaker than Cl-containing molecules. For the same reason Cl—Cl bond ($n = 3$) is stronger than the Br—Br bond with the Br—Cl bond (with one of each atom) being between. The Cl—F bond has the highest energy because the larger electronegativity difference of F and Cl atoms results in the stability that accompanies a polar bond.

9.61 ClO_2 $:O=\overset{..}{C}l=O: \longleftrightarrow :O=\overset{..}{C}l-\overset{..}{O}: \longleftrightarrow \overset{..}{.O}-\overset{..}{C}l=O:$

The Cl atom uses sp^3 hybridized orbitals to form σ bonds with the O atoms. The Cl atom uses 3d orbitals to form a π system extending over the 3 atoms. The compound is unusual because having an odd number of valence electrons, it has a lone electron in a π orbital.

9.63 Four bonds: H—O in H_2O, C=O in CO_2, H—C and $C\equiv N$ in HCN. The shortest bond is H—O. The longest bond is C=O.

H—O < H—C < $C\equiv N$ < C=O
96   109   116   120 pm

The small principal quantum number ($n = 1$) for H makes its bonds shorter. The greater electronegativity difference makes the H—O bond shorter than the H—C bond. The lengths of the other 2 bonds are longer because both atoms involved have $n = 2$. The $C\equiv N$ is shorter than the C=O because it has a triple bond versus a double bond.

9.65 $2 (-Si)_n$ + $n O_2$ → $2 (-O-Si)_n$
2 Si—Si bonds   1 O=O bond   4 Si—O bonds
$\Delta E = [(2 \text{ mol Si}-\text{Si})(220 \text{ kJ/mol}) + (1 \text{ mol } O_2)(493.6 \text{ kJ/mol})]$
$- [(4 \text{ mol Si}-\text{O})(450 \text{ kJ/mol})] = 934 \text{ kJ} - 1800 \text{ kJ} = -866 \text{ kJ/2 mol Si}$
$= -433 \text{ kJ/mol Si}$

$2 (-C)_n$ + $n O_2$ → $2 (-O-C)_n$
2 C—C bonds   1 O=O bond   4 C—O bonds
$\Delta E = [(2 \text{ mol C}-\text{C})(345 \text{ kJ/mol}) + (1 \text{ mol } O_2)(493.6 \text{ kJ/mol})]$
$- [(4 \text{ mol C}-\text{O})(360 \text{ kJ/mol})] = 1184 \text{ kJ} - 1440 \text{ kJ} = -256 \text{ kJ/2 mol C}$
$= -128 \text{ kJ/mol C}$

9.67   Increasing band gap: Bi < As < P
Bi: conductor   As: conductor   P: insulator
Both Bi and As have electrical resistivities that fall between what is usually defined
as conductors and semiconductors. As is the better conductor.

9.69   a) valence $e^- = 8(4 \text{ per C}) + 3(1 \text{ per H}) + 30 (6 \text{ per O}) + 5 (5 \text{ per N}) = 46 \ e^-$

46 e^- - 20 e^- (10 σ orbitals) - 18 e^- (lone pairs
on outer O) - 8 e^- (lone pairs on inner O) = 0

b) The left carbon has sp^3 hybridization. The other carbon has sp^2 hybridization. The
oxygen marked with an asterisk in the question has sp^3 hybridization. The nitrogen
has sp^2 hybridization.

c) The bond angles around the left carbon atom are 109.5°. The O=C—O bond
angle is 120°. The O=N—O bond angle is 120°.

d) There is a set of delocalized π orbitals over O=N—O.

9.71   The bond energy of C—Cl is 325 kJ/mol. The energy per bond is:
$$\left(\frac{325 \text{ kJ}}{\text{mol}}\right)\left(\frac{\text{mol}}{6.02 \times 10^{23}}\right) = 5.40 \times 10^{-22} \text{ kJ}$$

$E = h\nu$

$$\nu = \frac{E}{h} = \frac{(5.40 \times 10^{-22} \text{ kJ})(10^3 \text{ J / kJ})}{(6.63 \times 10^{-34} \text{ J s})} = 8.14 \times 10^{14} \ s^{-1} = 8.14 \times 10^{14} \text{ Hz}$$

9.73   The carbon-oxygen bond length in the carbonate ion is between the average length
for a C—O single bond and a C=O double bond. Delocalized electrons give the
C—O bonds in the carbonate ion properties that are intermediate between single and
double bonds.

9.75   a) For the N=N=N– portion of the molecule from left to right the first N uses 2p
        orbitals to the σ bond and the π bond and uses the 2s and a 2p orbital to hold the
        lone pairs of electrons. The middle N uses sp hybrid orbitals to form the 2 σ bonds
        and the remaining two 2p orbitals to form π bonds. The third N uses sp^2 hybrid
        orbitals to form 2 σ bonds and the lone pair of electrons. The remaining 2p orbital
        is used for the π bond. The 2 N's in the ring use sp^2 hybrid orbitals to each form 3
        σ bonds and a p orbital for the lone pairs of electrons. The lone pairs of electrons
        are part of a delocalized π bond system that involves the entire ring.
        b) The bond angle is approximately 109.5°.
        c) There are 6 C atoms with sp^3 hybrid orbitals. (4 in the 5-membered ring, 1
        attached to this ring and 1 attached to the 6-membered ring).
        d) There are 31 σ bonds in AZT.
        e) There are 5 π bonds in AZT.

9.77   a) Each carbon atom uses sp^2 hybrid orbitals to form 3 σ bonds and the remaining
        p orbital for π bonding.
        b) Five p orbitals contribute to the π bonding system to form 5 molecular orbitals.
        c) There are 4 delocalized electrons.
        d)

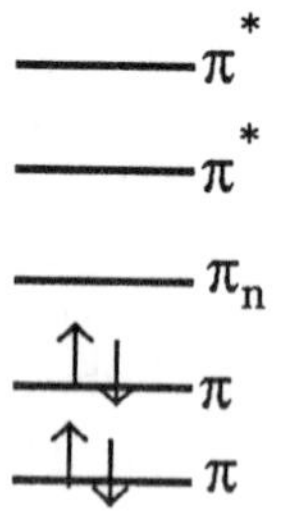

        ———— π^*

        ———— π^*

        ———— π_n

        ↑↓ —— π

        ↑↓ —— π

84

CHAPTER 10: EFFECTS OF INTERMOLECULAR FORCES

10.1 Your sketch should look like that in Figure 10-3 with the Br_2-Br_2 being replaced by carbon tetrachloride-carbon tetrachloride and the F_2-F_2 being replaced by methane-methane. Carbon tetrachloride has the stronger attraction of the two like Br_2 had the stronger attraction than F_2. The dashed line would remain very nearly at the same level in the two sketches.

10.3 a) Intermolecular attractions become less significant when a gas is expanded to a larger volume at constant temperature.
b) Intermolecular attractions become more significant when more gas is forced into the same volume at constant temperature.
c) Intermolecular attractions become more significant when the temperature of the gas is lowered at constant volume.

10.5 $\dfrac{PV}{nRT} = 1.00$ for an ideal gas.

a) $\dfrac{(19.7 \text{ atm})(1.20 \text{ L})}{(1.00 \text{ mol})(0.08206 \text{ L} \cdot \text{atm} / \text{mol} \cdot \text{K})(313.15 \text{ K})} = 0.92$

$\dfrac{1.00 - 0.92}{1.00} \times 100\% = 8.00\%$

b) $\dfrac{(2.00 \text{ atm})\left(189.18 \text{ cm}^3 \times 1 \text{ mL} / \text{cm}^3\right)(1.00 \text{ L} / 1000 \text{ mL})}{\left(\dfrac{3.00 \text{ g}}{2.016 \text{ g} / \text{mol}}\right)\left(0.08206 \dfrac{\text{L} \cdot \text{atm}}{\text{mol} \cdot \text{K}}\right)(273.15 \text{ K})} = 1.13$

$\dfrac{|1.00 - 1.13|}{1.00} \times 100\% = 13\%$

10.7 Arranged with easiest to liquefy first: $CCl_4 > CF_4 > CH_4$
The large Cl atoms in CCl_4 have large atomic orbitals and, therefore, high polarizability, giving CCl_4 strong intermolecular forces and ease of liquefaction (has a high boiling point). The small F atoms in CF_4 have small atomic orbitals, low polarizability, giving CF_4 weaker intermolecular forces and a lower boiling point. The H atoms in CH_4 have no lone pairs of electrons in their smallest atomic orbitals ($n = 1$) resulting in the weakest intermolecular forces and the lowest boiling point.

10.9 a) Mg^{2+} has a stronger interaction with water molecules because it has a greater positive charge.

b) Na^+ has the stronger interaction with water molecules because it has the greater charge density (same charge but smaller size).

c) SO_4^{2-} has the stronger interaction with water molecules because it has 4 O atoms with polarizable lone pairs of electrons versus only 3 O atoms with polarizable lone pairs of electrons on SO_3^{2-}.

10.11 propane < *n*-pentane < ethanol: boiling points
C_3H_8 C_5H_{12} C_2H_5OH
Dispersion forces are the only intermolecular forces for propane and *n*-pentane. The larger *n*-pentane is more polarizable resulting in stronger intermolecular forces and a higher boiling point than propane. Ethanol not only has dispersion forces but also dipole interactions because of the polar C—O and O—H bonds and the slightly stronger hydrogen bonding between the H atom on the O atom and the lone pairs of electrons on the O atom.

10.13 a) CH_2Cl_2 will not hydrogen bond.

b) H_2SO_4 will hydrogen bond with molecules of its own kind and other molecules such as H_2O.

c) H_3COCH_3 will hydrogen bond only with other molecules such as H_2O.

d) $H_2NCH_2CO_2H$ will hydrogen bond with molecules of its own kind and other molecules such as H_2O.

10.15 pentane < gasoline < fuel oil   The viscosity increases as the hydrocarbon chains get longer because the tangling between molecules increases and because the cohesive forces (dispersion forces) increase.

10.17 Flux must remove those substances to which metals will not adhere. Specifically, any oil or grease and any oxides or other forms of corrosion that prevent a good solder joint.

10.19 Sn - metallic; S_8 - molecular; Se - molecular; SiO_2 - covalent; Na_2SO_4 - ionic

10.21

The structure of carborundum has alternating carbon and silicon atoms that are covalently bonded. All atoms are bonded to four other atoms and have a tetrahedral geometry.

10.23 # of Ba atoms = 1/1(2 in interior) = 2 atoms
# of Cu atoms = 1/8(8 at corners) + 1/4(8 on edges) = 3 atoms
# of Y atoms = 1/1(1 in center) = 1 atom
# of O atoms = 1/4(12 on edges) + 1/2(8 on faces) = 7 atoms
$Ba_2Cu_3YO_7$ $YBa_2Cu_3O_7$

10.25 Water is not soluble in gasoline because the major intermolecular forces for water are hydrogen bonds and for gasoline are dispersion forces. Water should not be used to fight a gasoline fire because the gasoline will float on top of the water and continue to burn. Further, if the water flows anywhere, it will spread the fire.

10.27 Solids that form hydrogen bonds, ionic solids and polar solids would be expected to be soluble in liquid ammonia.

10.29 If the 1.10 atm pressure is the pressure of the CO_2:
$$C_i = (34 \times 10^{-3} \text{ M/atm})(1.10 \text{ atm}) = 0.037 \text{ M}$$
$$\left(0.037 \frac{\text{mol}}{\text{L}}\right)\left(\frac{250 \text{ mL}}{1000 \text{ mL} / \text{L}}\right)\left(\frac{44 \text{ g } CO_2}{1 \text{ mol}}\right) = 0.41 \text{ g } CO_2$$

10.31 The compound is hydrophobic. It will concentrate in the fatty tissue. This makes it dangerous because its concentration builds up in the fatty tissue of an organism through absorption by the organism or by consumption of organisms containing the DDT. Therefore, higher members of the food chain (which humans often eat) acquire the biggest concentration build up.

10.33 The drawing asked for in this question should look like one of the horizonal portions of the vesicle shown in Figure 10-37.

10.35 12% by mass ethanol: 12 g ethanol (CH_3CH_2OH) per 88 g H_2O

$$\text{mol of ethanol} = \frac{12 \text{ g}}{46 \text{ g} / \text{mol}} = 0.26 \text{ mol} \qquad \text{mol of } H_2O = \frac{88 \text{ g}}{18 \text{ g} / \text{mol}} = 4.89$$
$$\Delta T_f = K_f X_{solute} = (103.2°C)\left(\frac{0.26}{0.26 + 4.89}\right) = 5.2°C$$
$$0°C - 5.2°C = -5.2°C \text{ (freezing point of wine)}$$

10.37 a) $MM = \dfrac{mRT}{\Pi V} = \dfrac{(1.00 \text{ g})(0.08206 \text{ L} \cdot \text{atm} / \text{mol} \cdot \text{K})(298 \text{ K})}{(64.8 \text{ torr})\left(\dfrac{1 \text{ atm}}{760 \text{ torr}}\right)(1 \text{ L})} = 286.8 \text{ g} / \text{mol}$

(The Π has been corrected. If 17.8 torr is used one obtains the value of 1044 g/mol.)

b) MM of hydrocarbon portion $= \left(\dfrac{11 \text{ mol C}}{\text{mol}}\right)\left(\dfrac{12 \text{ g}}{\text{mol C}}\right) = \left(\dfrac{23 \text{ mol H}}{\text{mol}}\right)\left(\dfrac{1 \text{ g}}{\text{mol H}}\right)$
$$= 155 \text{ g} / \text{mol}$$
MM of polar portion = 287 g/mol - 155 g/mol = 132 g/mol

10.39 The impure diethyl ether is placed into a distilling flask and pieces of sodium metal are added. The water reacts with the sodium metal to form sodium hydroxide and hydrogen gas. Air is excluded from the distilling system to prevent recontamination and the ether is distilled from the sodium hydroxide.

10.41 95% of 250 g of sample = 237.5 g of $HgCl_2$
5% of 250 g of sample = 12.5 g of impurity
Minimum amount of H_2O at 100°C =
$$(237.5 \text{ g } HgCl_2)\left(\frac{1 \text{ L } H_2O}{380 \text{ g } HgCl_2}\right) = 0.625 \text{ L } H_2O = 625 \text{ mL } H_2O$$

Amount of $HgCl_2$ left in solution at 0°C $= (0.625 \text{ L})\left(\dfrac{30 \text{ g } HgCl_2}{1 \text{ L}}\right) = 18.75 \text{ g } HgCl_2$

Note: The impurity is more soluble than $HgCl_2$; all 12.5 g will remain in solution.
Amount of $HgCl_2$ precipitated at 0°C = 237.5 g - 18.75 g = 218.8 g
218.8 / 237.5 = 0.92 or 92%

10.43 Assuming that separation according to polarity means that least polar comes off first and most polar comes off last:

1st - heptane; 2nd - methylpentylether; 3rd (last) - 1-hexanol

Heptane is nonpolar, would have no interaction with the GC column and would come off first. The 1-hexanol with the C—O—H bonding would be the most polar, would interact with the GC column the most and would come off last. The methyl-pentylether with the C—O—C bonding would be intermediate in polarity and would come off between the other two.

Note: All three compounds have about the same size and molecular mass.

10.45 $MM = mRT/\Pi V$ (Equation 10-5)

$$MM = \frac{(1.00 \text{ g})(0.08206 \text{ L atm / mol K})(298 \text{ K})}{(1.36 \text{ atm})(0.100 \text{ L})} = 180 \text{ g / mol}$$

10.47 C_6H_6 < $C_5H_{11}OH$ < $HOCH_2CH(OH)CH(OH)CH_2OH$
benzene   pentanol   erythritol

Solubility in water will depend on the ability of the compound to form hydrogen bonds with H_2O. Benzene cannot form any hydrogen bonds with H_2O and is least soluble. Pentanol has only one OH group that can hydrogen bond with H_2O and a 5 carbon chain which would make it only slightly soluble in H_2O. Erythritol has 4 OH groups spread over a 4 carbon chain that allows it to form many hydrogen bonds with many water molecules and be quite soluble in water.

10.49 Glucose is in the form of molecules in solution. Therefore, a 1 M solution has 1 mole of molecules per liter and a 0.5 M solution has 0.5 moles of molecules per liter. In water, acetic acid is a weak acid that ionizes to a small degree; therefore, a 0.5 M acetic acid solution will have slightly more than 0.5 mole of particles (acetic acid molecules, hydronium ions and acetate ions) and the freezing point will be lower than 0.5 M glucose solution but higher than the freezing point of 1 M glucose (because the acetic acid does not completely ionize). A similar explanation holds for $MgSO_4$ which forms Mg^{2+} ions and SO_4^{2-} ions in aqueous solution. A 0.5 M solution of $MgSO_4$ should be quite close to having 1 mole of particles (Mg^{2+} ions and SO_4^{2-} ions). Because the freezing point is higher than 1 M glucose, there must be some Mg^{2+} - SO_4^{2-} ion pairs which are not separated in solution.

10.51 It is not practical to purify NaCl by recrystallization from water because upon cooling the solution from 100°C to 0°C only 2 g of every 28 g (~7 %) are recovered.

10.53 pentane < butanol < propane-1,3-diol
C_5H_{12} C_4H_9OH $HOCH_2CH_2CH_2OH$

The stronger the intermolecular forces between the molecules of a liquid, the more viscous the liquid. Pentane is least viscous because its intermolecular forces are only weak dispersion forces. Butanol is more viscous because it can hydrogen bond between molecules using the OH group (but only 1 OH per 4 C atoms). The propane-1,3-diol is most viscous because it also forms hydrogen bonds but many more than the butanol (2 OH per 3 C atoms).

10.55
a) slightly greater than 0.1 M,   a small amount of ionization

b) slightly less than 0.4 M, $FeCl_{3(aq)} \rightarrow Fe^{3+}_{(aq)} + 3Cl^{-}_{(aq)}$

c) approximately 0.2 M, $NaOH_{(aq)} \rightarrow Na^{+}_{(aq)} + OH^{-}_{(aq)}$

d) approximately 0.3 M, $(NH_4)_2CO_{3(aq)} \rightarrow 2NH_4^{+}_{(aq)} + CO_3^{2-}_{(aq)}$

10.57   0.050 M $KHSO_3$ 0.050 mol $KHSO_3$ per L of solution
Assuming that the density of the solution is 1.00 g/mL, 1 L of solution contains 6g $KHSO_3$ and 994 g H_2O.

$$994 \text{ g } H_2O = \frac{994 \text{ g } H_2O}{18.02 \text{ g / mol}} = 55.2 \text{ mol } H_2O$$

$$0.19°C = i(103.2°C)\left(\frac{0.050 \text{ mol}}{0.050 \text{ mol} + 55.2 \text{ mol}}\right)$$

$0.19°C = i \ (103.2°C)(0.000905)$ $i = 2.03$ $KHSO_{3(s)} \rightarrow K^+_{(aq)} + HSO_3^-_{(aq)}$

As can be seen from the calculation, the freezing point depression is about twice what it would be if the $KHSO_3$ did not dissociate. Therefore, it must dissociate into 2 particles (ions) as in equation (b).

10.59   Water has intermolecular forces of hydrogen bonding and dipolar forces. Carbon tetrachloride and iodine both have dispersion forces as their intermolecular forces. As a result of this, iodine is more soluble in CCl_4 than in H_2O. When a mixture of H_2O and CCl_4 is shaken and iodine is present, the iodine will collect in the solvent in which it is more soluble: carbon tetrachloride.

10.61 3.0×10^{-3} M glucose < 4.0×10^{-3} M glucose < 3.0×10^{-3} M KBr

$\Pi = MRT$; therefore, $\Pi \propto M$ (or concentration of particles.)

Glucose is present in aqueous solution as molecules. Therefore, 4.0×10^{-3} M glucose has a higher osmotic pressure than 3.0×10^{-3} M glucose because it has a third more molecules in solution (4.0×10^{-3} mol/L versus 3.0×10^{-3} mol/L). The 3.0×10^{-3} M KBr has the highest osmotic pressure because in aqueous solution the KBr dissociates into K^+ ions and Br^- ions. Therefore, the concentration of particles for KBr is approximately 6.0×10^{-3} mol of particles/L.

10.63   When a fish is placed in a strong salt solution, water is removed from the cells of the fish by osmosis. The fish does not spoil as quickly in this dehydrated condition.

10.65   **Compound**   **Intermolecular Forces**
a) NH_3 dispersion, dipolar, hydrogen bonding
b) Xe   dispersion
c) SF_4 dispersion, dipolar
d) CF_4 dispersion
e) CH_3CO_2H dispersion, dipolar, hydrogen bonding

10.67   1.1 kg NaCl = 1100 g NaCl = (1100 g NaCl)÷(58.44 g/mol) = 18.8 mol NaCl
7.25 kg of ice = 7250 g H_2O = (7250 g H_2O)÷(18.02 g/mol) = 402.3 mol H_2O

$$X_{NaCl} = \left(\frac{18.8 \text{ mol}}{18.8 \text{ mol} + 402.3 \text{ mol}}\right) = 0.0446$$

$\Delta T_f = i \ K_f X_{NaCl} = 2(103.2°C)(0.0446) = 9.2°C$ freezing point = -9.2°C

10.69 a) The CH_3OH has a higher boiling point than CH_3OCH_3 because CH_3OH has hydrogen bonding versus dipolar forces for CH_3OCH_3.
b) The SiO_2 has a higher boiling point than SO_2 because SiO_2 has covalent bonding versus dipolar forces for SO_2.
c) The HF has a higher boiling point than HCl because HF has hydrogen bonding versus dipolar forces (and, perhaps, some weak hydrogen bonding) for HCl.
d) The I_2 has a higher boiling point than Br_2. Although both have dispersion forces as their intermolecular forces, those for I_2 are greater because it is more polarizable due to its larger size.

10.71 The intermolecular (interatomic) forces involved here are dispersion forces. Dispersion forces depend on the polarizability of the electrons of the substance. The polarizability is easier for the electron cloud of an extended molecule than for more compact molecules or atoms. Therefore, H_2 is more polarizable than He and has a higher boiling point; methane is more polarizable than Ne and has a higher boiling point.

10.73

n-pentane   b.p. = 36°C

2,2-dimethylpropane
b.p. = -10°C

The more compact a molecule is the weaker its dispersion forces are. Therefore, a compact molecule will have a lower boiling point than a molecule of similar mass which is more extended.

10.75 Assuming the stationary phase is polar, the most polar dye would move the least distance from the original spot. The dye (black) left at the original spot would be most polar and the purple at the far right the least polar: black > yellow > blue green > orange > purple.

10.77 a) Cl_2 would deviate more from ideality than F_2 because it is larger and its dispersion forces are stronger.
b) SnH_4 would deviate more from ideality than CH_4 because it is larger and its dispersion forces are stronger.

10.79 # atoms Ti = 1/1(1 in center) = 1, # atoms Ca = 1/8(8 at corners) = 1, # atoms O = 1/2(6 face centers) = 3, $CaTiO_3$

10.81

cis
b.p. = 60°C

trans
b.p. = 47°C

Each has two polar C—Cl bonds. Because of the symmetry, the trans molecule is not a polar molecule. The cis form has a dipole moment and the dipole-dipole interactions between the molecules result in the cis form having a higher boiling point.

10.83

Compound	Forces to overcome to convert from liquid to gas
a) NH_3	dispersion, dipolar and hydrogen bonding
b) $CHCl_3$	dispersion and dipolar
c) CCl_4	dispersion
d) CO_2	dispersion

CHAPTER 11: MACROMOLECULES

11.1

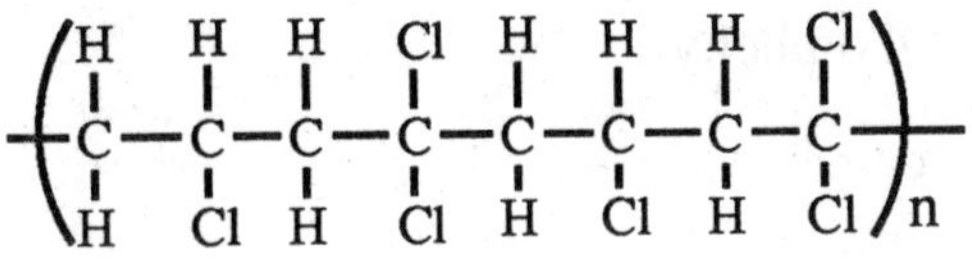

11.3 Initiation: Init.—Init. → 2 Init.•

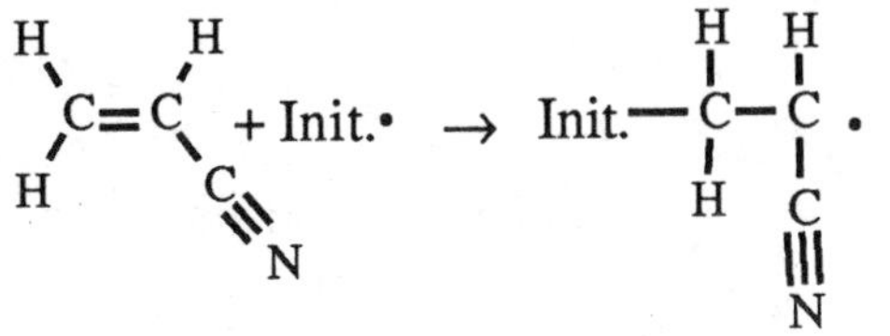

Propagation:

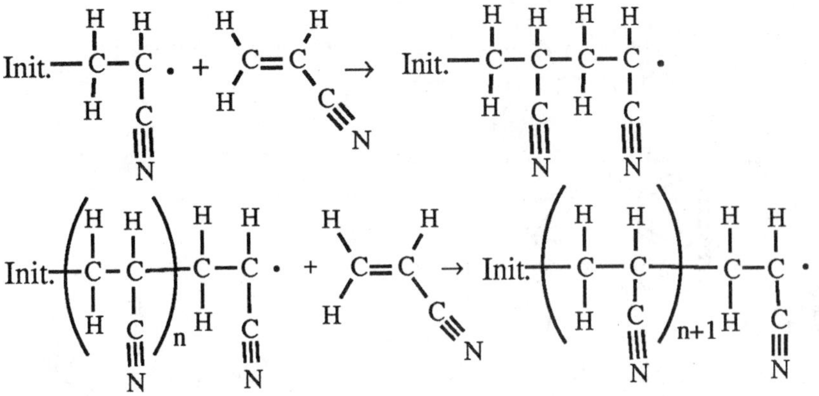

Termination:

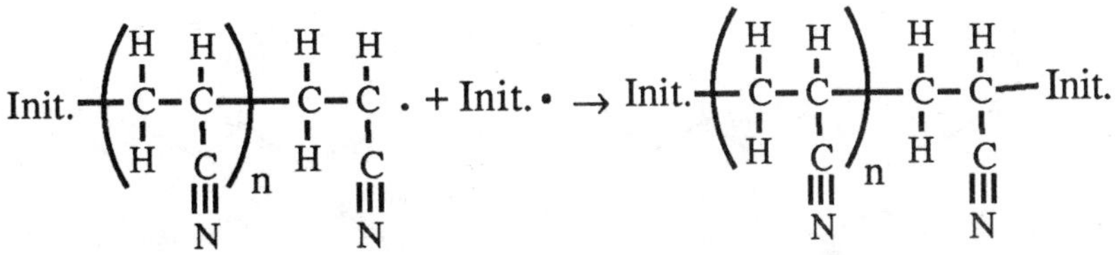

11.5 a)

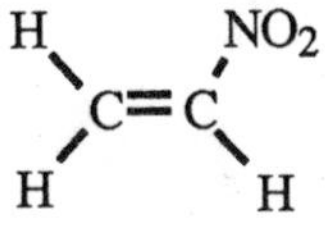

nitroethene or nitroethylene

b)

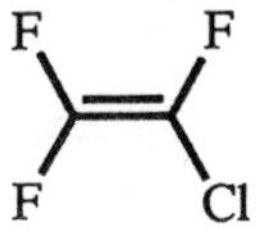

chlorotrifluoroethene or chlorotrifluoroethylene

11.7 a) 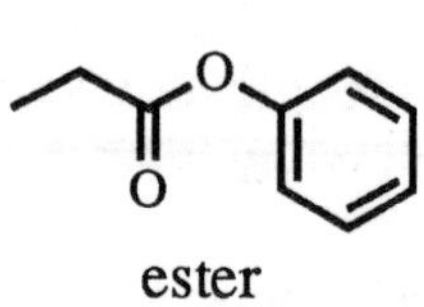
ester

b) 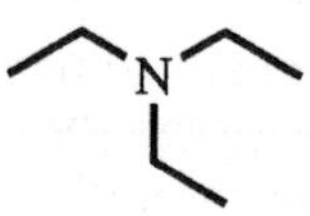
amine

c) 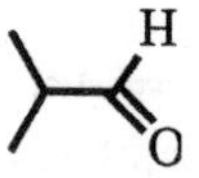
aldehyde

d) 
alkene
(also bromo or bromide)

d) 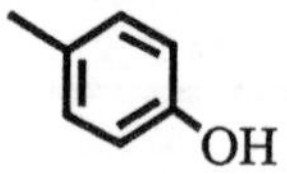
phenol
(and phenyl)

11.9 a) Other isomers are also possible.

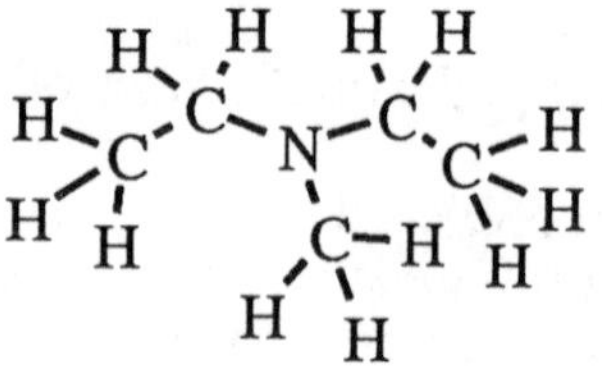

b) Other isomers are also possible.

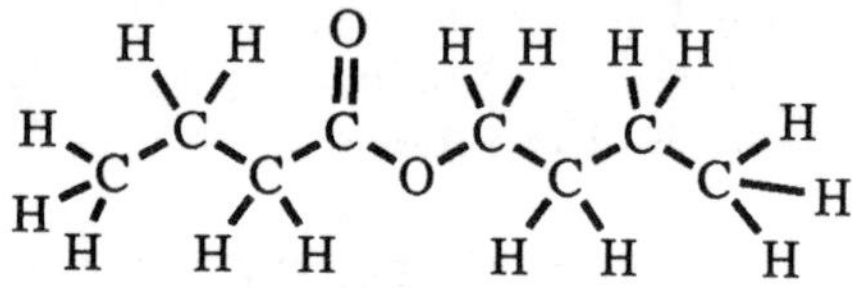

c) Others are also possible.

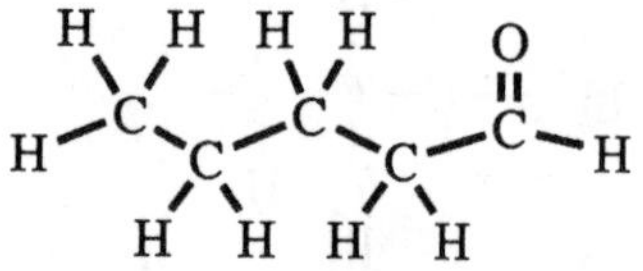

d) Others are also possible.

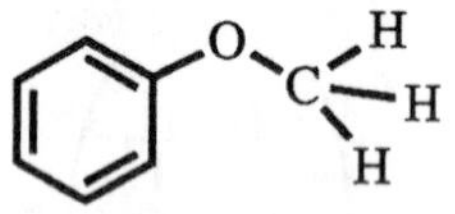

11.11 a)

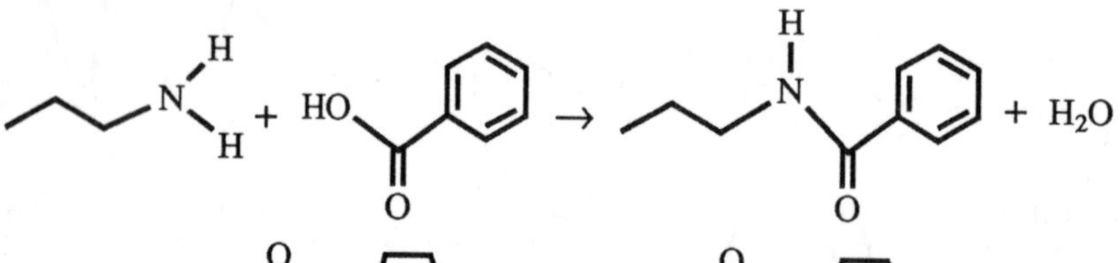

b)

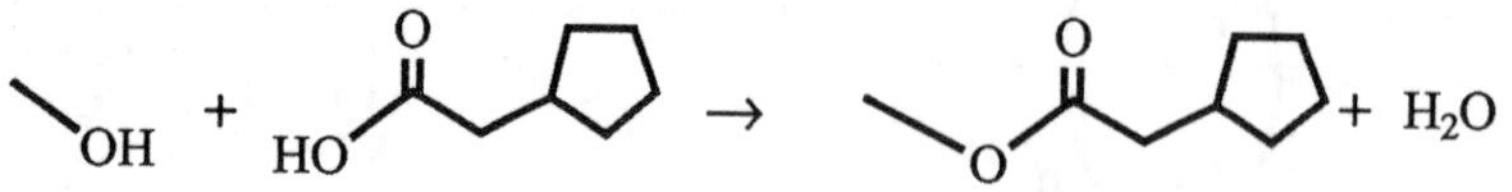

c)

11.13

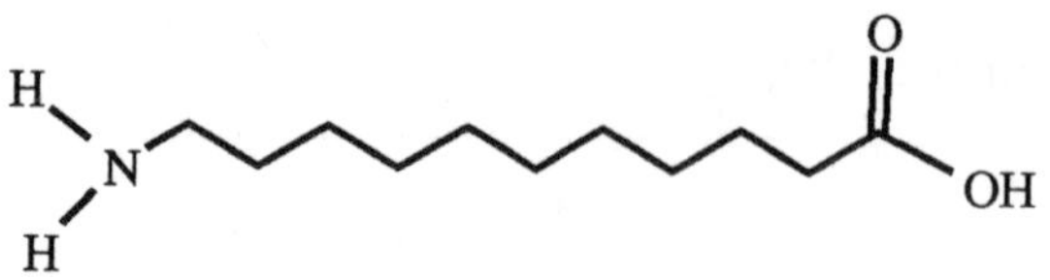

11.15 $-CH_2-CH_2-O-CH_2-CH_2-O-CH_2-CH_2-O-$ or $\left(CH_2-CH_2-O\right)_n$

11.17 The rubber in an automobile tire is more extensively cross-linked than the rubber in surgical gloves. Rubber for tires needs to have greater strength and not be very flexible compared to the rubber in surgical gloves. Cross-linking decreases the flexibility of rubber and increases its strength.

11.19

item	category of polymer
a) balloon	elastomer
b) rope	fiber
c) camera case	plastic

11.21 Dioctylphthalate is a common plasticizer that increases the flexibility of polymers when it is added to them.

11.23 Assigning each amino acid as hydrophobic or hydrophilic on the basis of its side chain:
a) Tyr - would be hydrophilic because of the hydrophilic side chain.
b) Phe - would be hydrophobic because of the hydrophobic side chain (ring).
c) Glu - would be hydrophilic because of the hydrophilic side chain.
d) His - would be hydrophilic because of the hydrophilic side chain (ring).
e) Leu - would be hydrophobic because of the hydrophobic side chain.
f) Pro - would be hydrophobic because of the hydrophobic side chain (ring).

11.25

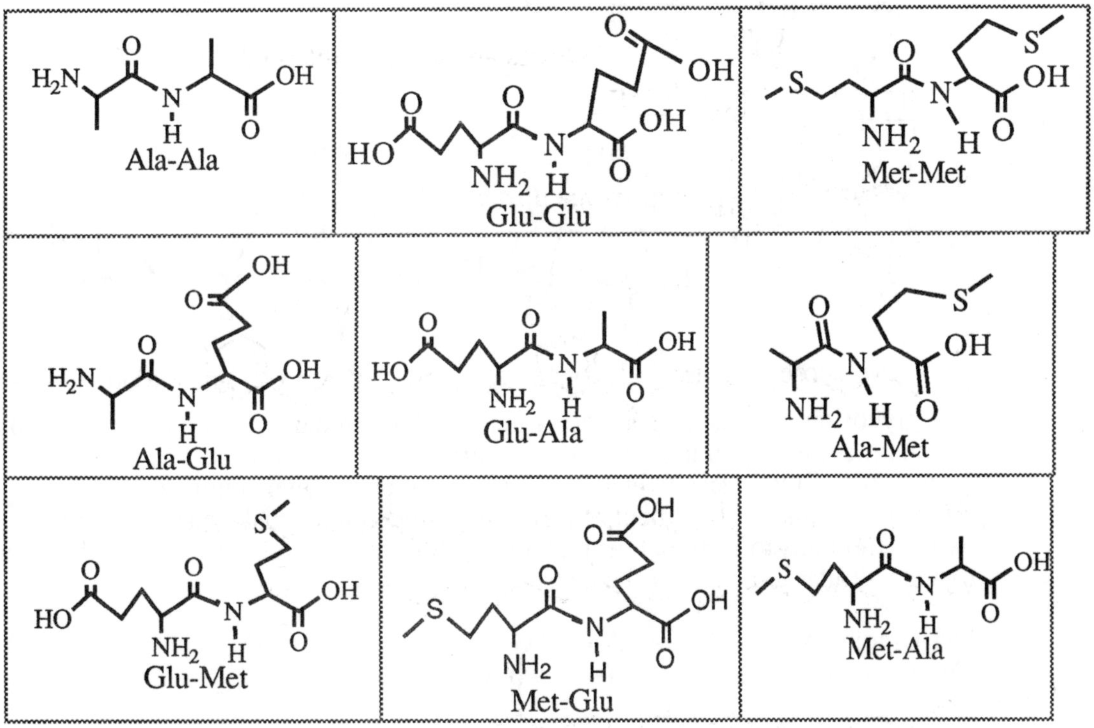

11.27

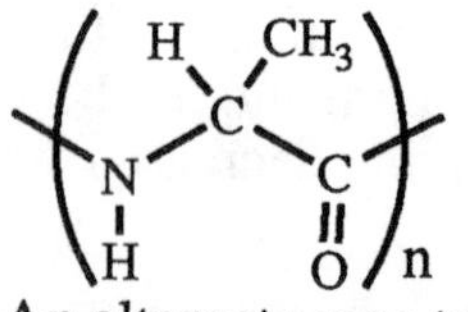

C_3H_5NO   MM = 71.08 g/empirical unit

$$\frac{1.20 \times 10^3 \text{ g / mol}}{71.08 \text{ g / unit}} = 16.9 \approx 17 \text{ units / mol or 17 amino acids}$$

An alternate way to obtain the same answer is:
MM of polypeptide - mass of end H - mass of end OH =

$1.20 \times 10^3 - 1 - 17 = 1.18 \times 10^3$ then $\frac{1.18 \times 10^3}{71} = 16.6 = 17$ units or amino acids

11.29

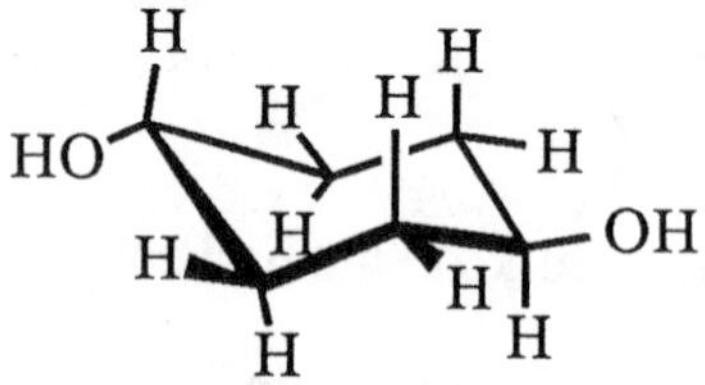

11.31

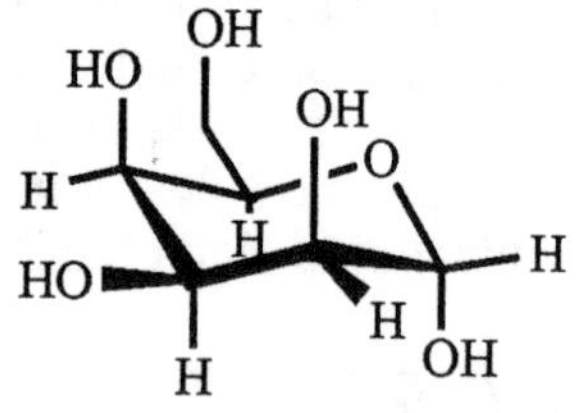

α-talose

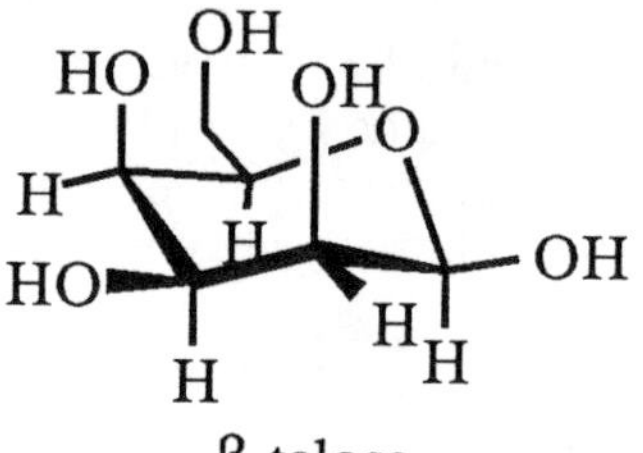

β-talose

11.33

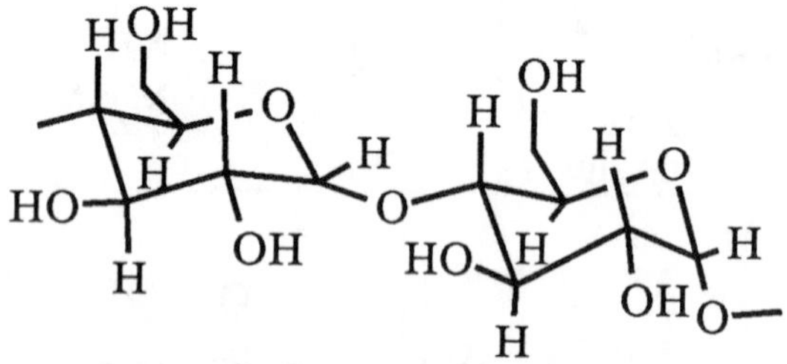

α-glucose linkage of glycogen

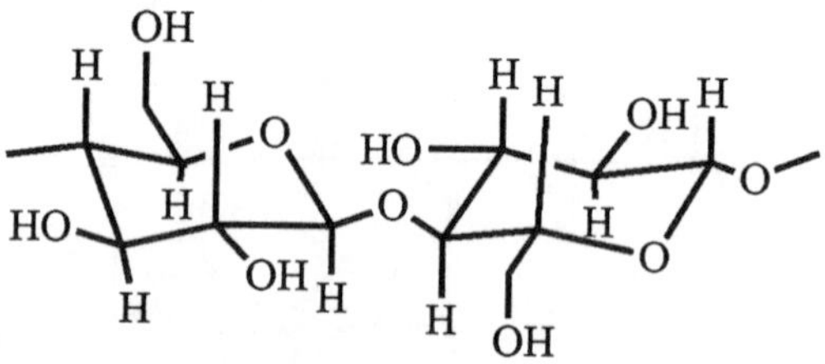

β-glucose linkage of cellulose

The glycosidic linkage of glycogen imparts a kink in the structure that can result in the chain coiling upon itself into the granular shape of deposits. The glycosidic linkage of cellulose results in a flat structure giving a long ribbon-like chain of units. The planar arrangement of monomers in cellulose makes it possible for hydrogen bonds to form between chains generating extended packages of ribbons that are similar to a pleated sheet structure.

11.35 DNA sequence: A-A-T-G-C-A-C-T-G
Complementary sequence: T-T-A-C-G-T-G-A-C

11.37

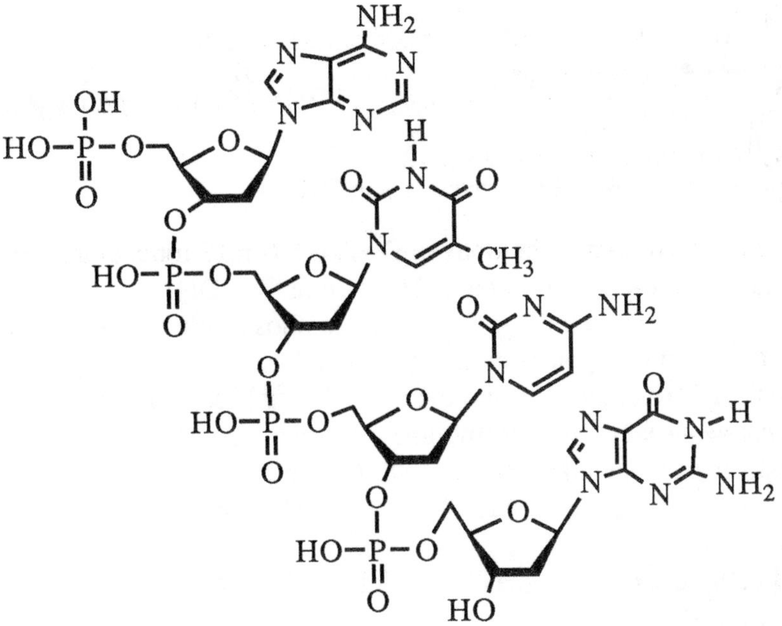

11.39   DNA sequence: A-T-C   Complementary sequence: T-A-G

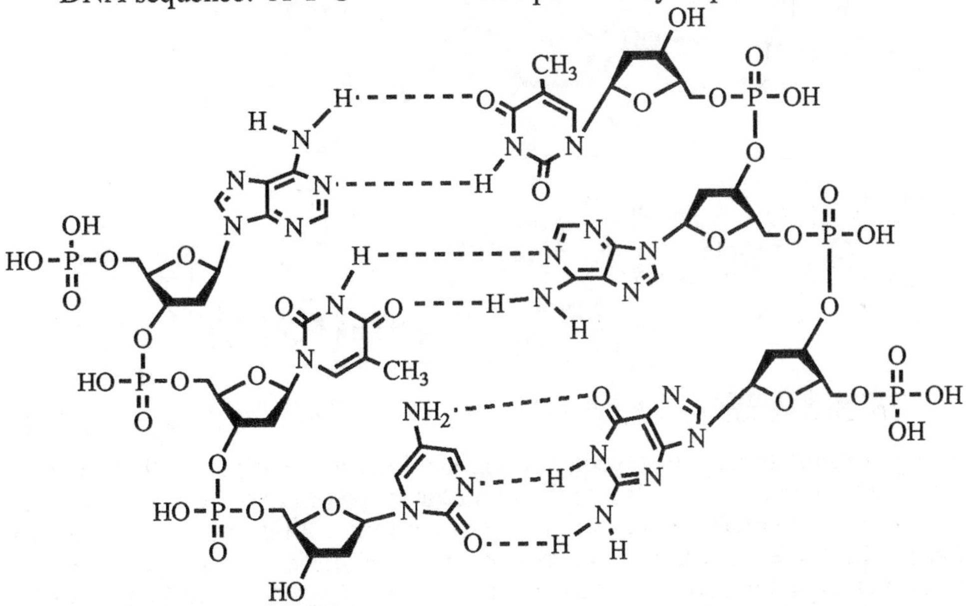

11.41 The nitrogen has sp^2 hybridization in a peptide linkage. By adopting a trigonal planar geometry using sp^2 hybridization, the nitrogen has a p orbital left that may form a delocalized π system with the carbonyl from the acid.

11.43

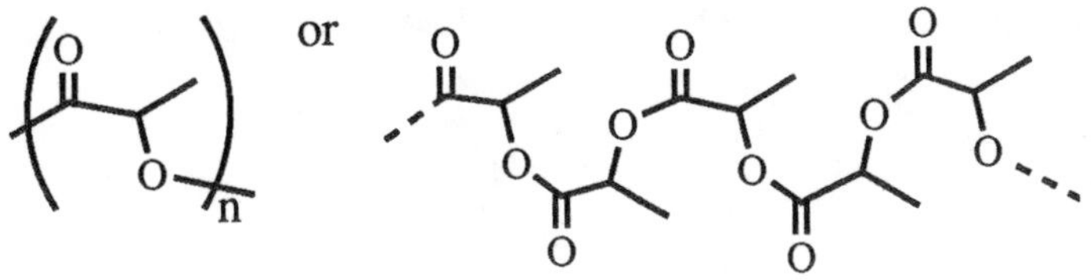

95

11.45

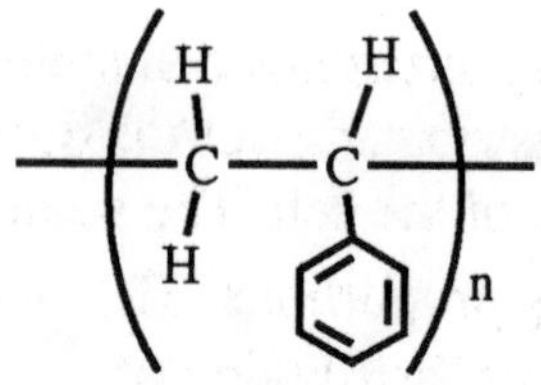

or $-\left(C_8H_8\right)_n-$ 452 (8 x 12.01 + 8 x 1.008) =
$$4.71 \times 10^4 \text{ g/mol}$$

11.47 There are 6 molecular building block units in DNA partners, 3 for each strand.
a) Each strand has a pentose sugar which is deoxyribose for DNA.
b) Each strand has a phosphate linkage derived from phosphoric acid.
c) Each strand contains a nitrogen-containing organic base. The possible bases for DNA are: thymine, adenine, guanine and cytosine. These bases are present in certain pairs: guanine with cytosine or thymine with adenine.
This question asks for a molecular picture of DNA partners. The picture in the answer to Question 11.39 is an example of 3 of these DNA partners.

11.49 base 1 = cytosine, base 2 = uracil, base 3 = adenine, base 4 = guanine

11.51

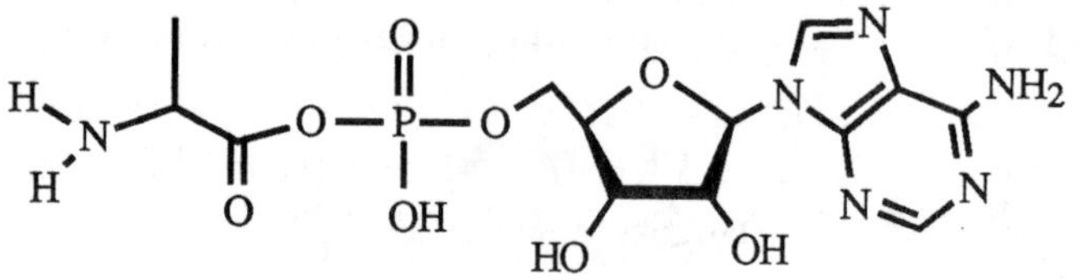

11.53 MM_{enzyme} = 64,000 g/mol

mass of Cu per mole = (64,000 g/mol)(0.0040) = 256 g Cu/mol enzyme

$$(256 \text{ g Cu / mol enzyme})\left(\frac{1 \text{ mol C}}{63.55 \text{ g Cu}}\right) = 4.03 \text{ mol Cu / mol enzyme}$$

Therefore, 4 atoms of Cu are in one molecule of fungal lactase.

11.55

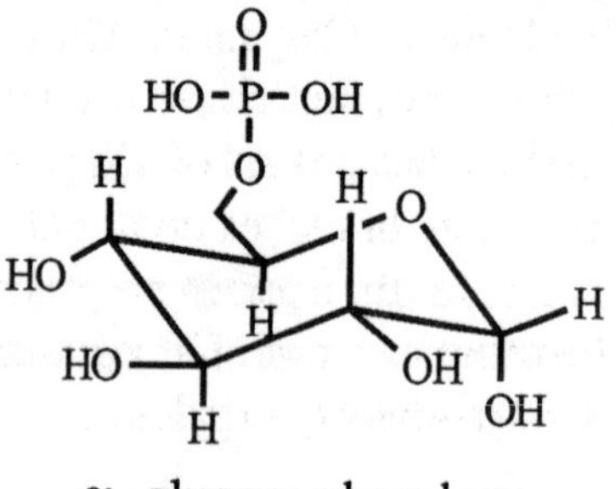

α–glucose phosphate

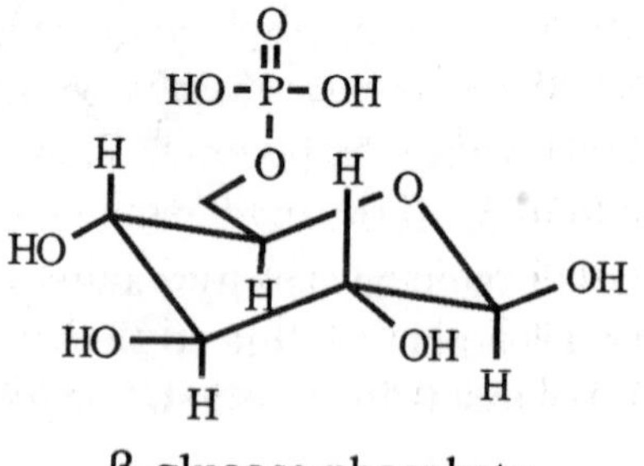

β-glucose phosphate

11.57 Globular proteins have compact, roughly spherical tertiary structures containing folds and grooves. Many have hydrophilic side chains distributed over the outer surface making them soluble in the aqueous environment of the cell. The secondary structures include α-helices and pleated sheets in varying proportions. The unique primary structure of each globular protein leads to a unique distribution of secondary structures and to a specific tertiary structure. Fibrous proteins are the structural components of cells and tissue. The α-helix is a prevalent secondary structure for some fibrous proteins. These molecules are long strands of helical protein that lie with their axes parallel to the axis of the fiber. Because of their compact helical chains, stretching the fibers will stretch and break the relatively weak hydrogen bonds of the α-helix, but as long as none of the stronger covalent bonds are broken, the protein will relax to its original length when released. Other fibrous proteins contain extensive regions of pleated sheets.

Globular proteins carry out most of the work done by cells, including synthesis, transport and energy production. Most globular proteins are enzymes which speed up biochemical reactions. Some are antibodies that protect us from disease. Some transport smaller molecules through the bloodstream such as hemoglobin which carries oxygen in the bloodstream. Some act as hormones. Others are bound in cell membranes where they facilitate passage of nutrients and ions into and out of cells. Fibrous proteins are the cables, girders, bricks, and mortar of organisms. One form called keratins includes wool, hair, skin, fingernails and fur.

11.59

11.61

11.63 High-density polyethylene is made up of straight chains of CH_2 units. These linear molecules maximize dispersion forces by lining up in rows that create crystalline regions within the polymer. Low-density polyethylene has chains of CH_2 groups that branch off the main backbone of the polymer. These branches do not allow the polymer molecules to pack close together resulting in an amorphous polymer. The close packed crystalline high-density ethylene therefore has more CH_2 groups per unit volume than the non-close packed amorphous low-density ethylene.

11.65

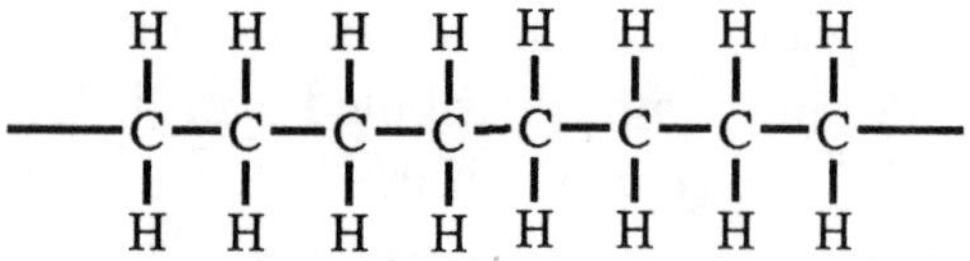

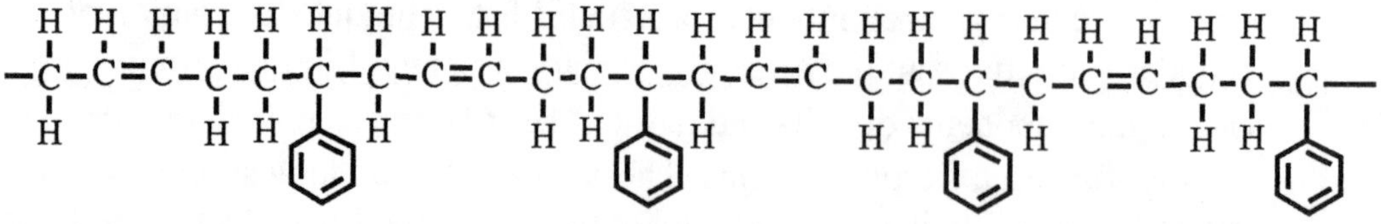

The polyethylene straight chain polymers can maximize dispersion forces by lining up in rows that create crystalline regions within the polymer. This gives strength and rigidity to the polymer. The butadienestyrene copolymer chain cannot line up closely to maximize dispersion forces because the benzene rings take up space and keep portions of the chains apart. Therefore, the butadienestyrene copolymer is less rigid.

11.67

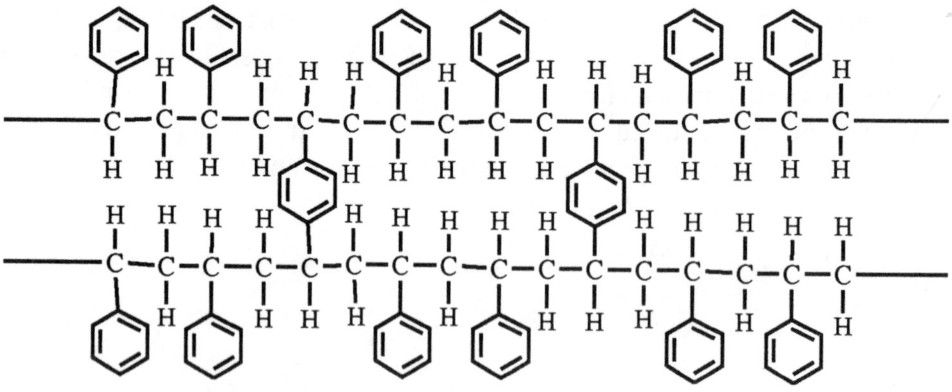

11.69 Amino acids with nonpolar hydrophobic side chains would most likely be found on the inside of a globular protein where they are tucked away from the aqueous environment of the cell. The amino acids with side chains that contain polar function groups are usually on the outer surface of a protein where their interaction with water molecules increases the solubility of the protein.

Amino Acid	Location	Polar or Nonpolar Side Chain	Polar Functional Group
a) Arg	outside	$-CH_2CH_2CH_2NH-C-NH_2$ with $\overset{\|}{NH}$ double bond	$-NH$, $-NH_2$, $=NH$
b) Val	inside	$CH_3-\overset{\|}{CH}-CH_3$	
c) Met	inside	$-CH_2CH_2SCH_3$	
d) Thr	outside	$CH_3-\overset{\|}{CH}-OH$	$-OH$
e) Asp	outside	$-CH_2CO_2H$	$-CO_2H$

CHAPTER 12: CHEMICAL ENERGIES

12.1   a) A human being is an open system.
b) Coffee in a thermos flask is an isolated system.
c) An ice-cube tray filled with water is a closed system.
d) A helium-filled balloon is an open system.

12.3   a) The temperature at which a solution is prepared is not a state variable for its present state.
b) The current temperature of the solution is a state variable.
c) The mass of NaCl in the solution (related to the number of moles) is a state variable.
d) The time when is was prepared is not a state variable.

12.5   The cylinder's contents, battery, cooling system, engine, the automobile and the human body are possible systems.

12.7   a) $(10.0 \text{ g Al})\left(\dfrac{1 \text{ mol Al}}{26.98 \text{ g Al}}\right) = 0.3706 \text{ mol Al}$

$\Delta T = \dfrac{q}{nC_p} = \dfrac{25.0 \text{ J}}{(0.3706 \text{ mol Al})(24.35 \text{ J / mol Al K})} = 2.77 \text{ K} = 2.77°C$
$T_f = 15.0°C + \Delta T = 15.0°C + 2.77°C = 17.8°C$

b) $(25.0 \text{ g Al})\left(\dfrac{1 \text{ mol Al}}{26.98 \text{ g Al}}\right) = 0.9266 \text{ mol Al}$

$\Delta T = \dfrac{25.0 \text{ J}}{(0.9266 \text{ mol Al})(24.35 \text{ J / mol Al K})} = 1.108 \text{ K}$
$T_f = 295 \text{ K} + \Delta T = 295 \text{ K} + 1.108 \text{ K} = 296 \text{ K}$

c) $(25.0 \text{ g Ag})\left(\dfrac{1 \text{ mol Ag}}{107.9 \text{ g Ag}}\right) = 0.2317 \text{ mol Ag}$

$\Delta T = \dfrac{25.0 \text{ J}}{(0.2317 \text{ mol Ag})(25.351 \text{ J / mol Ag K})} = 4.256 \text{ K}$
$T_f = 295 \text{ K} + \Delta T = 295 \text{ K} + 4.256 \text{ K} = 299 \text{ K}$

d) $(25.0 \text{ g H}_2\text{O})\left(\dfrac{1 \text{ mol H}_2\text{O}}{18.02 \text{ g H}_2\text{O}}\right) = 1.387 \text{ mol H}_2\text{O}$

$\Delta T = \dfrac{25.0 \text{ J}}{(1.387 \text{ mol H}_2\text{O})(75.291 \text{ J / mol H}_2\text{O K})} = 0.2394 \text{ K} = 0.3394°C$
$T_f = 22.0°C + 0.2394°C = 22.2°C$

12.9   Let the final temperature = t

$$(27.4 \text{ g Ag})\left(\frac{1 \text{ mol Ag}}{107.9 \text{ g Ag}}\right) = 0.2539 \text{ mol Ag}$$

$$(37.5 \text{ g H}_2\text{O})\left(\frac{1 \text{ mol H}_2\text{O}}{18.02 \text{ g H}_2\text{O}}\right) = 2.081 \text{ mol H}_2\text{O}$$

Heat lost by hot coin =
    (0.2539 mol)(25.351 J/mol K)(373 K - t) = 6.437 J/ K(373 K - t)
Heat gained by H_2O =
    (2.081 mol)(75.291 J/mol K)(t - 293.65 K) = 156.7 J/K(t - 293.65 K)
6.437 J/K(373 K - t) = 156.7 J/K(t - 293.65 K)
6.437 J/K(373 K) + 156.7 J/K(293.65 K) = 156.7 t J/K + 6.437 t J/K
2401 J + 46015 J = 163.1 t J/K   48416 J = 163.1 t J/K
t = 48416 J ÷ (163.1 J/K) = 296.8 K T_f = 296.8 K or 23.6°C

12.11   Assuming the atmospheric pressure is 1 atm and using Equation 12-3:
$w = -P\Delta V = -(1 \text{ atm})(2.5 \text{ L} - 0) = -2.5 \text{ atm L}$
$1 \text{ atm} = 1.013 \times 10^5 \text{ Pa} = 1.013 \times 10^5 \text{ kg m}^{-1} \text{ s}^{-2}$
$1 \text{ L} = 1000 \text{ cm}^3 = (1000 \text{ cm}^3)(1 \text{ m}/100 \text{ cm})^3 = 0.001000 \text{ m}^3$

$$w = (-2.5 \text{ atm L})\left(\frac{1.013 \times 10^5 \text{ kg m}^{-1} \text{ s}^{-2}}{1 \text{ atm}}\right)\left(\frac{0.001000 \text{ m}^3}{1 \text{ L}}\right)$$

$$= -253 \text{ kg m}^2 \text{ s}^{-2} = -2.5 \times 10^2 \text{ J}$$

12.13   a) Any change in the energy of a system must be counterbalanced by an opposite change in the energy of the surroundings.
b) The total energy of the universe remains unchanged.
c) $\Delta E_{surr} = -\Delta E_{sys}$ $q_{surr} + w_{surr} = -q_{sys} + (-w_{sys})$

12.15   q for burning of the glucose   q = (1.7500 g)(15.57 kJ/g) = 27.2475 kJ
ΔT = 23.34°C - 21.45°C = 1.89°C = 1.89 K

$$C_{cal} = \frac{q}{\Delta T} = \frac{27.2475 \text{ kJ}}{1.89 \text{ K}} = 14.4 \text{ kJ / K}$$

12.17   Assume that $C_{cal} \approx C_{H_2O} = n_{H_2O} C_{p,H_2O}$

$$n = \left(\frac{110.0 \text{ g H}_2\text{O}}{18.016 \text{ g / mol}}\right) = 6.1057 \text{ mol H}_2\text{O}$$

$C_{cal} = C_{H_2O}$ = (6.1057 mol)(75.291 J/mol K) = 459.7 J/K
ΔT = (29.7°C - 22.0°C) = 7.7°C = 7.7 K
q_{cal} = (459.7 J/K)(7.7 K) = 3.5 × 10³ J $q_{solution}$ = -q_{cal} = -3.5 × 10³ J

$$\left(\frac{-3.5 \times 10^3 \text{ J}}{4.75 \text{ g CaCl}_2}\right)\left(\frac{110.98 \text{ g CaCl}_2}{1 \text{ mol CaCl}_2}\right) = -8.2 \times 10^4 \text{ J / mol CaCl}_2$$

12.19 a) $2\ C_7H_6O_2 + 15\ O_2 \rightarrow 14\ CO_2 + 6\ H_2O$

b) $\left(\dfrac{35.61\ kJ}{1.350\ g\ benzoic\ acid}\right)\left(\dfrac{122.12\ g\ benzoic\ acid}{1\ mol\ benzoic\ acid}\right) = 3221\ kJ\,/\,mol\ benzoic\ acid$

c) $\left(\dfrac{3221\ kJ}{mol\ benzoic\ acid}\right)\left(\dfrac{2\ mol\ benzoic\ acid}{15\ mol\ O_2}\right) = 429.5\ kJ\,/\,mol\ O_2$

12.21 a) $C_2H_{4(g)} + 3\ O_{2(g)} \rightarrow 2\ CO_{2(g)} + 2\ H_2O_{(l)}$

$\Delta H_{rxn} = [2\ \Delta H_f^{\circ}(CO_2) + 2\ \Delta H_f^{\circ}(H_2O)] - [\Delta H_f^{\circ}(C_2H_4) + 3\ \Delta H_f^{\circ}(O_2)]$

$= \left[(2\ mol\ CO_2)\left(\dfrac{-393.5\ kJ}{mol\ CO_2}\right) + (2\ mol\ H_2O)\left(\dfrac{-285.8\ kJ}{mol\ H_2O}\right)\right]$

$- \left[(1\ mol\ C_2H_4)\left(\dfrac{52.26\ kJ}{mol\ C_2H_4}\right) + (3\ mol\ O_2)\left(\dfrac{0\ kJ}{mol\ O_2}\right)\right]$

$= [(-787.0\ kJ) + (-571.6\ kJ)] - [52.26\ kJ + 0\ kJ] = 1411\ kJ$

b) $2\ NH_{3(g)} \rightarrow N_{2(g)} + 3\ H_{2(g)}$

$\Delta H_{rxn} = [\Delta H^{\circ}{}_f(N_2) + 3\ \Delta H^{\circ}{}_f(H_2)] - [2\ \Delta H_f^{\circ}(NH_3)]$

$= \left[(1\ mol\ N_2)\left(\dfrac{0\ kJ}{mol\ N_2}\right) + (3\ mol\ H_2)\left(\dfrac{0\ kJ}{mol\ H_2}\right)\right] - \left[(2\ mol\ NH_3)\left(\dfrac{-46.1\ kJ}{mol\ NH_3}\right)\right]$

$= [0\ kJ + 0\ kJ] - [-92.2\ kJ] = 92.2\ kJ$

c) $5\ PbO_{2(s)} + 4\ P_{(s,\ white)} \rightarrow P_4O_{10(s)} + 5\ Pb_{(s)}$

$\Delta H_{rxn} = [\Delta H_f^{\circ}(P_4O_{10}) + 5\ \Delta H_f^{\circ}(Pb)] - [5\ \Delta H_f^{\circ}(PbO_2) + 4\ \Delta H_f^{\circ}(P)]$

$= [(1\ mol\ P_4O_{10})(-2984.0\ kJ/mol\ P_4O_{10}) + (5\ mol\ Pb)(0\ kJ/mol\ Pb)]$
$- [(5\ mol\ PbO_2)(-277.4\ kJ/mol\ PbO_2) + (4\ mol\ P)(0\ kJ/mol\ P)] = -1597\ kJ$

d) $SiCl_{4(l)} + 2\ H_2O_{(l)} \rightarrow SiO_{2(s)} + HCl_{(g)}$

$\Delta H_{rxn} = [\Delta H_f^{\circ}(SiO_2) + 4\ \Delta H_f^{\circ}(HCl)] - [\Delta H_f^{\circ}(SiCl_4) + 2\ \Delta H_f^{\circ}(H_2O)]$

$= [(1\ mol\ SiO_2)(-910.94\ kJ/mol\ SiO_2) + (4\ mol\ HCl)(-92.307\ kJ/mol\ HCl)]$
$- [(1\ mol\ SiCl_4)(-687.0\ kJ/mol\ SiCl_4) + (2\ mol\ H_2O)(-285.830\ kJ/mol\ H_2O)]$
$= -21.5\ kJ$

12.23 $\Delta E_{rxn} = \Delta H_{rxn} - \Delta(PV)_{rxn}$ \hspace{1cm} Refer to Problem 12.21 for ΔH_{rxn}.

$\Delta(PV) \cong 0$ for condensed phases; \hspace{1cm} $\Delta(PV) = \Delta(nRT)$ for gases;
$\Delta(PV) = RT\Delta n$ if T is constant.
$\Delta E_{rxn} = \Delta H_{rxn} - RT\Delta n$ if T constant \hspace{1cm} Assume T = 298 K

a) $C_2H_{4(g)} + 3\ O_{2(g)} \rightarrow 2\ CO_{2(g)} + 2\ H_2O_{(l)}$ \hspace{1cm} $\Delta n = 2 - 4 = -2$
$\Delta E_{rxn} = \Delta H_{rxn} - (8.314\ J/mol\ K)(298\ K)(-2)(kJ/10^3\ J) = 1411\ kJ + 4.96\ kJ$
$= 1416\ kJ$

b) $2\ NH_{3(g)} \rightarrow N_{2(g)} + 3\ H_{2(g)}$ \hspace{1cm} $\Delta n = 4 - 2 = +2$
$\Delta E_{rxn} = \Delta H_{rxn} - (8.314\ J/mol\ K)(298\ K)(+2)(kJ/10^3\ J) = 92.2\ kJ - 4.96\ kJ$
\hspace{6cm} $= 87.2\ kJ$

(continued)

    c) $5 PbO_{2(s)} + 4 P_{(s, white)} \rightarrow P_4O_{10(s)} + 5 Pb_{(s)}$ $\Delta n = 0 - 0 = 0$

    $\Delta E_{rxn} = \Delta H_{rxn} - 0 = -1597 \text{ kJ} - 0 = -1597 \text{ kJ}$

    d) $SiCl_{4(l)} + 2 H_2O_{(l)} \rightarrow SiO_{2(s)} + 4 HCl_{(g)}$

    $\Delta E_{rxn} = \Delta H_{rxn} = -(8.314 \text{ J/mol K})(298 \text{ K})(+4)(\text{kJ}/10^3 \text{ J})$

                         $= -21.5 \text{ kJ} - 9.9 \text{ kJ} = -31.4 \text{ kJ})$

12.25 $\Delta H_{rxn} - \Delta E_{rxn} = \Delta(PV)$ $\Delta(PV) = \Delta(nRT)$ for gases

    a) 0 because $\Delta(PV) \cong 0$ for condensed phases

    b) $\Delta(nRT) = \Delta n \, RT$ $\Delta n = 2 - 0 = 2$

    $= (2 \text{ mol})(8.314 \text{ J/mol K})(298 \text{ K})$ Assuming temperature = 298 K

    $= 4.96 \times 10^3 \text{ J}$

    c) $C_4H_9OH_{(l)} + 6 O_{2(g)} \rightarrow 4 CO_{2(g)} + 5 H_2O_{(l)}$ Assuming T = 298 K, $\Delta n = 4 - 6 = -2$

    $\Delta n \, RT = (-2 \text{ mol})(8.314 \text{ J/mol K})(298 \text{ K}) = -4.96 \times 10^3 \text{ J}$

12.27   a) $3 K_{(s)} + P_{(s, white)} + 2 O_{2(g)} \rightarrow K_3PO_{4(s)}$

    b) $2 C_{(s, graphite)} + O_{2(g)} + 2 H_{2(g)} \rightarrow CH_3CO_2H_{(l)}$

    c) $3 C_{(s, graphite)} + 9/2 H_{2(g)} + 1/2 N_{2(g)} \rightarrow (CH_3)_3N_{(g)}$

    d) $2 Al_{(s)} + 3/2 O_{2(g)} \rightarrow Al_2O_{3(s)}$

12.29   a) $4 NH_{3(g)} + 5 O_{2(g)} \rightarrow 4 NO_{(g)} + 6 H_2O_{(l)}$

    $\Delta H_{rxn} = [4 \, \Delta H_f^\circ(NO) + 6 \, \Delta H_f^\circ(H_2O)] - [(4 \, \Delta H_f^\circ(NH_3) + 5 \, \Delta H_f^\circ(O_2)]$

$$= \left[(4 \text{ mol NO})\left(\frac{90.25 \text{ kJ}}{\text{mol NO}}\right) + (6 \text{ mol } H_2O)\left(\frac{-285.8 \text{ kJ}}{\text{mol } H_2O}\right)\right]$$

$$-\left[(4 \text{ mol } NH_3)\left(\frac{-46.1 \text{ kJ}}{\text{mol } NH_3}\right) + (5 \text{ mol } O_2)\left(\frac{0 \text{ kJ}}{\text{mol } O_2}\right)\right]$$

    $= [361.0 \text{ kJ} + (-1714.8 \text{ kJ}] - [-184.4 \text{ kJ} + 0 \text{ kJ})] = -1169.4 \text{ kJ}$

    b) $4 NH_{3(g)} + 3 O_{2(g)} \rightarrow 2 N_{2(g)} + 6 H_2O_{(l)}$

    $\Delta H_{rxn} = [2 \, \Delta H_f^\circ(N_2) + 6 \, \Delta H_f^\circ(H_2O)] - [(4 \, \Delta H_f^\circ(NH_3) + 3 \, \Delta H_f^\circ(O_2)]$

$$= \left[(2 \text{ mol } N_2)\left(\frac{0 \text{ kJ}}{\text{mol } N_2}\right) + (6 \text{ mol } H_2O)\left(\frac{-285.8 \text{ kJ}}{\text{mol } H_2O}\right)\right]$$

$$-\left[(4 \text{ mol } NH_3)\left(\frac{-46.1 \text{ kJ}}{\text{mol } NH_3}\right) + (3 \text{ mol } O_2)\left(\frac{0 \text{ kJ}}{\text{mol } O_2}\right)\right]$$

    $= [0 \text{ kJ} + (-1714.8 \text{ kJ}] - [-184.4 \text{ kJ} + 0 \text{ kJ})] = -1530.4 \text{ kJ}$

12.31   a) Solution reaction: $MgSO_{4(s)} \xrightarrow{H_2O} Mg^{2+}_{(aq)} + SO_4^{2-}_{(aq)}$

    This is the reverse reaction of that given; therefore, ΔH is negative and when

    $MgSO_4$ dissolves in water, heat is released to the water and absorbed by the water.

(continued)

(12.31 continued)

b) $q_{H_2O} = -q_{salt} = -(n_{salt})(\Delta H_{soln})$

$$n_{salt} = \frac{2.55 \text{ g MgSO}_4}{120.371 \text{ g MgSO}_4 / \text{mol}} = 0.02118 \text{ mol}$$

$q_{H_2O} = -(0.02118 \text{ mol})(-91.2 \text{ kJ/mol}) = 1.932 \text{ kJ}$

c) moles of H_2O = $(500 \text{ mL})\left(\dfrac{1.00 \text{ g}}{1 \text{ mL}}\right)\left(\dfrac{1 \text{ mol H}_2O}{18.02 \text{ g}}\right) = 27.747 \text{ mol H}_2O$

$q_{H_2O} = (n_{H_2O})(C_{p, H_2O})(\Delta T)$

$$\Delta T = \frac{q_{H_2O}}{\left(n_{H_2O}\right)\left(C_{p, H_2O}\right)} = \frac{1.932 \text{ kJ} \times 10^3 \text{ J} / \text{kJ}}{(27.747 \text{ mol})(75.291 \text{ J} / \text{mol K})} = 0.925 \text{ K}$$

12.33   a) Ethane has stronger dispersion forces between molecules than methane has between its molecules. Therefore, it takes less energy to counteract these forces for methane and cause methane to vaporize resulting in a lower heat of vaporization for methane.
b) Ethanol forms hydrogen bonds between molecules which are much stronger than the dipolar forces between diethyl ether molecules. Therefore, ethanol has a much higher heat of vaporization than diethyl ether.
c) Argon and methane have dispersion forces as their interparticle forces. The polarizability for methane is larger than for argon but the shape of the argon atom allows the argon atoms to approach each other more closely than the methane molecules can approach each other. Dispersion forces decrease rapidly with an increase in distance. Therefore, the forces of attraction for argon are greater than for methane and the heat of fusion for argon is greater.

12.35   a) $\Delta H = q_p = nC_p\Delta T$ $\qquad n = \dfrac{2.50 \text{ g}}{18.02 \text{ g} / \text{mol}} = 0.1387 \text{ mol}$
$\Delta T = 100.0°C - 37.5°C = 62.5°C = 62.5 \text{ K}$
$\Delta H = q_p = (0.1387 \text{ mol})(75.291 \text{ J/mol K})(62.5 \text{ K}) = 653 \text{ J or } 0.653 \text{ kJ}$
b) $q_p = n\Delta H_{vap} + nC_p\Delta T$
$= (0.1387 \text{ mol})(40.79 \text{ kJ/mol})(10^3 \text{ J/kJ}) + 653 \text{ J}$
$= 5658 \text{ J} + 653 \text{ J} = 6.31 \times 10^3 \text{ J or } 6.31 \text{ kJ}$

12.37 $q_{Cu} + q_{ice} = 0$

$q_{Cu} = nC_p\Delta T = \left(\dfrac{12.7 \text{ g}}{63.55 \text{ g} / \text{mol}}\right)(24.435 \text{ J} / \text{mol K})(0°C - 200°C)\left(\dfrac{K}{°C}\right) = -976.6 \text{ J}$

$q_{ice} = -q_{Cu} = 976.6 \text{ J} = n\Delta H_{fus} = n \, (6.01 \text{ kJ/mol})(10^3 \text{ J/kJ})$

$n = \dfrac{976.6 \text{ J}}{(6.01 \text{ kJ} / \text{mol})\left(10^3 \text{ J/kJ}\right)} = 0.1625 \text{ mol}$

$m_{ice} = (0.1625 \text{ mol})(18.02 \text{ g/mol}) = 2.93 \text{ g of ice will melt}$

12.39 $C_{(s, \text{graphite})} + O_{2(g)} \rightarrow CO_{2(g)}$ $\qquad$ $\Delta H = -393.51$ kJ/mol

$C_{60} + 60\, O_{2(g)} \rightarrow 60\, CO_{2(g)}$

$\Delta H_f^{\circ}(C_{60}) = $ probably positive

Therefore, the ΔH for combustion of buckminsterfullerene equals 60 ΔH of combustion of graphite minus the positive ΔH_f° for C_{60}. Therefore, the value for buckminsterfullerene will be more negative. The heat of combustion per gram will be greater for buckminsterfullerene (by $(1/60)(1/12)$ of the ΔH_f° of buckminsterfullerene).

12.41 Any living organism will die if made into an isolated thermodynamic system. Even before dying, its performance will change from the normal.

12.43 Prefer classification as physical change. The process can be reversed by a physical process; that is, the water can be allowed to evaporate (or be boiled off) to recover the original substance, $NaCl_{(s)}$.

12.45 a) $C_6H_{12}O_{6(s)} + 6\, O_{2(g)} \rightarrow 6\, H_2O_{(l)} + 6\, CO_{2(g)}$

b) $\left(\dfrac{-15.7 \text{ kJ}}{1.00 \text{ g glucose}}\right)\left(\dfrac{180.156 \text{ g glucose}}{\text{mol glucose}}\right) = -2.83 \times 10^3$ kJ / mol glucose

c) $\Delta H_{rxn} = [6\, \Delta H_f^{\circ}(H_2O_{(l)}) + 6\, \Delta H_f^{\circ}(CO_2)] - [\Delta H_f^{\circ}(C_6H_{12}O_6) + 6\, \Delta H_f^{\circ}(O_2)]$

$-2.83 \times 10^3 \text{ kJ} = \left[(6 \text{ mol } H_2O_{(l)})\left(\dfrac{-285.8 \text{ kJ}}{\text{mol } H_2O_{(l)}}\right) + (6 \text{ mol } CO_2)\left(\dfrac{-393.5 \text{ kJ}}{\text{mol } CO_2}\right)\right]$

$-\left[(1 \text{ mol } C_6H_{12}O_6)\Delta H_f^{\circ}(C_6H_{12}O_6) + (6 \text{ mol } O_2)\left(\dfrac{0 \text{ kJ}}{\text{mol } O_2}\right)\right]$

$-2.83 \times 10^3 \text{ kJ} =$

$[-1714.8 \text{ kJ} - 2361.0 \text{ kJ}] - [(1 \text{ mol } C_6H_{12}O_6)\, \Delta H_f^{\circ}(C_6H_{12}O_6) + 0 \text{ kJ}]$

$-2.83 \times 10^3 \text{ kJ} + [4075.8 \text{ kJ}] = -(1 \text{ mol } C_6H_{12}O_6)\, \Delta H_f^{\circ}(C_6H_{12}O_6)$

$\Delta H_f^{\circ}(C_6H_{12}O_6) = -1.25 \times 10^3$ kJ/ mol $C_6H_{12}O_6$

12.47 $\Delta E = q + w = q - P\Delta V = nC_p\Delta T - P(V_f - V_i)$

$= \left(\dfrac{1.25 \text{ g}}{132.92 \text{ g / mol}}\right)\left(\dfrac{80.7 \text{ J}}{\text{mol K}}\right)(20^{\circ}C - 50^{\circ}C)\left(\dfrac{K}{^{\circ}C}\right)$

$-(1 \text{ atm})(248 \text{ mL} - 274 \text{ mL})\left(\dfrac{\text{cm}^3}{\text{mL}}\right)$

$= -22.767 \text{ J} + 26 \text{ atm cm}^3\left(\dfrac{1 \text{ m}}{10^2 \text{ cm}}\right)^3\left(\dfrac{1.013 \times 10^5 \text{ kg m}^{-1}\text{ s}^{-2}}{1 \text{ atm}}\right)$

$= -22.767 \text{ J} + 2.6338 \text{ kg m}^2\text{ s}^{-2} = -22.767 \text{ J} + 2.6338 \text{ J} = -20.1$ J

(continued)

(12.47 continued)

$\Delta H = \Delta E + \Delta(nRT); \ \Delta(nRT) = R\Delta(nT) = RT\Delta n + Rn\Delta T$ \qquad $\Delta n = 0$

$\Delta T = 20°C - 50°C = -30°C = -30 \text{ K}$

$= -20.1 \text{ J} + nR\Delta T = -20.1 \text{ J} + \left(\dfrac{1.25 \text{ g}}{132.92 \text{ g / mol}}\right)(8.314 \text{ J / mol K})(-30 \text{ K})$

$= -20.1 \text{ J} - 2.3 \text{ J} = -22.4 \text{ J}$

12.49 Let T = final temperature \qquad $q_{spoon} = -q_{coffee}$

$nC_{p,spoon}\Delta T = -nC_{pH_2O}\Delta T$

$\left(\dfrac{99 \text{ g Ag}}{107.9 \text{ Ag / mol}}\right)\left(\dfrac{25.351 \text{ J}}{\text{mol K}}\right)(T - 280 \text{ K})$

$= -(200 \text{ mL H}_2O)\left(\dfrac{1.00 \text{ g}}{\text{mL}}\right)\left(\dfrac{1 \text{ mol}}{18.02 \text{ g H}_2O}\right)\left(\dfrac{75.291 \text{ J}}{\text{mol K}}\right)(T - 350 \text{ K})$

$(23.26 \text{ J/K}) \text{ T} - 6512.8 \text{ J} = -(835.64 \text{ J/K}) \text{ T} + 292473.4 \text{ J}$

$2.98986 \times 10^5 \text{ J} = (858.9 \text{ J/K}) \text{ T}$

$T = 348.1 \text{ K}$

An aluminum spoon of the same mass (99 g) would be (99 g/27 g mol^{-1}) 3.7 moles while the silver spoon would be (99 g/107.9 g mol^{-1}) 0.92 moles. C_p times number of moles yields:

For Ag: (25.351 J/mol K)(0.92 mol) = 23 J/K
For Al: (24.35 J/mol K)(3.7 mol) = 90 J/K

Therefore, the final temperature of the coffee would be lower for the aluminum spoon.

12.51 $w_{sys} = -P_{ext}\Delta V_{sys}$

$w_{gas} = -(4.00 \text{ atm})(20.0 \text{ L} - 30.0 \text{ L})$

$= (+40.0 \text{ atm L})\left(\dfrac{1000 \text{ cm}^3}{\text{L}}\right)\left(\dfrac{1 \text{ m}}{10^2 \text{ cm}}\right)^3\left(\dfrac{1.013 \times 10^5 \text{ kg m}^{-1} \text{ s}^{-2}}{\text{atm}}\right)$

$= 4052 \text{ kg m}^2 \text{ s}^{-2} = 4052 \text{ J} \cong 4.05 \text{ kJ}$

$q_{gas} = nC_p\Delta T \quad \Delta T = 0 \therefore q_{gas} = 0$

$\Delta E_{gas} = q_{gas} + w_{gas} = 0 + 4.05 \text{ kJ} = 4.05 \text{ kJ or } 4.05 \times 10^3 \text{ J}$

12.53 a) $2 \text{ SO}_{2(g)} + \text{O}_{2(g)} \rightarrow 2 \text{ SO}_{3(g)}$

$\Delta H_{rxn} = [2 \ \Delta H_f^{\circ}(\text{SO}_3)] - [2 \ \Delta H_f^{\circ}(\text{SO}_2) + \Delta H_f^{\circ}(\text{O}_2)]$

$= \left[(2 \text{ mol SO}_3)\left(\dfrac{-454.51 \text{ kJ}}{\text{mol SO}_3}\right)\right] - \left[(2 \text{ mol SO}_2)\left(\dfrac{-296.83 \text{ kJ}}{\text{mol SO}_2}\right) + (1 \text{ mol O}_2)\left(\dfrac{0 \text{ kJ}}{\text{mol O}_2}\right)\right]$

$= [-909.02 \text{ kJ}] - [-593.66 \text{ kJ} + 0 \text{ kJ}] = -315.36 \text{ kJ}$

(continued)

(12.53 continued)

b) $2 \text{ NO}_{2(g)} \rightarrow \text{N}_2\text{O}_{4(g)}$

$\Delta H_{rxn} = [\Delta H_f^\circ(\text{N}_2\text{O}_4)] - [2 \; \Delta H_f^\circ(\text{NO}_2)]$

$$= \left[(1 \text{ mol N}_2\text{O}_4)\left(\frac{9.16 \text{ kJ}}{\text{mol N}_2\text{O}_4}\right) \right] - \left[(2 \text{ mol NO}_2)\left(\frac{33.18 \text{ kJ}}{\text{mol NO}_2}\right) \right]$$

$= [9.16 \text{ kJ}] - [66.36 \text{ kJ}] = -57.20 \text{ kJ}$

c) $\text{Fe}_2\text{O}_{3(s)} + 2 \text{ Al}_{(s)} \rightarrow \text{Al}_2\text{O}_{3(s)} + 2 \text{ Fe}_{(s)}$

$\Delta H_{rxn} = [\Delta H_f^\circ(\text{Al}_2\text{O}_3) + 2 \; \Delta H_f^\circ(\text{Fe})] - [\Delta H_f^\circ(\text{Fe}_2\text{O}_3) + 2 \; \Delta H_f^\circ(\text{Al})]$

$$= \left[(1 \text{ mol Al}_2\text{O}_3)\left(\frac{-1675.7 \text{ kJ}}{\text{mol Al}_2\text{O}_3}\right) + (2 \text{ mol Fe})\left(\frac{0 \text{ kJ}}{\text{mol Fe}}\right) \right] -$$

$$\left[(1 \text{ mol Fe}_2\text{O}_3)\left(\frac{-824.2 \text{ kJ}}{\text{mol NO}_2}\right) + (2 \text{ mol Al})\left(\frac{0 \text{ kJ}}{\text{mol Al}}\right) \right]$$

$= [-1675.7 \text{ kJ} + 0 \text{ kJ}] - [-824.2 \text{ kJ} + 0 \text{ kJ}] = -851.5 \text{ kJ}$

12.55 $(1 \text{ km})\left(\dfrac{1 \text{ L}}{6.0 \text{ km}}\right)\left(\dfrac{10^3 \text{ mL}}{1 \text{ L}}\right)\left(\dfrac{0.68 \text{ g}}{\text{mL}}\right)\left(\dfrac{48 \text{ kJ}}{\text{g}}\right) = 5.4 \times 10^3 \text{ kJ}$

12.57 a) $2 \text{ NH}_{3(g)} + 3 \text{ O}_{2(g)} + 2 \text{ CH}_{4(g)} \rightarrow 2 \text{ HCN}_{(g)} + 6 \text{ H}_2\text{O}_{(g)}$

$\Delta H_{rxn} = [2 \Delta H_f^\circ(\text{HCN}_{(g)}) + 6 \Delta H_f^\circ(\text{H}_2\text{O}_{(g)})]$

$\qquad - [2 \Delta H_f^\circ(\text{NH}_3) + 3 \Delta H_f^\circ(\text{O}_2) + 2 \Delta H_f^\circ(\text{CH}_4)]$

$$= \left[(2 \text{ mol HCN})\left(\frac{135.1 \text{ kJ}}{\text{mol HCN}}\right) + (6 \text{ mol H}_2\text{O})\left(\frac{-241.8 \text{ kJ}}{\text{mol H}_2\text{O}}\right) \right]$$

$$- \left[(2 \text{ mol NH}_3)\left(\frac{-46.1 \text{ kJ}}{\text{mol NH}_3}\right) + (3 \text{ mol O}_2)\left(\frac{0 \text{ kJ}}{\text{mol O}_2}\right) + (2 \text{ mol CH}_4)\left(\frac{-74.81 \text{ kJ}}{\text{mol CH}_4}\right) \right]$$

$= [270.2 \text{ kJ} + (-1450.8 \text{ kJ})] - [(-92.2 \text{ kJ}) + 0 \text{ kJ} + (-149.62 \text{ kJ})] = -938.8 \text{ kJ}$

b) $2 \text{ C}_2\text{H}_{2(g)} + 5 \text{ O}_{2(g)} \rightarrow 4 \text{ CO}_{2(g)} + 2 \text{ H}_2\text{O}_{(g)}$

$\Delta H_{rxn} = [4 \; \Delta H_f^\circ(\text{CO}_2) + 2 \; \Delta H_f^\circ(\text{H}_2\text{O}_{(g)})]$

$\qquad - [2 \; \Delta H_f^\circ(\text{C}_2\text{H}_2) + 5 \; \Delta H_f^\circ(\text{O}_2)]$

$$= \left[(4 \text{ mol CO}_2)\left(\frac{-393.5 \text{ kJ}}{\text{mol CO}_2}\right) + (2 \text{ mol H}_2\text{O})\left(\frac{-241.8 \text{ kJ}}{\text{mol H}_2\text{O}}\right) \right]$$

$$- \left[(2 \text{ mol C}_2\text{H}_2)\left(\frac{226.7 \text{ kJ}}{\text{mol C}_2\text{H}_2}\right) + (5 \text{ mol O}_2)\left(\frac{0 \text{ kJ}}{\text{mol O}_2}\right) \right]$$

$= [(-1574.0 \text{ kJ}) + (-483.6 \text{ kJ})] - [453.4 \text{ kJ} + 0 \text{ kJ}] = -2511.0 \text{ kJ}$

c) $\text{C}_2\text{H}_{4(g)} + \text{O}_{3(g)} \rightarrow \text{CH}_3\text{CHO}_{(g)} + \text{O}_{2(g)}$

(continued)

(12.57 continued)

$$\Delta H_{rxn} = [\Delta H_f^\circ(CH_3CHO) + \Delta H_f^\circ(O_2)] - [\Delta H_f^\circ(C_2H_4) + \Delta H_f^\circ(O_3)]$$

$$= \left[(1 \text{ mol } CH_3CHO)\left(\frac{-166.2 \text{ kJ}}{\text{mol } CH_3CHO}\right) + (1 \text{ mol } O_2)\left(\frac{0 \text{ kJ}}{\text{mol } O_2}\right)\right]$$

$$- \left[(1 \text{ mol } C_2H_4)\left(\frac{52.3 \text{ kJ}}{\text{mol } C_2H_4}\right) + (1 \text{ mol } O_3)\left(\frac{142.7 \text{ kJ}}{\text{mol } O_3}\right)\right]$$

$$= [-166.2 \text{ kJ} + 0 \text{ kJ}] - [52.3 \text{ kJ} + 142.7 \text{ kJ}] = -361.2 \text{ kJ}$$

12.59 Heat to convert $H_2O = nC_{p,H_2O}\Delta T + n\Delta H_{vap}$ $\Delta T = (100°C - 25°C) = 75°C = 75 \text{ K}$

$$q = \left(\frac{250 \text{ g } H_2O}{18.02 \text{ g } H_2O/\text{mol}}\right)\left(\frac{75.291 \text{ J}}{\text{mol K}}\right)(75 \text{ K}) + \left(\frac{250 \text{ g } H_2O}{18.02 \text{ g } H_2O/\text{mol}}\right)\left(\frac{40.79 \text{ kJ}}{\text{mol}}\right)$$

$$= 7.834 \times 10^4 \text{ J} + 565.9 \text{ kJ} = 7.834 \times 10^1 \text{ kJ} + 565.9 \text{ kJ} = 644 \text{ kJ}$$

$$\text{mass of } CH_4 = (644 \text{ kJ})\left(\frac{\text{mol } CH_4}{803 \text{ kJ}}\right)\left(\frac{16.043 \text{ g } CH_4}{\text{mol } CH_4}\right) = 12.9 \text{ g } CH_4$$

12.61 $H_2O_{(l, 298 \text{ K})} \rightarrow H_2O_{(g, 298 \text{ K})}$ $\Delta H_{rxn} = [\Delta H_f^\circ(H_2O_{(g)})] - [\Delta H_f^\circ(H_2O_{(l)})] =$

$$\left[(1 \text{ mol } H_2O_{(g)})\left(\frac{-241.8 \text{ kJ}}{\text{mol } H_2O_{(g)}}\right)\right] - \left[(1 \text{ mol } H_2O_{(l)})\left(\frac{-285.8 \text{ kJ}}{\text{mol } H_2O_{(l)}}\right)\right]$$

$$= -241.8 \text{ kJ} + 285.8 \text{ kJ} = 44.0 \text{ kJ} \qquad \Delta H_{vap} = 40.79 \text{ kJ/mol}$$

$44.0 \text{ kJ/mol} \cong 5.65 \text{ kJ/mol} + 40.79 \text{ kJ/mol} - 2.74 \text{ kJ/mol} = 43.7 \text{ kJ/mol}$
The ΔH_{vap} is for the steam and water at 100°C (373 K) and 1 atm. The ΔH_{rxn} is for steam and water at standard conditions of 25°C (298 K) and 1 atm. Note that the 2 paths for the change give the same enthalpy change to two significant digits.

12.63 $$C_{p,Pb} = \frac{100.0 \text{ J}}{(52.5 \text{ g})(299.6 \text{ K} - 280.0 \text{ K})}\left(\frac{207.2 \text{ g Pb}}{1 \text{ mol Pb}}\right) = 20.1 \text{ J/mol K}$$

12.65 $P_4S_{3(s)} + 8\ O_{2(g)} \rightarrow P_4O_{10(s)} + 3\ SO_{2(g)}$ $\Delta H_{rxn} = -3677$ kJ

-3677 kJ $= \Delta H_{rxn} = [\Delta H_f^\circ(P_4O_{10}) + 3\ \Delta H_f^\circ(SO_2)] - [\Delta H_f^\circ(P_4S_3) + 8\ \Delta H_f^\circ(O_2)]$

$$= \left[(1\ \text{mol } P_4O_{10})\left(\frac{-2984.0\ \text{kJ}}{\text{mol } P_4O_{10}}\right) + (3\ \text{mol } SO_2)\left(\frac{-296.8\ \text{kJ}}{\text{mol } SO_2}\right)\right]$$

$$-\left[\Delta H_f^\circ(P_4S_3) + (8\ \text{mol } O_2)\left(\frac{0\ \text{kJ}}{\text{mol } O_2}\right)\right]$$

-3677 kJ $= -2984.0$ kJ $+ (-890.4$ kJ$) - \Delta H_f^\circ(P_4S_3) + 0$ $\Delta H_f^\circ(P_4S_3) = -197$ kJ

12.67 The process would appear to be deposition (the opposite of sublimation).
a) ΔH_{sys} would be (-) as heat is released (exothermic process) as a gas is transformed into a solid.
b) ΔE_{surr} would be (+) as it would have the opposite sign of ΔE_{sys} which would be approximately the same value as ΔH_{sys}.
c) $\Delta E_{universe}$ would equal zero (0); the total energy of the universe is conserved.

12.69 The process is an expansion at constant temperature.
a) w_{sys} is (-) for an expansion: [$w_{sys} = -P_{ext}\Delta V_{sys}$, ΔV_{sys} is (+)].
b) ΔE_{surr} would be (+) because its sign is opposite ΔE_{sys} which is (-).
c) $q_{sys} = 0$ because this is a constant temperature process.

12.71 $C_{20}H_{32}O_{2(s)} + 27\ O_{2(g)} \rightarrow 20\ CO_{2(g)} + 16\ H_2O_{(l)}$

$\Delta H_{combustion} = [20\ \Delta H_f^\circ(CO_2) + 16\ \Delta H_f^\circ(H_2O_{(l)})]$

$- [\Delta H_f^\circ(C_{20}H_{32}O_2) + 27\ \Delta H_f^\circ(O_2)]$

$$= \left[(20\ \text{mol } CO_2)\left(\frac{-393.5\ \text{kJ}}{\text{mol } CO_2}\right) + (16\ \text{mol } H_2O)\left(\frac{-285.8\ \text{kJ}}{\text{mol } H_2O}\right)\right]$$

$$-\left[(1\ \text{mol } C_{20}H_{32}O_2)\left(\frac{-636\ \text{kJ}}{\text{mol } C_{20}H_{32}O_2}\right) + (27\ \text{mol } O_2)\left(\frac{0\ \text{kJ}}{\text{mol } O_2}\right)\right]$$

$= -1.18 \times 10^4$ kJ

heat needed to warm bear flesh =

$(500\ \text{kg})(10^3\ \text{g/kg})(4.18\ \text{J g}^{-1}\ \text{K}^{-1})(25°C - 5°C)(K/°C) = 4.18 \times 10^7$ J

$(4.18 \times 10^7\ \text{J})\left(\frac{\text{kJ}}{10^3\ \text{J}}\right)\left(\frac{1\ \text{mol } C_{20}H_{32}O_2}{1.18 \times 10^4\ \text{kJ}}\right)\left(\frac{304.46\ \text{g } C_{20}H_{32}O_2}{1\ \text{mol } C_{20}H_{32}O_2}\right)$

$= 1080\ \text{g } C_{20}H_{32}O_2$ or 1.08 kg

12.73 a) The drawing of this figure after work has been done on the system would need to reflect a $-\Delta V$. Therefore, the piston would need to be moved in to keep pressure constant.

b) Pressure will increase and the piston will move out as work is done by the system.

CHAPTER 13: SPONTANEITY OF CHEMICAL PROCESSES

13.1   a) The ordered arrangement of sand particles in the sand castle is destroyed as the ocean waves wash them away into a disordered state.

b) The secretary expends energy as she orders (straightens) the objects on the boss' desk. Her body becomes disordered as it produces CO_2 and H_2O from carbohydrates and/or fat and the CO_2 is exhaled.

c) The sticks become disordered as they drop to the floor and land in a disorganized manner.

d) The engine is ordered as the skilled mechanic works to reassemble the engine. The mechanic's body becomes disordered as it produces energy and reaction products from the metabolism of carbohydrates and/or fat.

13.3   The organized molecules of ink in the drop become disorganized in the beaker of water. After the drop enters the water, collisions with H_2O molecules cause the ink molecules to become separated and to become homogeneously distributed throughout the beaker of water.

13.5   a) The air confined inside the tire has some order. Upon puncturing, the air molecules escape and become disordered as they spread throughout the atmosphere.

b) The ordered molecules of perfume in the bottle escape out into the air of the room and become disordered as they spread throughout the room.

13.7   a) As the dry ice sublimes (changes from solid CO_2 to gaseous CO_2), its entropy increases and some of the water freezes to become ice with a decrease in its entropy.

b) $q_{CO_2} = +(n_{CO_2})(\Delta H_{subl}) = \left(\dfrac{12.5 \text{ g } CO_2}{44.01 \text{ g } CO_2/\text{mole}} \right)\left(\dfrac{25.2 \text{ k J}}{\text{mol}} \right) = 7.157 \text{ kJ}$

$\Delta S_{CO_2} = \left(\dfrac{7.157 \text{ kJ}}{195 \text{ K}} \right)\left(\dfrac{10^3 \text{ J}}{\text{kJ}} \right) = 36.7 \text{ J/K}$

$q_{H_2O} = -q_{CO_2} = -7.157 \text{ kJ}$

$\Delta S_{H_2O} = \dfrac{-7.157 \text{ kJ}}{273.15 \text{ K}}\left(\dfrac{10^3 \text{ J}}{\text{kJ}} \right) = -26.2 \text{ J/K}$

$\Delta S_{overall} = \Delta S_{CO_2} + \Delta S_{H_2O} = 36.7 \text{ J/K} - 26.2 \text{ J/K} = 10.5 \text{ J/K}$

13.9 $q_{H_2O} = -n_{H_2O}\Delta H_{vap} = -\left(\dfrac{15.5 \text{ g } H_2O}{18.02 \text{ g } H_2O/mol}\right)\left(\dfrac{40.79 \text{ kJ}}{mol}\right) = -35.09 \text{ kJ}$

$$\Delta S_{H_2O} = \dfrac{-35.09 \text{ kJ}}{373.15 \text{ K}}\left(\dfrac{10^3 \text{ J}}{kJ}\right) = -94.0 \text{ J/K}$$

The entropy change of the surroundings will be positive and greater in value than 94.0 J/K because $\Delta S_{universe} > 0$.

13.11 a) The soft drink is the system; ΔS is negative (cooling decreases disorder).

b) The air is the system; ΔS is negative (concentration is increased).

c) The juice concentrate is the system; ΔS is positive (concentration is decreased).

13.13 a) HgS would have the larger molar entropy because HgO and HgS have similar structures but HgS has the larger molar mass.

b) The $MgCl_2$ in aqueous solution would have the larger molar entropy because it dissociates into 3 ions in solution while NaCl only dissociates into 2 ions in solution.

c) Br_2 would have the larger molar entropy because it is a liquid while I_2 is a solid.

13.15 $S^{\circ}_{He} = 126.15 \text{ J/mol K}$ Per mole of atoms = 126.15 J/mol K

$S^{\circ}_{H_2} = 130.684 \text{ J/mol K}$ Per mole of atoms =

$$\left(\dfrac{130.684 \text{ J}}{mol \text{ K}}\right)\left(\dfrac{1 \text{ molecule}}{2 \text{ atoms}}\right) = 65.34 \text{ J/mol atoms K}$$

$S^{\circ}_{CH_4} = 186.26 \text{ J/mol K}$ Per mole of atoms =

$$\left(\dfrac{186.26 \text{ J}}{mol \text{ K}}\right)\left(\dfrac{1 \text{ molecule}}{5 \text{ atoms}}\right) = 37.25 \text{ J/mol atoms K}$$

$S^{\circ}_{C_2H_6} = 229.60 \text{ J/mol K}$ Per mole of atoms =

$$\left(\dfrac{229.60 \text{ J}}{mol \text{ K}}\right)\left(\dfrac{1 \text{ molecule}}{8 \text{ atoms}}\right) = 28.70 \text{ J/mol atoms K}$$

$S^{\circ}_{C_3H_6} = 226.9 \text{ J/mol K}$ Per mole of atoms =

$$\left(\dfrac{226.9 \text{ J}}{mol \text{ K}}\right)\left(\dfrac{1 \text{ molecule}}{9 \text{ atoms}}\right) = 25.21 \text{ J/mol atoms K}$$

The tabulated values of S° do not include a value for C_3H_8. The trend for CH_4 and C_2H_6 and C_3H_6 suggest the S° for C_3H_8 would be 260 - 270 J/mol K, resulting in 23 - 24.5 J/mol atoms K. These values show the trend that as the number of bonded atoms increases the entropy increases per mole of molecules. This is to be expected since larger molecules have more ways to arrange their atoms in space. Thus, they have greater intermolecular disorder. However, the above data also shows that as atoms are bonded the order increases for each mole of atoms. This is to be expected since free atoms have less order than bonded atoms.

13.17 a) $N_{2(g)} + 3 H_{2(g)} \rightarrow 2 NH_{3(g)}$

$\Delta S^{\circ}_{rxn} = [2\ S^{\circ}(NH_3)] - [S^{\circ}(N_2) + 3\ S^{\circ}(H_2)]$

$= \left[(2\ mol\ NH_3)\left(\dfrac{192.45\ J/K}{mol\ NH_3} \right) \right]$

$- \left[(1\ mol\ N_2)\left(\dfrac{191.61\ J/K}{mol\ N_2} \right) + (3\ mol\ H_2)\left(\dfrac{130.684\ J/K}{mol\ H_2} \right) \right] = -198.76\ J/K$

b) $3\ O_{2(g)} \rightarrow 2\ O_{3(g)}$

$\Delta S^{\circ}_{rxn} = [2\ S^{\circ}(O_3)] - [3\ S^{\circ}(O_2)]$

$= \left[(2\ mol\ O_3)\left(\dfrac{238.93\ J/K}{mol\ O_3} \right) \right] - \left[(3\ mol\ O_2)\left(\dfrac{205.138\ J/K}{mol\ O_2} \right) \right] = -137.55\ J/K$

c) $PbO_{2(s)} + 2\ Ni_{(s)} \rightarrow Pb_{(s)} + 2\ NiO_{(s)}$

$\Delta S^{\circ}_{rxn} = [S^{\circ}(Pb) + 2\ S^{\circ}(NiO)] - [S^{\circ}(PbO_2) + 2\ S^{\circ}(Ni)]$

$= \left[(1\ mol\ Pb)\left(\dfrac{64.81\ J/K}{mol\ Pb} \right) + (2\ mol\ NiO)\left(\dfrac{37.99\ J/K}{mol\ NiO} \right) \right]$

$- \left[(1\ mol\ PbO_2)\left(\dfrac{68.6\ J/K}{mol\ PO_2} \right) + (2\ mol\ Ni)\left(\dfrac{29.87\ J/K}{mol\ Ni} \right) \right] = 12.5\ J/K$

d) $C_2H_{4(g)} + 3\ O_{2(g)} \rightarrow 2\ CO_{2(g)} + 2\ H_2O_{(l)}$

$\Delta S^{\circ}_{rxn} = [2\ S^{\circ}(CO_2) + 2\ S^{\circ}(H_2O_{(l)})] - [S^{\circ}(C_2H_4) + 3\ S^{\circ}(O_2)]$

$= \left[(2\ mol\ CO_2)\left(\dfrac{213.74\ J/K}{mol\ CO_2} \right) + (2\ mol\ H_2O)\left(\dfrac{69.91\ J/K}{mol\ H_2O} \right) \right]$

$- \left[(1\ mol\ C_2H_4)\left(\dfrac{219.56\ J/K}{mol\ C_2H_4} \right) + (3\ mol\ O_2)\left(\dfrac{205.14\ J/K}{mol\ O_2} \right) \right] = -267.68\ J/K$

13.19 a) Disorder decreases due to increased number of bonds. The magnitude reflects the increased bonds.

b) Disorder decreases due to increased number of bonds. The magnitude reflects the increased bonds.

c) Disorder increases slightly due to the formation of the smaller NiO compared to PbO_2.

d) Disorder increases by a considerable amount due to the large amount of intermolecular interactions present in water.

111

13.21 a) 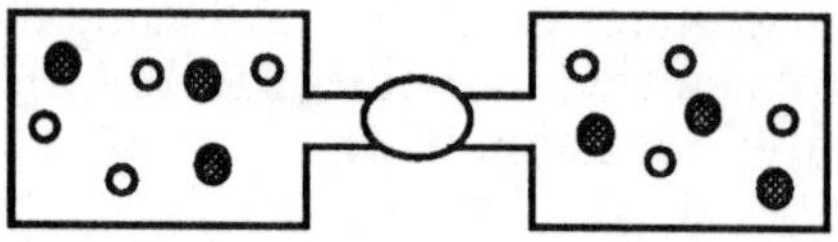 b)

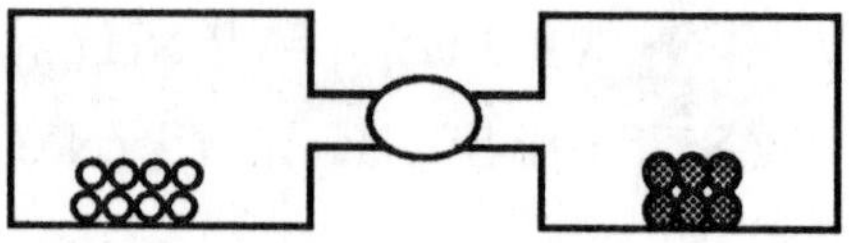

13.23 a) $2\,H_{2(g)} + O_{2(g)} \rightarrow 2\,H_2O_{(l)}$

$\Delta G^{\circ}_{rxn} = [2\,\Delta G^{\circ}(H_2O_{(l)})] - [\,2\,\Delta G^{\circ}(H_2) + \Delta G^{\circ}(O_2)]$

$= [(2\text{ mol }H_2O_{(l)})(-237.13\text{ kJ/mol }H_2O)] -$
$\qquad [(2\text{ mol }H_2)(0.00\text{ kJ/mol }H_2)] + (1\text{ mol }O_2)(0.00\text{ kJ/mol }O_2)]$

$$= -474.26\text{ kJ}$$

b) $C_{(s)} + 2\,H_{2(g)} \rightarrow CH_{4(g)}$ \qquad Assuming $C_{(s)}$ is graphite form:

$\Delta G^{\circ}_{rxn} = [\Delta G^{\circ}(CH_4)] - [\Delta G^{\circ}(C_{(s,\,graphite)}) + 2\,\Delta G^{\circ}(H_2)]$

$$= \left[(1\text{ mol }CH_4)\left(\frac{-50.72\text{ kJ}}{\text{mol }CH_4}\right)\right]$$

$$-\left[(1\text{ mol }C_{(s,\,graphite)})\left(\frac{0.00\text{ kJ}}{\text{mol }C_{(s)}}\right) + (2\text{ mol }H_2)\left(\frac{0.00\text{ kJ}}{\text{mol }H_2}\right)\right] = -50.72\text{ kJ}$$

c) $C_2H_5OH_{(l)} + 3\,O_{2(g)} \rightarrow 2\,CO_{2(g)} + 3\,H_2O_{(l)}$

$\Delta G^{\circ}_{rxn} = [2\,\Delta G^{\circ}(CO_2) + 3\,\Delta G^{\circ}(H_2O_{(l)})] - [\Delta G^{\circ}(C_2H_5OH) + 3\,\Delta G^{\circ}(O_2)]$

$$= \left[(2\text{ mol }CO_2)\left(\frac{-394.36\text{ kJ}}{\text{mol }CO_2}\right) + \left(3\text{ mol }H_2O_{(l)}\right)\left(\frac{-237.13\text{ kJ}}{\text{mol }H_2O}\right)\right]$$

$$-\left[(1\text{ mol }C_2H_5OH)\left(\frac{-174.78\text{ kJ}}{\text{mol }C_2H_5OH}\right) + (3\text{ mol }O_2)\left(\frac{0.00\text{ kJ}}{\text{mol }O_2}\right)\right] = -1,325.3\text{ kJ}$$

d) $Fe_2O_{3(s)} + 2\,Al_{(s)} \rightarrow Al_2O_{3(s)} + 2\,Fe_{(s)}$

$\Delta G^{\circ}_{rxn} = [\Delta G^{\circ}(Al_2O_3) + 2\,\Delta G^{\circ}(Fe)] - [\Delta G^{\circ}(Fe_2O_3) + 2\,\Delta G^{\circ}(Al)]$

$$= \left[(1\text{ mol }Al_2O_3)\left(\frac{-1582.3\text{ kJ}}{\text{mol }Al_2O_3}\right) + (2\text{ mol }Fe)\left(\frac{0.0\text{ kJ}}{\text{mol }Fe}\right)\right]$$

$$-\left[(1\text{ mol }Fe_2O_3)\left(\frac{-742.2\text{ kJ}}{\text{mol }Fe_2O_3}\right) + (2\text{ mol }Al)\left(\frac{0.00\text{ kJ}}{\text{mol }Al}\right)\right] = -840.1\text{ kJ}$$

13.25 We can estimate free energy changes at temperatures other than 298 using the equation $\Delta G_{rxn} = \Delta H° + T\Delta S°$. $\Delta S°$'s were calculated in Problem 13.17. $\Delta H°$'s will need to be calculated before we can answer this question.

a) $N_{2(g)} + 3\,H_{2(g)} \rightarrow 2\,NH_{3(g)}$

$\Delta H°_{rxn} = [2\,\Delta H°(NH_3)] - [\Delta H°(N_2) + 3\,\Delta H°(H_2)]$

$$= \left[(2\text{ mol }NH_3)\left(\frac{-46.11\text{ kJ}}{\text{mol }NH_3} \right) \right]$$

$$- \left[(1\text{ mol }N_2)\left(\frac{0.00\text{ kJ}}{\text{mol }N_2} \right) - (3\text{ mol }H_2)\left(\frac{0.00\text{ kJ}}{\text{mol }H_2} \right) \right] = -92.22\text{ kJ}$$

$\Delta G°_{rxn,425} = \Delta H°_{rxn} - T\Delta S°_{rxn} = -92.22\text{ kJ} - [425\text{ K }(-198.76\text{ J/K})(10^{-3}\text{ kJ/J})]$
$$= -7.7\text{ kJ}$$

b) $3\,O_{2(g)} \rightarrow 2\,O_{3(g)}$

$\Delta H°_{rxn} = [2\,\Delta H°(O_3)] - [3\,\Delta H°(O_2)]$

$$= \left[(2\text{ mol }O_3)\left(\frac{142.7\text{ kJ}}{\text{mol }O_3} \right) \right] - \left[(3\text{ mol }O_2)\left(\frac{0.00\text{ kJ}}{\text{mol }O_2} \right) \right] = 285.4\text{ kJ}$$

$\Delta G°_{rxn,425} = \Delta H°_{rxn} - T\Delta S°_{rxn} = 285.4\text{ kJ} - [425\text{ K }(-137.55\text{ J/K})(10^{-3}\text{ kJ/J})]$
$$= 343.9\text{ kJ}$$

c) $PbO_{2(s)} + 2\,Ni_{(s)} \rightarrow Pb_{(s)} + 2\,NiO_{(s)}$

$\Delta H°_{rxn} = [\Delta H°(Pb) + 2\,\Delta H°(NiO)] - [\Delta H°(PbO_2) + 2\,\Delta H°(Ni)]$

$$= \left[(1\text{ mol }Pb)\left(\frac{0.00\text{ kJ}}{\text{mol }Pb} \right) + (2\text{ mol }NiO)\left(\frac{-239.7\text{ kJ}}{\text{mol }NiO} \right) \right]$$

$$- \left[(1\text{ mol }PbO_2)\left(\frac{-277.4\text{ kJ}}{\text{mol }PO_2} \right) + (2\text{ mol }Ni)\left(\frac{0.00\text{ kJ}}{\text{mol }Ni} \right) \right] = -202.0\text{ kJ}$$

$\Delta G°_{rxn,425} = \Delta H°_{rxn} - T\Delta S°_{rxn} = -202.0\text{ kJ} - [425\text{ K }(12.5\text{ J/K})(10^{-3}\text{ kJ/J})]$
$$= -207.3\text{ kJ}$$

d) $C_2H_{4(g)} + 3\,O_{2(g)} \rightarrow 2\,CO_{2(g)} + 2\,H_2O_{(l)}$

$\Delta H°_{rxn} = [2\,\Delta H°(CO_2) + 2\,\Delta H°(H_2O_{(l)})] - [\Delta H°(C_2H_4) + 3\,\Delta H°(O_2)]$

$$= \left[(2\text{ mol }CO_2)\left(\frac{-393.509\text{ kJ}}{\text{mol }CO_2} \right) + (2\text{ mol }H_2O)\left(\frac{-285.83\text{ kJ}}{\text{mol }H_2O} \right) \right]$$

$$- \left[(1\text{ mol }C_2H_4)\left(\frac{52.26\text{ kJ}}{\text{mol }C_2H_4} \right) + (3\text{ mol }O_2)\left(\frac{0.00\text{ kJ}}{\text{mol }O_2} \right) \right] = -1410.94\text{ kJ}$$

$\Delta G°_{rxn,425} = \Delta H°_{rxn} - T\Delta S°_{rxn} = -1410.94\text{ kJ} - [425\text{ K }(-267.68\text{ J/K})(10^{-3}\text{ kJ/J})]$
$$= -1297\text{ kJ}$$

13.27 $2 NH_{3(g)} + 2 O_{2(g)} \rightarrow NH_4NO_{3(s)} + H_2O_{(l)}$

$\Delta H^{\circ}_{rxn} = [\Delta H^{\circ}_f(NH_4NO_3) + \Delta H^{\circ}_f(H_2O_{(l)})] - [2 \Delta H^{\circ}_f(NH_3) + 2 \Delta H^{\circ}_f(O_2)]$

$= \left[(1 \text{ mol } NH_4NO_3)\left(\dfrac{-365.56 \text{ kJ}}{\text{mol } NH_4NO_3}\right) + (1 \text{ mol } H_2O)\left(\dfrac{-285.83 \text{ kJ}}{\text{mol } H_2O}\right) \right]$

$- \left[(2 \text{ mol } NH_3)\left(\dfrac{-46.11 \text{ kJ}}{\text{mol } NH_3}\right) + (2 \text{ mol } O_2)\left(\dfrac{0 \text{ kJ}}{\text{mol } O_2}\right) \right] = -559.17 \text{ kJ}$

$\Delta S^{\circ}_{rxn} = [S^{\circ}(NH_4NO_3) + S^{\circ}(H_2O_{(l)})] - [2 S^{\circ}(NH_3) + 2 S^{\circ}(O_2)]$

$= \left[(1 \text{ mol } NH_4NO_3)\left(\dfrac{151.08 \text{ J/K}}{\text{mol } NH_4NO_3}\right) + (1 \text{ mol } H_2O_{(l)})\left(\dfrac{69.91 \text{ J/K}}{\text{mol } H_2O}\right) \right]$

$- \left[(2 \text{ mol } NH_3)\left(\dfrac{192.45 \text{ J/K}}{\text{mol } NH_3}\right) + (2 \text{ mol } O_2)\left(\dfrac{205.14 \text{ J/K}}{\text{mol } O_2}\right) \right] = -574.19 \text{ J / K}$

$\Delta G^{\circ}_{rxn} = [\Delta G^{\circ}_f(NH_4NO_3) + \Delta G^{\circ}_f(H_2O_{(l)})] - [2 \Delta G^{\circ}_f(NH_3) + 2 \Delta G^{\circ}_f(O_2)]$

$= \left[(1 \text{ mol } NH_4NO_3)\left(\dfrac{-183.87 \text{ kJ}}{\text{mol } NH_4NO_3}\right) + (1 \text{ mol } H_2O_{(l)})\left(\dfrac{-237.13 \text{ kJ}}{\text{mol } H_2O}\right) \right]$

$- \left[(2 \text{ mol } NH_3)\left(\dfrac{-16.45 \text{ kJ}}{\text{mol } NH_3}\right) + (2 \text{ mol } O_2)\left(\dfrac{0.00 \text{ kJ}}{\text{mol } O_2}\right) \right] = -388.10 \text{ kJ}$

or $\Delta G^{\circ}_{rxn} = \Delta H^{\circ}_{rxn} - T \Delta S^{\circ}_{rxn} = -559.17 \text{ kJ} - (298 \text{ K})(-574.19 \text{ J/K})(10^{-3} \text{ kJ/J})$

$= -388.06 \text{ kJ}$

13.29 Ammonia is a gas at standard conditions while both urea and ammonium nitrate are solids. All three are quite water soluble. Some of the gaseous ammonia can be lost during placement in the field and after placement before use by the plants. An aqueous solution of ammonia is a weak base. A solution with too high a concentration of ammonia not only increases the percentage lost as gas but raises the pH of the soil. Neither the dissolved molecular urea nor the ionic ammonium nitrate raises the pH of the soil.

13.31 a) As the CO_2 gas is compressed from 1.00 to 50.0 atm at a constant T of 298 K, it liquefies at ~6 - 6.5 atm and remains a liquid during the rest of the compression to 50.0 atm.

b) As the N_2 gas is compressed from 1.00 to 50.0 atm at constant T of 298 K, it remains a gas and the volume decreases to ~1/50 of original volume.

c) As the N_2 is cooled from 298 to 50 K at a constant P of 1.00 atm, it liquefies at 77 K and then freezes at 63 K.

d) As the CO_2 is cooled from 298 to 50 K at a constant P of 1.00 atm, it condenses directly to the solid phase at 195 K.

114

13.33 At cellular conditions, the pressure is not 1 atm and the concentrations are not 1 M.

13.35 Reactions

a) acetyl phosphate + H_2O → acetic acid + H_3PO_4 $\quad \Delta G° = -41.8$ kJ

b) glutamic acid + NH_3 → glutamine + H_2O $\quad\quad \Delta G° = +14$ kJ

Net: glutamic acid + NH_3 + acetyl phosphate → glutamine + acetic acid + H_3PO_4

$$\Delta G°_{rxn} = \Delta G°_{rxn(a)} + \Delta G°_{rxn(b)} = -41.8 \text{ kJ} + (+14. \text{ kJ}) = -28 \text{ kJ}$$

13.37 No, this will not work. Moving heat from the cooler refrigerator to the warmer surroundings (the kitchen) is a nonspontaneous process. Therefore, not only is the kitchen made warmer by the heat removed from the refrigerator, but according to the second law of thermodynamics, excess heat must be generated by the system while driving this nonspontaneous process.

13.39 a) $\Delta E_{universe} = 0$ $\quad\quad$ b) $\Delta E_{teaspoon}$ is positive $\quad\quad$ c) $\Delta S_{universe}$ is positive

d) ΔS_{water} is negative $\quad$ e) $q_{teaspoon}$ increases.

13.41 0.50 mol H_2O (liquid, 298 K) < 1.0 mol H_2O (liquid, 298 K) < 1.0 mol H_2O (liquid, 373 K) < 1.0 mol H_2O (gas, 373 K, 1 atm) < 1.0 mol H_2O (gas, 373 K, 0.1 atm)

13.43 a) $q_{H_2O} = (n_{H_2O})(\Delta H_{fusion}) = \left(\dfrac{150 \text{ g } H_2O}{18.02 \text{ g } H_2O/mol}\right)\left(-6.01\dfrac{kJ}{mol}\right) = -50.03$ kJ

$\Delta S_{H_2O} = \left(\dfrac{-50.03 \text{ kJ}}{273 \text{ K}}\right)\left(\dfrac{10^3 \text{ J}}{kJ}\right) = -183$ J/K

b) $q_{freezer} = -q_{H_2O} = 50.03$ kJ $\quad\quad \Delta S_{freezer} = \left(\dfrac{50.03 \text{ kJ}}{253 \text{ K}}\right)\left(\dfrac{10^3 \text{ J}}{kJ}\right) = 198$ J/K

$\Delta S_{universe} = \Delta S_{H_2O} + \Delta S_{freezer} = -183$ J/K $+ 198$ J/K $= 15$ J/K

c) Yes, an additional entropy change would occur for the universe as the tray of ice was cooled to -20°C because the freezer would generate excess heat and, therefore, more disorder.

13.45 $Al_2O_3 + 3 H_2 \rightarrow 2 Al + 3 H_2O_{(l)}$

$\Delta H°_{rxn} = [2 \Delta H°(Al) + 3 \Delta H°(H_2O_{(l)})] - [\Delta H°(Al_2O_3) + 3 \Delta H°(H_2)]$

$\Delta H°_{rxn} = [0.00 + 3(-285.830 \text{ kJ})] - [-1675.7 \text{ kJ} + 0.00] = 818.2$ kJ

$\Delta G°_{rxn} = [2 \Delta G°(Al) + 3 \Delta G°(H_2O_{(l)})] - [\Delta G°(Al_2O_3) + 3 \Delta G°(H_2)]$

$\Delta G°_{rxn} = [0.00 + 3 (-237.129 \text{ kJ})] - [-1582.3 \text{ kJ} + 0.00] = 870.9$ kJ

$\Delta S°_{rxn} = [2 S°(Al) + 3 S°(H_2O_{(l)})] - [S°(Al_2O_3) + 3 S°(H_2)]$

$\Delta S°_{rxn} = [(2 \times 28.33 \text{ J/K}) + 3(69.91)] - [+50.92 \text{ J/K} + 3(130.684 \text{ J/K})] = -176.58$ J/K

(continued)

(13.45 continued)

a) As indicated by the positive value for ΔG°_{rxn}, the reaction is not spontaneous.

b) As indicated by the positive value for ΔH°_{rxn}, the reaction absorbs heat.

c) As indicated by the negative value for ΔS°_{rxn}, the products are more ordered than the reactants.

13.47 $C_{16}H_{32}O_2 + 23\ O_2 + 130\ ADP + 130\ H_3PO_4 \rightarrow 16\ CO_2 + 146\ H_2O + 130\ ATP$

$C_{16}H_{32}O_2 + 23\ O_2 \rightarrow 16\ CO_2 + 16\ H_2O \qquad \Delta G^\circ = -9790\ kJ$

$130\ ADP + 130\ H_3PO_4 \rightarrow 130\ H_2O + 130\ ATP$
$\Delta G^\circ = (130\ mol\ ATP)(+30.6\ kJ/mol) = +3978\ kJ$

$\Delta G^\circ_{overall} = -9790\ kJ + 3978\ kJ = -5812\ kJ$

$\text{efficiency} = \left(\dfrac{3978\ kJ}{3978\ kJ + 9790\ kJ}\right) \times 100\% = 40.6\%$

13.49 $N_2 + O_2 + \text{Lightning} \rightarrow 2\ NO \qquad \Delta G$ would be positive because lightning is needed to supply the energy to drive this nonspontaneous reaction.

Luciferin $\rightarrow$ Dehydroluciferin + Light $\qquad \Delta G$ could be negative because an enzyme catalyzes this reaction; could be positive if energy from metabolism drives the reaction; probably positive because firefly exerts control over "lightning".

13.51 a) $2\ ClO_2^-{}_{(aq)} + O_{2(g)} \rightarrow 2\ ClO_3^-{}_{(aq)}$

$\Delta S^\circ_{rxn} = [2\ S^\circ(ClO_3^-{}_{(aq)})] - [2\ S^\circ(ClO_2^-{}_{(aq)}) + S^\circ(O_2)]$

$= \left[(2\ mol\ ClO_3^-{}_{(aq)})\left(\dfrac{162\ J/K}{mol\ ClO_3^-}\right)\right]$

$-\left[(2\ mol\ ClO_2^-)\left(\dfrac{101\ J/K}{mol\ ClO_2^-}\right) + (1\ mol\ O_2)\left(\dfrac{205.138\ J/K}{mol\ O_2}\right)\right] = -83\ J/K$

b) $4\ FeCl_{3(s)} + 3\ O_{2(g)} \rightarrow 2\ Fe_2O_{3(s)} + 6\ Cl_{2(g)}$

$\Delta S^\circ_{rxn} = [2\ S^\circ(Fe_2O_3) + 6\ S^\circ(Cl_2)] - [4\ S^\circ(FeCl_3) + 3\ S^\circ(O_2)]$

$= \left[(2\ mol\ Fe_2O_3)\left(\dfrac{87.40\ J/K}{mol\ Fe_2O_3}\right) + (6\ mol\ Cl_2)\left(\dfrac{223.066\ J/K}{mol\ Cl_2}\right)\right]$

$-\left[(4\ mol\ FeCl_3)\left(\dfrac{142.3\ J/K}{mol\ FeCl_3}\right) + (3\ mol\ O_2)\left(\dfrac{205.138\ J/K}{mol\ O_2}\right)\right] = 328.6\ J/K$

(continued)

(13.51 continued)

c) $3 N_2H_{4(l)} + 4 O_{3(g)} \rightarrow 6 NO_{(g)} + 6 H_2O_{(l)}$

$\Delta S^{\circ}_{rxn} = [6 S^{\circ}(NO) + 6 S^{\circ}(H_2O_{(l)})] - [3 S^{\circ}(N_2H_4) + 4 S^{\circ}(O_3)]$

$= \left[(6 \text{ mol NO})\left(\dfrac{210.761 \text{ J/K}}{\text{mol NO}} \right) + (6 \text{ mol } H_2O_{(l)})\left(\dfrac{69.91 \text{ J/K}}{\text{mol } H_2O} \right) \right]$

$- \left[(3 \text{ mol } N_2H_4)\left(\dfrac{121.21 \text{ J/K}}{\text{mol } N_2H_4} \right) + (4 \text{ mol } O_3)\left(\dfrac{238.93 \text{ J/K}}{\text{mol } O_3} \right) \right] = -478.37 \text{ J/K}$

13.53 a) $2 ClO_2^-{}_{(aq)} + O_{2(g)} \rightarrow 2 ClO_3^-{}_{(aq)}$

$\Delta H^{\circ}_{rxn} = [2 \Delta H^{\circ}(ClO_3^-{}_{(aq)})] - [2 \Delta H^{\circ}(ClO_2^-{}_{(aq)}) + \Delta H^{\circ}(O_2)]$

$= [2(-104)] - [2(-67) + (0.00)] = -74 \text{ kJ}$

$\Delta G^{\circ}_{350} = \Delta H^{\circ} - T\Delta S^{\circ}$ (Use ΔS° from Problem 13.51)

$= -74 \text{ kJ} - [(350 \text{ K})(-83 \text{ J/K})(10^{-3} \text{ kJ/J})] = -45 \text{ kJ}$

b) $4 FeCl_{3(s)} + 3 O_{2(g)} \rightarrow 2 Fe_2O_{3(s)} + 6 Cl_{2(g)}$

$\Delta H^{\circ}_{rxn} = [2 \Delta H^{\circ}(Fe_2O_3) + 6 \Delta H^{\circ}(Cl_2)] - [4 \Delta H^{\circ}(FeCl_3) + 3 \Delta H^{\circ}(O_2)]$

$= [2(-824.2) + 6(0.00)] - [4(-399.49) + 3(0.00)] = -50.44 \text{ kJ}$

$\Delta G^{\circ}_{350} = \Delta H^{\circ} - T\Delta S^{\circ} = -50.44 \text{ kJ} - (350 \text{ K})(328.6 \text{ J/K})(10^{-3} \text{ kJ/J}) = -165 \text{ kJ}$

c) $3 N_2H_{4(l)} + 4 O_{3(g)} \rightarrow 6 NO_{(g)} + 6 H_2O_{(l)}$

$\Delta H^{\circ}_{rxn} = [6 \Delta H^{\circ}(NO) + 6 \Delta H^{\circ}(H_2O_{(l)})] - [3 \Delta H^{\circ}(N_2H_4) + 4 \Delta H^{\circ}(O_3)]$

$= [6(90.25) + 6(-285.83)] - [3(50.63) + 4(142.7)] = -1896 \text{ kJ}$

$\Delta G^{\circ}_{350} = \Delta H^{\circ} - T\Delta S^{\circ} = -1896 \text{ kJ} - (350 \text{ K})(-478.37 \text{ J/K})(10^{-3} \text{ kJ/J}) = -1729 \text{ kJ}$

13.55 a) W_{sys} is 0 because the system is constant volume.

b) q_{sys} is positive because system requires energy to separate the molecules into their constituent atoms (ΔS is positive).

c) ΔS_{surr} is negative because the surroundings supply energy to the system.

13.57 a) $P_{4\beta}$ has a more ordered crystalline structure; q is negative; therefore, ΔS is negative.

b) ΔS is negative for this process because the process results in a more ordered system. ΔH is negative because the process is spontaneous (ΔG is negative) at -77°C.

13.59 a) The sign of ΔH is - and of ΔG is -. The NO_2 molecule contains an unpaired electron and readily dimerizes to form N_2O_4; therefore, ΔG is negative. $\Delta G = \Delta H - T\Delta S$. For the reaction $2\ NO_2 \rightarrow N_2O_4$, ΔS is negative. Therefore, in order that ΔG be negative, ΔH must be negative and of greater magnitude than $T\Delta S$.

b) The sign of ΔH is + and of ΔG is +. The gold requires an input of energy (+ q) to be melted; therefore, ΔH is positive. The melting is **nonspontaneous**; therefore, ΔG is positive.

c) The sign of ΔH is - and of ΔG is -. This is a typical **exothermic** combustion that is **spontaneous**; therefore, ΔH is negative and ΔG is negative.

13.61 The process of breaking the bonds in oxygen (or in sulfur) and forming oxide (or sulfide) bonds with other elements are spontaneous (ΔG negative) because the oxide (or sulfide) bonds are stronger than the bonds in oxygen (or sulfur). The bond in nitrogen is stronger than a nitride bond and, therefore, nitrogen exists as a pure element rather than being bonded to other elements.

13.63 The K^+ amd Ca^{2+} ions are similar in size and mass. The Cl^- ion is both larger and of greater mass than the O^{2-} ion which could account for KCl having a greater disorder.

13.65 a) The water flow is spontaneous. The turning of the turbine and the generating of the electricity is nonspontaneous.

b) The burning of the gasoline is spontaneous. The movement of the engine and the movement of the water uphill are nonspontaneous.

13.67 $3\ NO_{2(g)} + H_2O_{(l)} \rightarrow 2\ HNO_{3(g)} + NO_{(g)}$
a) $\Delta G^\circ = [\Delta G^\circ(HNO_3) + 1\ \Delta G^\circ(NO)] - [3\ \Delta G^\circ(NO_2) + 1\ \Delta G^\circ(H_2O)]$
   $= [2(-74.72\ kJ) + 1(86.55\ kJ)] - [3(51.31\ kJ) + 1(-237.129\ kJ)]$
   $= 20.31\ kJ$ Not thermodynamically feasible at standard conditions.
b) $\Delta G = \Delta G^\circ + RT \ln Q$

$$\Delta G = (20.31\ kJ) + (8.314\ J/mol\ K)(298K)(10^{-3}\ kJ\ /\ J)\ln \frac{(10^{-6})^3}{(10^{-6})^3}$$

$= 20.31\ kJ + 2.478\ kJ \ln 1$

$= 20.31\ kJ + 0 = 20.31\ kJ$ Not thermodynamically feasible.

If $H_2O_{(g)}$ were used instead of $H_2O_{(l)}$ in this problem:

a) $\Delta G^\circ = 11.90\ kJ$ b) $\Delta G^\circ = 46.13\ kJ$

It is still not thermodynamically feasible but at higher pressures (~123 atm) it could be.

13.69 a) The sample will condense to a liquid at 331.9 K and freeze at 265.9 K if maintained at 1 atm pressure. Therefore, at 250 K and 1.00 atm the sample will be a solid. This can be shown by drawing a horizontal line at 1 atm between T = 400 K and T = 250 K.

(continued)

(13.69 continued)

b) The temperature of 265.8 K is between the temperature of the triple point and the normal freezing point. A vertical line drawn at 265.8 K between the pressures of 1.00×10^{-3} atm and 1.00×10^3 atm will start in the vapor phase, pass through the liquid phase, and terminate in the solid phase.

c) A horizontal line drawn at $P = 2.00 \times 10^{-2}$ atm will pass below the triple point. Therefore, a sample heated from 250 to 400 K at a constant $P = 2.00 \times 10^{-2}$ atm will start as a solid and pass directly into the vapor phase (sublime).

13.71   Methane would be expected to have the highest $S°$ because it has the most atoms (5 vs. 4 for ammonia and 3 for water). Methane may have a lower absolute entropy because it has 4 hydrogen atoms located symmetrically around the carbon atom. The other molecules have hydrogen atoms and lone pairs of electrons located around the central atom.

13.73 $CCl_{4(l)} + 5\ O_{2(g)} \rightarrow CO_{2(g)} + 4\ ClO_{2(g)}$

$\Delta H°_{rxn} = (1\ \text{mol}\ CO_2)\ \Delta H°(CO_2) + (4\ \text{mol}\ ClO_2)\ \Delta H°ClO_2 - [(1\ \text{mol}\ CCl_4)$

$\Delta H°(CCl_4) + (5\ \text{mol}\ O_2)\ \Delta H°(O_2)] = (1\ \text{mol}\ CO_2)(-393.509\ \text{kJ/mol}) + (4\ \text{mol}\ ClO_2)$

$(102.5\ \text{kJ/mol}) - [(1\ \text{mol}\ CCl_4)(-135.44\ \text{kJ/mol}) + (5\ \text{mol}\ O_2)(0.0\ \text{kJ/mol})]$

$= -393.509\ \text{kJ} + 410.0\ \text{kJ} + 135.44\ \text{kJ} + 0.0\ \text{kJ} = 151.9\ \text{kJ}$

$\Delta G°_{rxn} = (1\ \text{mol}\ CO_2)\ \Delta G°(CO_2) + (4\ \text{mol}\ ClO_2)\ \Delta G°ClO_2 - [(1\ \text{mol}\ CCl_4)$

$\Delta G°(CCl_4) + (5\ \text{mol}\ O_2)\ \Delta G°(O_2)]$

$= (1\ \text{mol}\ CO_2)(-394.359\ \text{kJ/mol}) + (4\ \text{mol}\ ClO_2)(120.5\ \text{kJ/mol}) - [(1\ \text{mol}\ CCl_4)$

$(-65.21\ \text{kJ/mol}) + (5\ \text{mol}\ O_2)(0\ \text{kJ/mol})]$

$= -394.359\ \text{kJ} + 482.0\ \text{kJ} + 65.21\ \text{kJ} + 0.0\ \text{kJ} = 152.9\ \text{kJ}$

$CS_{2(l)} + 3\ O_{2(g)} \rightarrow CO_{2(g)} + 2\ SO_{2(g)}$

$\Delta H°_{rxn} = (1\ \text{mol}\ CO_2)\ \Delta H°(CO_2) + (2\ \text{mol}\ SO_2)\ \Delta H°(SO_2) - [(1\ \text{mol}\ CS_2)\ \Delta H°(CS_2)$

$+ (3\ \text{mol}\ O_2)\ \Delta H°(O_2)] = (1\ \text{mol}\ CO_2)(-393.509\ \text{kJ/mol})$

$+ (2\ \text{mol}\ SO_2)(-296.830\ \text{kJ/mol}) - [(1\ \text{mol}\ CS_2)(89.70\ \text{kJ/mol}) + (3\ \text{mol}\ O_2)(0)]$

$= -393.509\ \text{kJ} - 593.66\ \text{kJ} - 89.70\ \text{kJ} + 0\ \text{kJ} = -1076.87\ \text{kJ}$

$\Delta G°_{rxn} = (1\ \text{mol}\ CO_2)\ \Delta G°(CO_2) + (2\ \text{mol}\ SO_2)\ \Delta G°(SO_2) - [(1\ \text{mol}\ CS_2)\ \Delta G°(CS_2)$

$+ (3\ \text{mol}\ O_2)\ \Delta G°(O_2)]$

$= (1\ \text{mol}\ CO_2)(-394.359\ \text{kJ/mol}) + (2\ \text{mol}\ SO_2)(-300.194\ \text{kJ/mol}) - [(1\ \text{mol}\ CS_2)$

$(65.27\ \text{kJ/mol}) + (3\ \text{mol}\ O_2)(0)]$

$= -394.359\ \text{kJ} - 600.388\ \text{kJ} - 65.27\ \text{kJ} + 0.0\ \text{kJ} = -1060.02\ \text{kJ}$

Special precautions against fires would be recommended for industrial plants using carbon disulfide. The combustion reaction is spontaneous ($\Delta G° = -1060$ kJ) and a great amount of heat is produced ($\Delta H° = -1077$ kJ) which would keep the fire going and might well ignite other flammable substances. The combustion of carbon tetrachloride is nonspontaneous and is endothermic.

13.75   a) $q_v = \Delta E$ b) $q_p = \Delta H$ c) $q_T = T\Delta S$

13.77 $NaCl_{(s)} \rightarrow Na^+_{(aq)} + Cl^-_{(aq)}$

a) $\Delta G^\circ_{rxn} =$

(1 mol $Na^+_{(aq)}$) $\Delta G^\circ_f(Na^+_{(aq)})$ + (1 mol $Cl^-_{(aq)}$) $\Delta G^\circ_f(Cl^-_{(aq)})$ - (1 mol $NaCl_{(s)}$)

$\Delta G^\circ_f(NaCl_{(s)})$ = (1 mol)(-261.905 kJ/mol) + (1 mol)(-131.228 kJ/mol) -
(1 mol)(-384.138 kJ/mol) = -8.995 kJ

This reaction is spontaneous; ΔG is negative.

b) ΔH°_{rxn} = (1 mol $Na^+_{(aq)}$) $\Delta H^\circ_f(Na^+_{(aq)})$ + (1 mol $Cl^-_{(aq)}$) $\Delta H^\circ_f(Cl^-_{(aq)})$ -

(1 mol $NaCl_{(s)}$) $\Delta H^\circ_f(NaCl_{(s)})$

= (1 mol)(-240.12 kJ/mol) + (1 mol)(-167.59 kJ/mol) - (1 mol)(-411.153 kJ/mol)
= +3.443 kJ

This reaction does not release energy; it is endothermic (ΔH is positive.)

c) $\Delta S^\circ_{rxn} =$

(1 mol $Na^+_{(aq)}$) $S^\circ(Na^+_{(aq)})$ + (1 mol $Cl^-_{(aq)}$) $S^\circ(Cl^-_{(aq)})$ - (1 mol $NaCl_{(s)}$) $S^\circ(NaCl_{(s)})$
= (1 mol)(59.0 J/K) + (1 mol)(56.5 J/K) - (1 mol)(72.13 J/K) = 43.4 J/K

No, the amount of disorder of the system increases; ΔS is positive.

CHAPTER 14: MECHANISMS OF CHEMICAL REACTIONS

14.1 a) Filling each cup with coffee from the urn.
b) The ringing up of the items by the checker.
c) Making payment.
d) The passing through the airplane door.

14.3 a) $\underset{Cl_2}{\bigcirc\!\!\bigcirc} \xrightarrow{h\nu} \underset{\text{activated}}{[\bigcirc\!\!\bigcirc]} \rightarrow \bigcirc + \bigcirc$
    $\underset{2\,Cl}{\qquad\qquad}$

b) $\underset{NO\ +\ Cl_2}{\bigcirc\!\!\bullet + \bigcirc\!\!\bigcirc}\ \underset{\text{approach collision}}{\longrightarrow}\ [\bigcirc\!\!\bullet\!\!\bigcirc\!\!\bigcirc]\ \underset{\text{separation}}{\longrightarrow}\ \underset{NOCl\ +\ Cl}{\bigcirc\!\!\bullet\!\!\bigcirc + \bigcirc}$

c) $\underset{NO\ +\ Cl_2\ +\ NO}{\bigcirc\!\!\bullet + \bigcirc\!\!\bigcirc + \bullet\!\!\bigcirc}\ \underset{\text{approach collision}}{\rightarrow}\ [\bigcirc\!\!\bullet\!\!\bigcirc\!\!\bigcirc\!\!\bullet\!\!\bigcirc]\ \underset{\text{separation}}{\rightarrow}\ \underset{2\ NOCl}{2\,\bigcirc\!\!\bullet\!\!\bigcirc}$

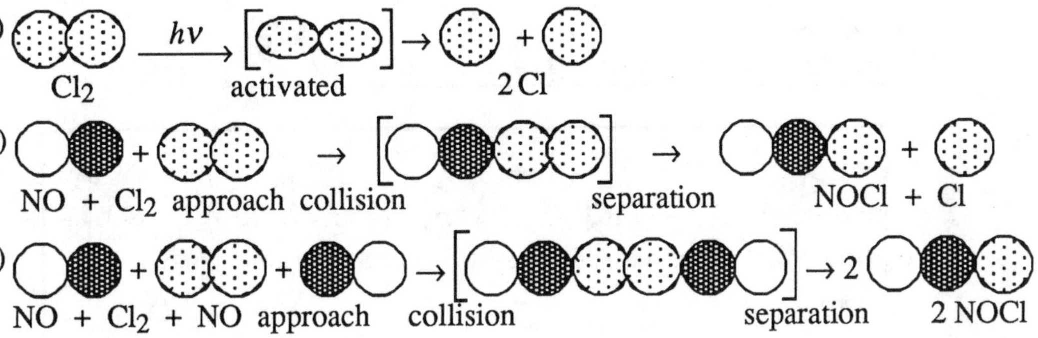

14.5 The intermediate for the decompositon of ozone is oxygen atoms (O).

14.7 $2\,NO + Cl_2 \rightarrow 2\,NOCl$

a) Rate $= -\dfrac{\Delta[Cl_2]}{\Delta t}$

b) $\dfrac{1}{2}\dfrac{\Delta[NOCl]}{\Delta t} = -\dfrac{\Delta[Cl_2]}{\Delta t}$, i.e., the rate of NOCl formation is twice the rate of Cl_2 disappearance.

c) The rate of NOCl formation = 2 x rate of Cl_2 disappearance = 2 x 47 M s^{-1}
    = 94 M s^{-1}

14.9 $2\,NO + Cl_2 \rightarrow 2\,NOCl$

a)

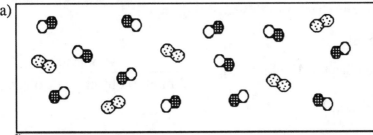

(continued)

121

(14.9 continued)

b)

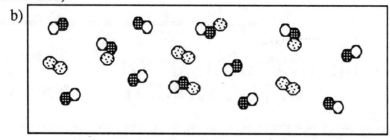

c)

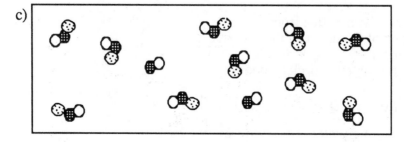

14.11 a)

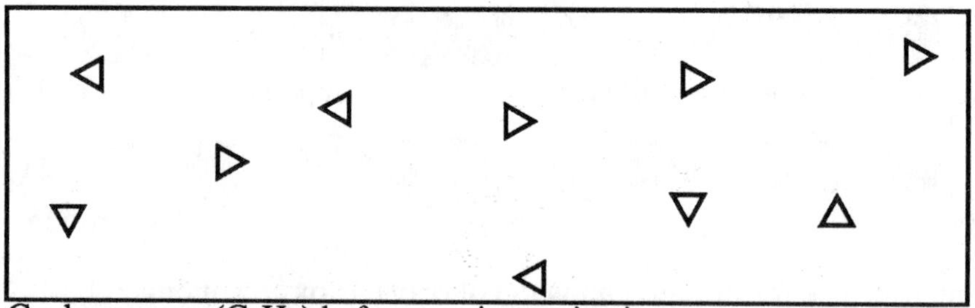

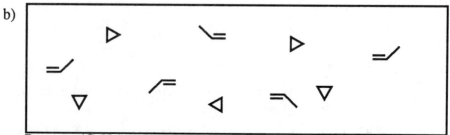

Cyclopropane (C_3H_6) before any isomerization.

b)

Propene (C_3H_6) and cyclopropane (C_3H_6) after the sample in part a has isomerized for 20 minutes.

14.13 A single-step elementary reaction would require 4 molecules to collide while having proper orientation. The chances of this occurring are nil.

14.15 Rate = k [red][white]

$$\frac{4 \text{ objects}}{2 \text{ min}} = k\ [20 \text{ objects}_r][10 \text{ objects}_w];\ k = \frac{4 \text{ objects}}{2 \text{ min}(20 \text{ objects}_r)(10 \text{ objects}_w)}$$

$$= 0.01 \text{ objects}^{-1} \text{ min}^{-1}$$

(continued)

(14.15 continued)

a) rate $= (0.01$ objects^{-1} min^{-1})[10 objects$_r$][10 objects$_w$]

$= 1$ object/min $= 2$ objects/2 min

The concentration of red balls was halved; therefore, half as many pairs are formed: (4 pairs/2 min)(1/2) = 2 pairs/2 min.

b) rate $= (0.01$ objects^{-1} min^{-1})[20 objects$_r$][15 objects$_w$]

$= 3$ objects/min $= 6$ objects/2 min

The concentration of red balls was kept the same (compared to original) and the concentration of white balls was increased by a factor of 1.5: (4 pairs/2 min)(1.5) = 6 pairs/2min

c) rate $= (0.01$ objects^{-1} min^{-1})[10 objects$_r$][20 objects$_w$]

$= 2$ objects/min $= 4$ objects/2 min

The concentration of red balls was halved and the concentration of white balls was doubled (compared to original): (4 pairs/2 min)(1/2)(2) = 4 pairs/2 min

d) rate $= (0.01$ object^{-1} min^{-1})[40 objects$_r$][20 objects$_w$]

$= 8$ objects/min $= 16$ objects/2 min

The concentration of red balls was doubled and the concentration of white balls was doubled (compared to original): (4 pairs/2 min)(2)(2) = 16 pairs/2 min

This assumes that cramming 60 balls in the machine does not exceed the capacity of the machine to keep all the balls in motion. If all of the balls are not kept in motion (some lying in the bottom of the machine), the rate would be less.

e) When the number of balls is increased (in the same volume) there will be more collisions and, therefore, more collisions with the correct orientation for reaction. The reaction rate will be greater. If the number of balls is decreased, the opposite is true and the rate will be smaller.

14.17 a) (concentration)(time)$^{-1}$ = k (concentration)^2(concentration)

$$k = \frac{(\text{concentration})(\text{time})^{-1}}{(\text{concentration})^2 (\text{concentration})} = (\text{concentration})^{-2}(\text{time})^{-1}$$

b) (concentration)(time)$^{-1}$ = k (concentration)^2

$$k = \frac{(\text{concentration})(\text{time})^{-1}}{(\text{concentration})^2} = (\text{concentration})^{-1}(\text{time})^{-1}$$

c) (concentration)(time)$^{-1}$ = k (concentration)(concentration)

$$k = \frac{(\text{concentration})(\text{time})^{-1}}{(\text{concentration})(\text{concentration})} = (\text{concentration})^{-1}(\text{time})^{-1}$$

d) (concentration)(time)$^{-1}$ = k (concentration)

$$k = \frac{(\text{concentration})(\text{time})^{-1}}{(\text{concentration})} = \text{time}^{-1}$$

14.19 a) rate $= k[H_2]^2$ rate ratio $= \dfrac{k\,[3\,x]^2}{k\,[x]^2} = 9$ The rate would increase 9 times.

b) rate $= k[H_2]^0$ rate ratio $= \dfrac{k\,[3\,x]^0}{k\,[x]^0} = 1$

The rate would not change because it is independent of the concentration of H_2.

c) rate $= k[H_2]^{3/2}$ rate ratio $= \dfrac{k\,[3\,x]^{3/2}}{k\,[x]^{3/2}} = [3]^{3/2} = 5.196$

The rate would increase by 5.196 times.

14.21 $2\,N_2O_{5(g)} \rightarrow 4\,NO_{2(g)} + O_{2(g)}$

Time (s)	0	200	400	600	800
$[N_2O_5]$(atm)	2.50	2.22	1.96	1.73	1.53
$\ln[N_2O_5]$	0.916	0.798	0.673	0.548	0.425
$1/[N_2O_5]$	0.400	0.450	0.510	0.578	0.654

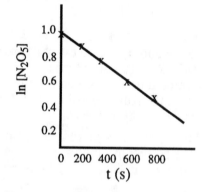
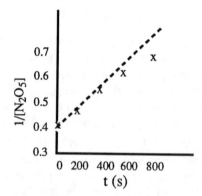

The rate is first order with respect to N_2O_5; rate $= k[N_2O_5]$. The rate constant, k, equals the − slope for the graph of $\ln[N_2O_5]$ vs. t.

From graph:− slope $= -\dfrac{\Delta y}{\Delta x} = -\dfrac{0.673 - 0.890}{400\ s - 0\ s} = +5.43 \times 10^{-4}\ s^{-1}$

14.23 Rate $= k\,[NOBr]^2 = 25\ L\ mol^{-1}\ min^{-1}\ [NOBr]^2$

a) $\dfrac{1}{[A]} - \dfrac{1}{[A]_0} = kt = 25\ L\ mol^{-1}\ min^{-1}\ t$

$\dfrac{1}{0.010\ M} - \dfrac{1}{0.025\ M} = 25\ L\ mol^{-1}\ min^{-1}\ t$

$60\ M^{-1} = 60\ L\ mol^{-1} = 25\ L\ mol^{-1}\ min^{-1}\ t$

$t = \dfrac{60\ L\ mol^{-1}\ min^{-1}}{25\ L\ mol^{-1}\ min^{-1}} = 2.4\ min$

b) $\dfrac{1}{[A]} = kt + \dfrac{1}{[A]_0} = (25\ L\ mol^{-1}\ min^{-1})(100\ min) + \dfrac{1}{0.025\ M}$

$1/[A] = 2500\ L\ mol^{-1} + 40\ M^{-1} = 2540\ M^{-1}$

$[A] = 0.00039\ M$

14.25 $C_2H_6N_2 \rightarrow N_2 + C_2H_6$

Time (s)	0	100	150	200	250	300
Azomethane (mM)	7.94	6.15	5.40	4.75	4.20	3.69
ln [Azomethane]	2.07	1.82	1.69	1.56	1.44	1.31
1/[Azomethane]	0.126	0.163	0.185	0.211	0.238	0.271

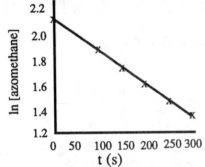 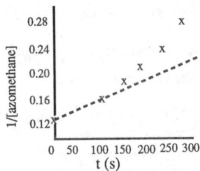

The rate of reaction is first order.

$k = -\text{slope} = -\Delta y/\Delta x = -(1.31 - 1.82)/(300\text{ s} - 100\text{ s}) = 0.00255\text{ s}^{-1}$

14.27 In a system where one of the products of a fast, reversible reaction is being used as a reactant for a second reaction, the forward and reverse rates for the fast, reversible reaction will not ever be exactly equal. The forward and reverse rates are equal only when the concentrations of the reactant(s) and product(s) fulfill a certain relationship. Because one of the products is constantly being used by the second reaction, this relationship will be deficient in the concentration of that product. Therefore, the forward rate will be slightly faster as it tries to replace the product being used by the second reaction.

14.29 $O_3 + O_2 \rightleftarrows O_5$ (fast, reversible)

$O_5 \rightarrow 2\,O_2 + O$ (slow, rate determining)

a) $O_3 + O \rightarrow 2\,O_2$ (fast)

b) $k_1[O_3][O_2] = k_{-1}[O_5]$

$[O_5] = k_1/k_{-1}[O_3][O_2]$

$\text{rate} = k_2[O_5] = k_2\dfrac{k_1}{k_{-1}}[O_3][O_2] = k[O_3][O_2]$

c) The mechanism requires O_5 to decompose by two different processes; a fast one that returns it to the starting materials and a slow one that leads to the product.

14.31 a) $Cl_2 + h\nu \underset{k_{-1}}{\overset{k_1}{\rightleftharpoons}} 2\,Cl$ (fast, reversible)

$NO + Cl \xrightarrow{\ k_2\ } NOCl$ (rate determining)

$k_1[Cl_2] = k_{-1}[Cl]^2$ $[Cl]^2 = \left(\dfrac{k_1}{k_{-1}}\right)[Cl_2]$ $[Cl] = \left(\dfrac{k_1}{k_{-1}}\right)^{1/2}[Cl_2]^{1/2}$

(continued)

125

$$\text{rate} = k_2[\text{NO}][\text{Cl}] = k_2\left(\frac{k_1}{k_{-1}}\right)^{1/2}[\text{NO}][\text{Cl}_2]^{1/2} = k[\text{NO}][\text{Cl}_2]^{1/2}$$

b) $\text{NO} + \text{Cl}_2 \underset{k_{-1}}{\overset{k_1}{\rightleftarrows}} \text{NOCl} + \text{Cl}$ (rev.) $\text{NO} + \text{Cl} \xrightarrow{k_2} \text{NOCl}$ (rate determining)

$$k_1[\text{NO}][\text{Cl}_2] = k_{-1}[\text{NOCl}][\text{Cl}] \qquad [\text{Cl}] = \left(\frac{k_1}{k_{-1}}\right)\frac{[\text{NO}][\text{Cl}_2]}{[\text{NOCl}]}$$

$$\text{rate} = k_2[\text{NO}][\text{Cl}] = k_2\left(\frac{k_1}{k_{-1}}\right)[\text{NO}]\frac{[\text{NO}][\text{Cl}_2]}{[\text{NOCl}]} = k\frac{[\text{NO}]^2[\text{Cl}_2]}{[\text{NOCl}]}$$

14.33 If the activation energy (E_a) is equal to zero, the rate constant is independent of temperature (i.e., it does not change when the temperature changes).
$k = A\,e^{-E_a/RT} = A\,e^{-0/RT} = A$
A zero E_a means that no bonds are distorted or broken during the reaction.

14.35 a) $\Delta E_{rxn} = A - C$; false; $\Delta E_{rxn} = C - A$ (Reaction is exothermic.)

b) $\Delta E_{rxn} = B - C$; false

c) E_a (forward) = E_a (reverse); false; E_a (forward) = $B - A$ and E_a (reverse) = $B - C$
d) true

e) E_a (forward) = $B - C$; false
f) true

14.37

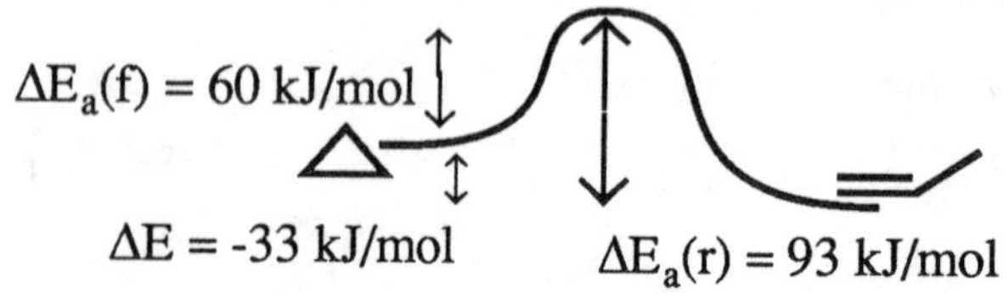

$\Delta E_a(f) = 60$ kJ/mol

$\Delta E = -33$ kJ/mol $\Delta E_a(r) = 93$ kJ/mol

14.39 An intermediate appears as a product of an early step and is used as a reactant in a later step. A catalyst is used as a reactant in an early step and is regenerated as a product in a later step of the mechanism.

14.41 a) A and B are O_2 and Br^- (or Br^- and O_2)
b) $2\,H_2O_{2(aq)} \rightarrow 2\,H_2O_{(l)} + O_{2(g)}$

$$\Delta H_{rxn} = \left(2 \text{ mol } H_2O_{(l)}\right)\left(\frac{-285.830 \text{ kJ}}{\text{mol } H_2O_{(l)}}\right) + (1 \text{ mol } O_2)\left(\frac{0.0 \text{ kJ}}{\text{mol } O_2}\right)$$

$$-\left[\left(2 \text{ mol } H_2O_{2(aq)}\right)\left(\frac{-187.78 \text{ kJ}}{\text{mol } H_2O_{2(aq)}}\right)\right] = -196.1 \text{ kJ}$$

(continued)

(14.41 continued)
Note: $\Delta E = \Delta H$ assumed and $\Delta H_{H_2O_{2(l)}}$ used for $\Delta H_{H_2O_{2(aq)}}$

$H_2O_{2(aq)} + Br^-_{(aq)} \rightarrow H_2O_{(l)} + BrO^-_{(aq)}$ (First step in catalyzed mechanism.)

$$\Delta H_{int} = \left(1 \text{ mol } H_2O_{(l)}\right)\left(\frac{-285.830 \text{ kJ}}{\text{mol } H_2O_{(l)}}\right) + \left(1 \text{ mol } BrO^-_{(aq)}\right)\left(\frac{-94.1 \text{ kJ}}{\text{mol } BrO^-_{(aq)}}\right)$$

$$-\left[\left(1 \text{ mol } H_2O_{2(aq)}\right)\left(\frac{-187.78 \text{ kJ}}{\text{mol } H_2O_{2(aq)}}\right) + \left(1 \text{ mol } Br^-_{(aq)}\right)\left(\frac{-121.55 \text{ kJ}}{\text{mol } Br^-_{(aq)}}\right)\right] = -70.60 \text{ kJ}$$

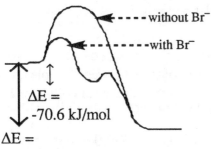

- - - - without Br^-

- - - - with Br^-

$\Delta E =$
-70.6 kJ/mol

$\Delta E =$
-196.1 kJ/mol

14.43 Enzymes are proteins with many different shapes. They have different grooves or pockets which have shape and size similar to the shape and size of the specific reactant molecule that they catalyze. Only that molecule tends to fit in the groove or pocket. During binding of the reactant molecule in the enzyme's groove or pocket, the enzyme often changes shape to match the reactant molecule's shape even more closely in order to bind it more tightly.
Your diagrams of a hypothetical enzyme need to show the shape of a groove or pocket that will fit the shape of the square but not the triangle.

14.45 Rate = k [CO][NO$_2$]

a) Rate $= k\left[\dfrac{0.5 \text{ mol}}{2.0 \text{ L}}\right]\left[\dfrac{0.5 \text{ mol}}{2.0 \text{ L}}\right] = 0.06 \text{ k mol}^2 \text{ L}^{-2}$

b) Rate $= k\left[\dfrac{0.5 \text{ mol}}{1.0 \text{ L}}\right]\left[\dfrac{0.5 \text{ mol}}{1.0 \text{ L}}\right] = 0.25 \text{ k mol}^2 \text{ L}^{-2}$

c) Rate $= k\left[\dfrac{0.1 \text{ mol}}{1.0 \text{ L}}\right]\left[\dfrac{2.0 \text{ mol}}{1.0 \text{ L}}\right] = 0.20 \text{ k mol}^2 \text{ L}^{-2}$

The fastest rate would be for the conditions for (b). The rate for (b) is faster than the rate for (a) because the concentrations for (b) are each twice the concentrations for (a). The rate for (b) is faster than the rate for (c) because the concentration of CO is 5 times greater for (b) than for (c) and the concentration of NO$_2$ for (b) is 1/4 that for (c). Therefore, (b) is 5/4 faster than (c).

14.47 a) Rate $= \dfrac{\Delta[C_6H_6]}{\Delta T} = -\dfrac{1}{3}\dfrac{\Delta[C_2H_2]}{\Delta T}$

(continued)

(14.47 continued)
     b) Not enough information is given to find the rate law. Experiments would need to be performed in which the starting concentrations of acetylene are varied and either the formation of benzene vs. time or the disappearance of acetylene vs. time is measured. Comparison of the change of starting concentrations of acetylene with the changes of rates of reaction would determine the order of the reaction with respect to acetylene. Then substitution of the starting concentration of acetylene and the rate (for any of these experiments) into the rate expression allows solving for k.

14.49 $A \rightarrow B + C$ rate $= k[A]^n$

rate of flask 1 $= rate_1 = k[\,5\,A]^n$

rate of flask 2 $= rate_2 = 4\,rate_1 = k[10\,A]^n$

$$\frac{rate_2}{rate_1} = \frac{4\,rate_1}{rate_1} = 4 = \frac{k[10\,A]^n}{k[5\,A]^n} = [2]^n \qquad n = 2$$

The rate-determining step depends on a collision between 2 molecules of A. Doubling the concentration of A increases the number of collisions by $[2]^2 = 4$ times; that increases the reaction rate by 4 times.

14.51

Time (s)	0	10.0	20.0	30.0
[A](M)	0.64	0.52	0.40	0.28
ln [A]	-0.45	-0.65	-0.92	-1.27
1/[A]	1.6	1.9	2.5	3.6

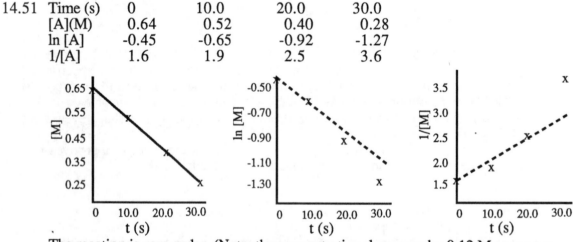

The reaction is zero-order. (Note: the concentration decreases by 0.12 M every ten seconds; therefore, the rate is independent of concentration.)

14.53 $k = Ae^{-E_a/RT}$ (Arrhenius equation) $21°C = 294\,K$

$rate_{urease} = A\,e^{-(46\,kJ/mol)/RT}$ $rate_{uncatalyzed} = A\,e^{-(125\,kJ/mol)/RT}$

$$\frac{rate_{urease}}{rate_{uncat.}} = \frac{A\,e^{-(46\,kJ/mol)/RT}}{A\,e^{-(125\,kJ/mol)/RT}} = e^{\left(\frac{-46\,kJ/mol}{RT} + \frac{125\,kJ/mol}{RT}\right)} = e^{\frac{79\,kJ/mol}{RT}}$$

$$= e^{\dfrac{79\,kJ/mol}{(8.314\,J/mol\,K)\left(\dfrac{kJ}{10^3\,J}\right)(294\,K)}} = e^{32.32} = 1.1 \times 10^{14}$$

14.55 a) A catalyst lowers the activation energy for the reaction. This speeds up the reaction because more molecules possess enough energy to react.

b) An increase in temperature increases the average energy of the molecules that in turn increases the proportion of molecules with enough energy to exceed the activation energy.

c) An increase in concentration increases the number of collisions that have the correct orientation and enough energy to react.

14.57 a)

Time (hr)	0	2	3	5	7	9
conc (M)	0.500	0.300	0.250	0.188	0.150	0.125
ln [ ]	-0.693	-1.20	-1.39	-1.67	-1.90	-2.08
1/[ ]	2.00	3.33	4.00	5.32	6.67	8.00

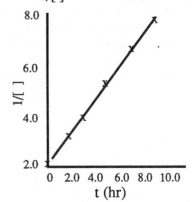

$$Rate = k[NH_4NCO]^2$$

b) Rate constant (k) at 50°C equals the slope of line;

$$k_2 = \frac{\Delta y}{\Delta x} = \frac{8.00 \ M^{-1} - 2.00 \ M^{-1}}{9 \ hr - 0 \ hr} = \frac{6.00}{9} M^{-1} \ hr^{-1} = 0.67 \ M^{-1} \ hr^{-1} \ (at \ 50°C)$$

c) 25°C $\quad k_1 = \dfrac{(1/0.300 \ M) - (1/0.500 \ M)}{6 \ hr} = \dfrac{3.33 \ M^{-1} - 2.00 \ M^{-1}}{6 \ hr} = 0.22 \ M^{-1} \ hr^{-1}$

$$E_a = \frac{R \ln\left(\dfrac{k_2}{k_1}\right)}{\left(\dfrac{1}{T_1} - \dfrac{1}{T_2}\right)} = \frac{(8.314 \ J \ mol^{-1} \ K^{-1})\left[\ln\left(\dfrac{0.67 \ M^{-1} \ hr^{-1}}{0.22 \ M^{-1} \ hr^{-1}}\right)\right]}{\left(\dfrac{1}{298 \ K} - \dfrac{1}{323 \ K}\right)}$$

$$= \frac{(8.314 \ J \ mol^{-1} \ K^{-1})[\ln(3.05)]}{(3.356 \times 10^{-3} \ K^{-1} - 3.096 \times 10^{-3} \ K^{-1})} = \frac{9.271 \ J \ mol^{-1} \ K^{-1}}{2.60 \times 10^{-4} \ K^{-1}}$$

$$= 3.57 \times 10^4 \ J \ mol^{-1} \ = 35.7 \ kJ \ mol^{-1}$$

14.59 a) false (The reaction is first-order, not fourth-order.)
b) false (One needs the rate constant for 2 different temperatures to evaluate the activation energy.)
c) true (This assumes that A, the geometric factor, does not change over this temperature range. If A changes, the truth of the statement depends on how A changes.)
d) true (This step is first-order.)

14.61

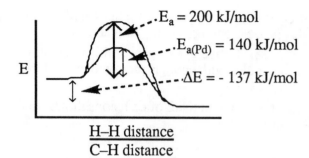

$E_a = 200$ kJ/mol

$E_{a(Pd)} = 140$ kJ/mol

$\Delta E = -137$ kJ/mol

E

$\underline{\text{H–H distance}}$
C–H distance

14.63 a) $NO_2 + O_3 \rightarrow NO_3 + O_2$ (rate determining)

$NO_2 + NO_3 \rightarrow N_2O_5$

b) Your drawing of a molecular picture should be similar to Figure 14-12 for the first step except you have NO_2 instead of NO. The drawing of a molecular picture for the second step would involve the combining of NO_2 and NO_3 to form N_2O_5.

c)

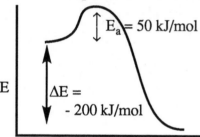

$E_a = 50$ kJ/mol

E

$\Delta E = -200$ kJ/mol

14.65

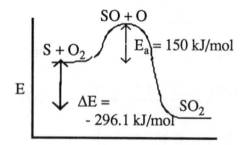

SO + O

S + O_2

$E_a = 150$ kJ/mol

E

$\Delta E = -296.1$ kJ/mol

SO_2

14.67 $\ln\left[\dfrac{A}{A_0}\right] = -kt$ $t = \dfrac{\ln\left[\dfrac{A}{A_0}\right]}{-k} = \dfrac{\ln\left[\dfrac{A}{A_0}\right]}{-5.5 \times 10^{-4} \text{ s}^{-1}}$

10% decomposed
A = 0.90 A_0

$t = \dfrac{\ln\left(\dfrac{0.90\ A_0}{A_0}\right)}{-5.5 \times 10^{-4} \text{ s}^{-1}} = 1.9 \times 10^2$ s   (or 3.2 min)

50% decomposed
A = 0.50 A_0

$t = \dfrac{\ln\left(\dfrac{0.50\ A_0}{A_0}\right)}{-5.5 \times 10^{-4} \text{ s}^{-1}} = 1.3 \times 10^3$ s   (or 21 min)

(Note: this is $t_{1/2} = \dfrac{\ln 2}{5.5 \times 10^{-4} \text{ s}^{-1}} = 1.26 \times 10^3$ s)

(continued)

130

(14.67 continued)
99.9% decomposed
A = 0.001 A_0

$$t = \frac{\ln\left(\dfrac{0.001\ A_0}{A_0}\right)}{-5.5 \times 10^{-4}\ s^{-1}} = 1.3 \times 10^4\ s$$

(or 209 min) (or 3.5 hr)

14.69 a) Rate = $k[H_2][X_2]$

b) Rate = $k[X_2]$

c) $X_2 \underset{k_{-1}}{\overset{k_1}{\rightleftarrows}} X + X$

$$k_1[X_2] = k_{-1}[X]^2 \qquad [X] = \left(\frac{k_1}{k_{-1}}\right)^{1/2}[X_2]^{1/2}$$

$$\text{Rate} = k_2[X][H_2] = k_2\left(\frac{k_1}{k_{-1}}\right)^{1/2}[X_2]^{1/2}[H_2] = k[X_2]^{1/2}[H_2]$$

d) $X_2 + X_2 \underset{k_{-1}}{\overset{k_1}{\rightleftarrows}} X_3 + X$

$$k_1[X_2]^2 = k_{-1}[X_3][X] \qquad [X] = \frac{k_1}{k_{-1}}\frac{[X_2]^2}{[X_3]}$$

$$\text{Rate} = k_2[X][H_2] = k_2\left(\frac{k_1}{k_{-1}}\right)\frac{[X_2]^2}{[X_3]} = k\frac{[X_2]^2}{[X_3]}$$

This rate law cannot be tested experimentally because it contains the intermediate X_3.

14.71 a) Assume the rate (and, therefore, the rate constant) at 25°C (298 K) is double the

$$E_a = \frac{R \ln\left(\dfrac{k_2}{k_1}\right)}{\left(\dfrac{1}{T_1} - \dfrac{1}{T_2}\right)} = \frac{8.314\ J\,/\,mol\ K\ \ln\ (2)}{\left(\dfrac{1}{288\ K} - \dfrac{1}{298\ K}\right)} = \frac{5.7628\ J\,/\,mol\ K}{\left(1.165 \times 10^{-4}\ K^{-1}\right)}$$

= 4.95×10^4 J/mol or 49.5 kJ/mol

(continued)

131

(14.71 continued)

b) $\ln\left(\dfrac{k_2}{k_1}\right) = \dfrac{E_a}{R}\left(\dfrac{1}{T_1} - \dfrac{1}{T_2}\right)$ $T_1 = 20°C = 293\ K$ $T_2 = 25°C = 298\ K$

$k_1 = 1/10 = 0.1$ $k_2 = 1/t$

$\ln\left(\dfrac{1/t}{0.1\ min^{-1}}\right) = \dfrac{4.95 \times 10^4\ J/mol}{8.314\ J/mol\ K}\left(\dfrac{1}{293\ K} - \dfrac{1}{298\ K}\right)$

$= 5.9538 \times 10^3\ K\ (5.726 \times 10^{-5}\ K^{-1}) = 0.3409$

$\ln\left(\dfrac{1\ min}{t}\right) - \ln(0.1) = 0.3409$

$\ln\left(\dfrac{1\ min}{t}\right) = 0.3409 + \ln(0.1) = (0.3409) - 2.3026 = -1.9617$

$\left(\dfrac{1\ min}{t}\right) = 0.1406$ $t = 7.1\ min$

14.73 Initially the rate of formation of product increases with the increase of the concentration of reactant. As the concentration of reactant is increased, more and more active sites on the enzyme are occupied and more and more product is formed. At some point all active sites are occupied and the rate of formation of product becomes constant as reactant molecules can only bind to the active sites after reaction occurs and the product is released.

14.75 $CF_2Cl_2 \xrightarrow{\ h\nu\ } CF_2Cl + Cl, \qquad O_3 + Cl \rightarrow O_2 + OCl, \qquad OCl + O \rightarrow O_2 + Cl$

14.77 Rate $= k[\text{\o}]^{1/2}[\bullet]$

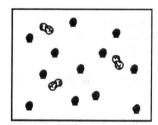

14.79 The time required for 6 atoms of ^{32}P to decompose to 3 atoms of ^{32}P (i.e., the half-life) will be the same for both flasks. The half-life ($t_{1/2}$) of a first-order process is independent of concentration and equals: $t_{1/2} = \dfrac{\ln 2}{k} = \dfrac{0.6931}{k}$

14.81 Rate $= k[X_2]$

In the first pair of drawings A has 5 X_2's and B has 10 X_2's. The number of Y's is constant. The rate will double with the doubling of the X_2's since the reaction is first order with respect to X_2.
In the second pair of drawings A has 5 X_2's and 8 Y's B has 5 X_2's and 16 Y's. The rate will not change with the doubling of the Y's since the reaction is zero order with respect to Y.

132

CHAPTER 15: PRINCIPLES OF CHEMICAL EQUILIBRIUM

15.1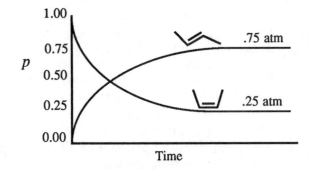

15.3 $ClO^-_{(aq)} + ClO^-_{(aq)} \xrightarrow{k_1} Cl^-_{(aq)} + ClO_2^-_{(aq)}$

$ClO^-_{(aq)} + ClO_2^-_{(aq)} \xrightarrow{k_2} Cl^-_{(aq)} + ClO_3^-_{(aq)}$

a) Every elementary reaction is a molecular rearrangement that goes both in the forward and the reverse direction. We are given elementary reactions in the forward direction, so we need to write elementary reactions in the reverse direction:

$Cl^-_{(aq)} + ClO_2^-_{(aq)} \xrightarrow{k_{-1}} 2\ ClO^-_{(aq)}$

$Cl^-_{(aq)} + ClO_3^-_{(aq)} \xrightarrow{k_{-2}} ClO^-_{(aq)} + ClO_2^-_{(aq)}$

b) $K_{eq} = \dfrac{\left[Cl^-\right]^2_{eq}\left[ClO_3^-\right]_{eq}}{\left[ClO^-\right]^3_{eq}}$

c) At equilibrium, each step in the mechanism has a forward rate that equals its reverse rate:

$2\ ClO^- \underset{k_{-1}}{\overset{k_1}{\rightleftharpoons}} Cl^- + ClO_2^-$ and $ClO^- + ClO_2^- \underset{k_{-2}}{\overset{k_2}{\rightleftharpoons}} Cl^- + ClO_3^-$

$k_1\left[ClO^-\right]^2_{eq} = k_{-1}\left[Cl^-\right]_{eq}\left[ClO_2^-\right]_{eq} \Rightarrow \dfrac{k_1}{k_{-1}} = \dfrac{\left[Cl^-\right]_{eq}\left[ClO_2^-\right]_{eq}}{\left[ClO^-\right]^2_{eq}}$

$k_2\left[ClO^-\right]_{eq}\left[ClO_2^-\right]_{eq} = k_{-2}\left[Cl^-\right]_{eq}\left[ClO_3^-\right]_{eq} \Rightarrow \dfrac{k_2}{k_{-2}} = \dfrac{\left[Cl^-\right]_{eq}\left[ClO_3^-\right]_{eq}}{\left[ClO^-\right]_{eq}\left[ClO_2^-\right]_{eq}}$

(continued)

(15.3 continued)

$$\frac{k_1k_2}{k_{-1}k_{-2}} = \frac{\left[Cl^-\right]_{eq}^2\left[ClO_3^-\right]_{eq}}{\left[ClO^-\right]_{eq}^3}$$

$$K_{eq} = \frac{k_{formation}}{k_{decomp.}} = \frac{k_1k_2}{k_{-1}k_{-2}}$$

15.5   Using Figure 15.1 as a guide, your molecular picture should include two hypochlorite ions approaching each other, colliding, and fragmenting to form Cl^- and ClO_2^-. In a second picture another hypochlorite and a ClO_2^- approach each other, collide, and fragment to form chloride and chlorate ions. The reversibility of the reaction can be shown either by drawing molecular pictures of the reaction going in the opposite direction or by showing that the reactions are reversible by using arrows showing the process occurring in both directions. See Figure 15-1.

15.7 ClO^- ions are placed in the reaction vessel, brought to reaction conditions and allowed to react. Analysis of the contents will show the presence of Cl^- and ClO_3^- Cl^- and ClO_3^- ions are placed in the reaction vessel, brought to reaction conditions and allowed to react. Analysis of the contents will show the presence of ClO^- ions.

15.9   a) $K_{eq} = \dfrac{\left(p_{PCl_5}\right)_{eq}}{\left(p_{Cl_2}\right)_{eq}\left(p_{PCl_3}\right)_{eq}}$; units = atm^{-1}

b) $K_{eq} = \dfrac{\left(X_{P_4O_{10}}\right)_{eq}}{\left(X_{P_4}\right)_{eq}\left(p_{O_2}\right)_{eq}^5} = \dfrac{1}{\left(p_{O_2}\right)_{eq}^5}$; units = atm^{-5}

c) $K_{eq} = \dfrac{\left(X_{BaO}\right)_{eq}\left(p_{CO}\right)_{eq}^2}{\left(X_{BaCO_3}\right)_{eq}\left(X_C\right)_{eq}} = \left(p_{CO}\right)_{eq}^2$; units = atm^2

d) $K_{eq} = \dfrac{\left(X_{CH_3OH}\right)_{eq}}{\left(p_{CO}\right)_{eq}\left(p_{H_2}\right)_{eq}^2} = \dfrac{1}{\left(p_{CO}\right)_{eq}\left(p_{H_2}\right)_{eq}^2}$; units = atm^{-3}

e) $K_{eq} = \dfrac{\left[PO_4^{3-}\right]_{eq}\left[H_3O^+\right]_{eq}^3}{\left[H_3PO_4\right]_{eq}\left(X_{H_2O}\right)_{eq}^3} = \dfrac{\left[PO_4^{3-}\right]_{eq}\left[H_3O^+\right]_{eq}^3}{\left[H_3PO_4\right]_{eq}}$; units = M^3

15.11 a) $K_{eq} = \dfrac{\left(X_{P_4}\right)_{eq}\left(p_{O_2}\right)_{eq}^5}{\left[X_{P_4O_{10}}\right]_{eq}} = \left(p_{O_2}\right)_{eq}^5$; units = atm^5

 b) $K_{eq} = \dfrac{\left(X_{BaCO_3}\right)_{eq}\left(X_C\right)_{eq}}{\left(X_{BaO}\right)_{eq}\left(p_{CO}\right)_{eq}^2} = \dfrac{1}{\left(p_{CO}\right)_{eq}^2}$; units = atm^{-2}

 c) $K_{eq} = \dfrac{\left[H_3PO_4\right]_{eq}\left(X_{H_2O}\right)_{eq}^3}{\left[PO_4^{3-}\right]_{eq}\left[H_3O^+\right]_{eq}^3} = \dfrac{\left[H_3PO_4\right]_{eq}}{\left[PO_4^{3-}\right]_{eq}\left[H_3O^+\right]_{eq}^3}$; units = M^{-3}

15.13 a) $K_{eq} = \dfrac{\left(X_{I_2}\right)_{eq}}{\left(p_{I_2}\right)_{eq}} = \dfrac{1}{\left(p_{I_2}\right)_{eq}} = \dfrac{1}{\text{vapor pressure}}$

 b) $K_{eq} = \dfrac{\left[I_{2(\text{solution})}\right]_{eq}}{\left(p_{I_2}\right)_{eq}} = K_H$?

 c) $K_{eq} = \dfrac{\left(X_{PbI_2}\right)_{eq}}{\left[Pb^{2+}\right]_{eq}\left[I^-\right]_{eq}^2} = \dfrac{1}{\left[Pb^{2+}\right]_{eq}\left[I^-\right]_{eq}^2} = \dfrac{1}{K_{sp}}$

 d) $K_{eq} = \dfrac{\left[CN^-\right]_{eq}\left[H_3O^+\right]_{eq}}{\left[HCN\right]_{eq}\left(X_{H_2O}\right)_{eq}} = \dfrac{\left[CN^-\right]_{eq}\left[H_3O^+\right]_{eq}}{\left[HCN\right]_{eq}} = K_a$

 e) $K_{eq} = \dfrac{\left[Pt^{2+}\right]_{eq}\left[Cl^-\right]_{eq}^4}{\left[PtCl_4^{2-}\right]_{eq}} = \dfrac{1}{K_f}$?

15.15 You may wish to draw two pictures. The first picture would be like Figure 15.4 except it should show more molecules of gas escaping and more being dissolved. The second picture would have reduced pressure (fewer molecules) above the liquid. In that second picture you need to show that the number of molecules escaping the liquid is the same as that in the first picture but that the number of molecules being dissolved is less.

15.17 Initial species: $H_{2(g)}$ and $CO_{2(g)}$

Chemical reaction: $H_{2(g)} + CO_{2(g)} \rightleftharpoons H_2O_{(g)} + CO_{(g)}$

$$K_{eq} = \frac{(p_{H_2O})_{eq}(p_{CO})_{eq}}{(p_{H_2})_{eq}(p_{CO_2})_{eq}} = \frac{[H_2O]_{eq}[CO]_{eq}}{[H_2]_{eq}[CO_2]_{eq}}$$

$(p_{CO})_{eq} = 0.49$ mol $\Rightarrow (p_{H_2O})_{eq} = (p_{CO})_{eq} = 0.49$ mol

Reaction:	$H_{2(g)}$ +	$CO_{2(g)}$	$\rightleftharpoons$ $H_2O_{(g)}$ +	$CO_{(g)}$
Init. (M)	1.0	1.0	0	0
Change (M)	-0.49	-0.49	+0.49	+0.49
Equilibrium (M)	0.51	0.51	0.49	0.49

$$K_{eq} = \frac{(0.49)^2}{(0.51)^2} = 0.92$$

15.19 $C_2H_5CO_2H_{(aq)} + H_2O_{(l)} \rightleftharpoons C_2H_5CO_2^-{}_{(aq)} + H_3O^+{}_{(aq)}$

$$K_{eq} = \frac{[C_2H_5CO_2^-]_{eq}[H_3O^+]_{eq}}{[C_2H_5CO_2H]_{eq}} = K_a$$

Reaction:	$C_2H_5CO_2H_{(aq)}$ +	$H_2O_{(l)} \rightleftharpoons C_2H_5CO_2^-{}_{(aq)}$ +	$H_3O^+{}_{(aq)}$
Init. (M)	$0.050/.500 = 1.00 \times 10^{-1}$	0	0
Change (M)	-1.15×10^{-3}	$+1.15 \times 10^{-3}$	$+1.15 \times 10^{-3}$
Equilibrium (M)	9.89×10^{-2}	1.15×10^{-3}	1.15×10^{-3}

$$K_a = \frac{(1.15 \times 10^{-3})^2}{(9.89 \times 10^{-2})} = 1.3 \times 10^{-5} \text{ M}$$

15.21

Reaction:	$PbF_{2(s)}$ + $H_2O_{(l)} \rightleftharpoons$	$Pb^{2+}{}_{(aq)}$ +	$2\ F^-{}_{(aq)}$
Initial		0	0
Change		$+ 1.9 \times 10^{-3}$	$+3.8 \times 10^{-3}$
Equilibrium		$+ 1.9 \times 10^{-3}$	$+3.8 \times 10^{-3}$

$$K_{sp} = K_{eq} = \frac{[Pb^{2+}]_{eq}[F^-]_{eq}^2}{[X_{PbF_2}]_{eq}[X_{H_2O}]} = [Pb^{2+}]_{eq}[F^-]_{eq}^2$$

$$K_{sp} = (1.9 \times 10^{-3}\text{M})(3.8 \times 10^{-3}\text{M})^2 = 2.7 \times 10^{-8} \text{ M}^3$$

15.23 $H_{2(g)} + Br_{2(g)} \rightleftharpoons 2\,HBr_{(g)}$ $K_{eq} = \dfrac{(p_{HBr})^2_{eq}}{(p_{H_2})_{eq}(p_{Br_2})_{eq}} = 1.6 \times 10^5$

Reaction:	$H_{2(g)}$	$+\ Br_{2(g)}$	$\rightleftharpoons$	$2\,HBr_{(g)}$
Initial (atm)	0	0		10.0
Change (atm)	+x	+x		-2x
Equilibrium (atm)	+x	+x		10.0 - 2x

$1.6 \times 10^5 = \dfrac{(10-2x)^2}{x^2} = \dfrac{100 - 40x + 4x^2}{x^2}$

$0 = 100 - 40x + 4x^2 - 1.6 \times 10^5 x^2 = -1.6 \times 10^5 x^2 - 40x + 100$

$x = \dfrac{-b \pm \sqrt{b^2 - 4ac}}{2a} = \dfrac{40 \pm \sqrt{1600 - 4(-1.6 \times 10^5)(100)}}{-2(1.6 \times 10^5)}$

$= \dfrac{40 \pm \sqrt{64001600}}{-3.2 \times 10^5} = \dfrac{40 \pm 8000.10}{-3.2 \times 10^5} = 2.49 \times 10^{-2}$ or -2.51×10^{-2}

$(p_{H_2})_{eq} = (p_{Br_2})_{eq} = 2.5 \times 10^{-2}$ atm $(p_{HBr})_{eq} = 10.0 - 2(2.5 \times 10^{-2}) = 10$ atm

where did 4x² go

15.25 $K_{eq} = \dfrac{(p_{COCl_2})_{eq}}{(p_{CO})_{eq}(p_{Cl_2})_{eq}} = 1.5 \times 10^8$ atm^{-1}

Reaction:	$CO_{(g)}$	$+\ Cl_{2(g)}$	$\rightleftharpoons$	$COCl_{2(g)}$
Init. (atm)	0	0		0.250 atm
Change (atm)	+x	+x		-x
Equil. (atm)	x	x		0.250 - x

$1.5 \times 10^8 = \dfrac{0.250 - x}{x^2}$ $0 = 1.5 \times 10^8 x^2 + x - 0.250$

$x = \dfrac{-b \pm \sqrt{b^2 - 4ac}}{2a} = \dfrac{-1 \pm \sqrt{1 - 4(1.5 \times 10^8)(-0.250)}}{2(1.5 \times 10^8)}$

$= \dfrac{-1 \pm 1.22 \times 10^4}{3.0 \times 10^8} = 4.1 \times 10^{-5}$ $(p_{CO})_{eq} = (p_{Cl_2})_{eq} = 4.1 \times 10^{-5}$ atm

$(p_{COCl_2})_{eq} = 0.250 - 4.1 \times 10^{-5} = 2.50 \times 10^{-1}$ atm

15.27 $PbCl_{2(s)} + H_2O_{(l)} \rightleftharpoons Pb^{2+}_{(aq)} + 2\,Cl^-_{(aq)}$ $K_{sp} = [Pb^{2+}]_{eq}[Cl^-]^2_{eq} = 2 \times 10^{-5}$ M^3

$Q = [Pb^{2+}][Cl^-]^2$

$[Pb^{2+}] = \dfrac{0.50\ g\ /\ 278.1\ g\ mol^{-1} \times 1\ mol\ Pb^{2+}\ /\ mol\ PbCl_2}{0.300\ L} = 6.0 \times 10^{-3}$ M

$[Cl^-] = 2[Pb^{2+}] = 1.2 \times 10^{-2}$ M $Q = (6.0 \times 10^{-3}\ M)(1.2 \times 10^{-2}\ M)^2 = 8.6 \times 10^{-7}$ M^3

$Q < K_{sp}$ so all of the solid dissolves.

15.29 a) $2 \text{SO}_{2(g)} + \text{O}_{2(g)} \rightleftharpoons 2 \text{SO}_{3(g)}$

$\Delta G^\circ = -RT \ln K_{eq}$ $\ln K_{eq} = -\Delta G^\circ / RT$ $R = 8.314$ J/K $T = 298$ K

$\Delta G^\circ_{rxn} = \Sigma(\text{coeff}) \Delta G^\circ_f (\text{products}) - \Sigma(\text{coeff}) \Delta G^\circ_f (\text{reactants})$

$= 2 \Delta G^\circ_f (\text{SO}_3) - 2 \Delta G^\circ_f (\text{SO}_2) - \Delta G^\circ_f (\text{O}_2)$

$= 2 (-371.06 \text{ kJ mol}^{-1}) - 2(-300.194 \text{ kJ mol}^{-1}) - 0 = -141.73$ kJ/mol

$\ln K_{eq} = \dfrac{141.73 \text{ kJ / mol}}{(8.314 \text{ J / mol K})(298 \text{ K})} 1000 \text{ J / kJ} = 57.21$

$K_{eq} = e^{57.21} = 7.0 \times 10^{24} \text{ atm}^{-1}$

b) $2 \text{CO}_{(g)} + \text{O}_{2(g)} \rightleftharpoons 2 \text{CO}_{2(g)}$

$\Delta G^\circ_{rxn} = 2 \Delta G^\circ_f (\text{CO}_2) - \Delta G^\circ_f (\text{O}_2) - 2 \Delta G^\circ_f (\text{CO})$

$= 2(-394.359) - 0 - 2(-137.168) = -514.38$ kJ/mol

$\ln K_{eq} = \dfrac{-\Delta G}{RT} = \dfrac{(514.38)10^3}{(8.314)(298)} = 207.62$ $K_{eq} = 1.5 \times 10^{90} \text{ atm}^{-1}$

c) $\text{BaCO}_{3(s)} + \text{C}_{(s)} \rightleftharpoons \text{BaO}_{(s)} + 2 \text{CO}_{(g)}$

$\Delta G^\circ_{rxn} = \Delta G^\circ_f (\text{BaO}) + 2 \Delta G^\circ_f (\text{CO}) - \Delta G^\circ_f (\text{BaCO}_3) - \Delta G^\circ_f (\text{C})$

$= -525.1 + 2(-137.168) + 1137.6 - 0 = 338.16$ kJ/mol

$\ln K_{eq} = \dfrac{-(338.16)(1000)}{(8.314)(298)} = -136.49$ $K_{eq} = 5.3 \times 10^{-60} \text{ atm}^2$

15.31 a) $\text{CO}_{(g)} + \text{H}_2\text{O}_{(g)} \rightleftharpoons \text{CO}_{2(g)} + \text{H}_{2(g)}$

$\Delta G^\circ = \Delta G^\circ(\text{CO}_2) + \Delta G^\circ(\text{H}_2) - \Delta G^\circ(\text{CO}) - \Delta G^\circ(\text{H}_2\text{O})$

$= -394.359 + 0 + 137.168 + 228.72 = -28.47$ kJ/mol

$\ln K_{eq} = \dfrac{-\Delta G^\circ}{RT} = \dfrac{(28.47)(1000)}{(8.314)(298)} = 11.49$ $K_{eq} = 9.8 \times 10^4$

b) $\text{CH}_{4(g)} + \text{H}_2\text{O}_{(g)} \rightleftharpoons \text{CO}_{(g)} + 3 \text{H}_{2(g)}$

$\Delta G^\circ = -137.168 + 3(0) - 228.72 + 50.72 = -315.17$ kJ/mol

$\ln K_{eq} = \dfrac{-(-315.17)(1000)}{(8.314)(298)} = 127.21$ $K_{eq} = 1.8 \times 10^{55} \text{ atm}^2$

c) $\text{SnO}_{2(s)} + 2 \text{H}_{2(g)} \rightleftharpoons \text{Sn}_{(s)} + 2 \text{H}_2\text{O}_{(g)}$

$\Delta G^\circ = 0 + 2(-228.72) + 519.6 - 2(0) = 62.16$ kJ/mol

$\ln K_{eq} = \dfrac{-(62.16)(1000)}{(8.314)(298)} = -25.09$ $K_{eq} = 1.3 \times 10^{-11}$

d) $4 \text{NH}_{3(g)} + 5 \text{O}_{2(g)} \rightleftharpoons 4 \text{NO}_{(g)} + 6 \text{H}_2\text{O}_{(g)}$

$\Delta G^\circ = 4(86.55) + 6(-228.72) - 4(-16.45) - 5(0) = -960.32$ kJ/mol

$\ln K_{eq} = \dfrac{-(-960.32)(1000)}{(8.314)(298)} = 387.61$ $K_{eq} = 2.2 \times 10^{168} \text{ atm}$

(continued)

(15.31 continued)

e) $3 Fe_{(s)} + 4 H_2O_{(g)} \rightleftarrows Fe_3O_{4(s)} + 2 H_{2(g)}$

$\Delta G° = -1015.4 + 4(0) - 3(0) - 4(-228.72) = -100.52 \text{ kJ/mol}$

$\ln K_{eq} = \dfrac{-(-100.52)(1000)}{(8.314)(298)} = 40.57$ $\qquad K_{eq} = 4.2 \times 10^{17} \text{ atm}^{-2}$

15.33 a) $CO_{(g)} + H_2O_{(g)} \rightleftarrows CO_{2(g)} + H_{2(g)}$; T = 700 K

$\ln K_{eq} = \dfrac{-\Delta H°}{RT} + \dfrac{\Delta S°}{R}$

$\Delta H° = -393.509 + 0 - (-110.525) - (-241.818) = -41.166 \text{ kJ/mol}$

$\Delta S° = 213.74 + 130.684 - 197.674 - 188.825 = -42.075 \text{ J/mol K}$

$\ln K_{eq} = \dfrac{-(41.166 \text{ kJ / mol})(1000 \text{ J / kJ})}{(8.314 \text{ J / K})(700 \text{ K})} + \dfrac{(-42.075 \text{ J / mol K})}{(8.314 \text{ J / K})} = 2.013$

$K_{eq} = e^{2.013} = 7.5$

b) $CH_{4(g)} + H_2O_{(g)} \rightleftarrows CO_{(g)} + 3 H_{2(g)}$

$\Delta H \approx \Delta H° = -110.525 + 3(0) - (-74.81) - (-241.818) = 206.103 \text{ kJ/mol}$

$\Delta S \approx \Delta S° = 197.674 + 3(130.684) - 186.264 - 188.825 = 214.637 \text{ J/mol K}$

$\ln K_{eq} = \dfrac{-\Delta H°}{RT} + \dfrac{\Delta S°}{R} = \dfrac{-(206.103)(1000)}{(8.314)(700 \text{ K})} + \dfrac{214.637}{8.314} = -9.598$

$K_{eq} = 6.8 \times 10^{-5} \text{ atm}^2$

c) $SnO_{2(s)} + 2 H_{2(g)} \rightleftarrows Sn_{(s)} + 2 H_2O_{(g)}$

$\Delta H \approx \Delta H° = 0 + 2(-241.818) - (-580.7) - 2(0) = 97.064 \text{ kJ/mol}$

$\Delta S \approx \Delta S° = 51.55 + 2(188.825) - 52.3 - 2(130.684) = 115.532 \text{ J/mol K}$

$\ln K_{eq} = \dfrac{-\Delta H°}{RT} + \dfrac{\Delta S°}{R} = -2.782$; $K_{eq} = 6.2 \times 10^{-2}$

d) $4 NH_{3(g)} + 5 O_{2(g)} \rightleftarrows 4 NO_{(g)} + 6 H_2O_{(g)}$

$\Delta H \approx \Delta H° = 4(90.25) + 6(-241.818) - 4(-46.11) - 5(0) = -905.468 \text{ kJ/mol}$

$\Delta S \approx \Delta S° = 4(210.761) + 6(188.825) - 4(192.45) - 5(205.138) = 180.504 \text{ J/mol K}$

$\ln K_{eq} = 177.29$; $K_{eq} = 9.96 \times 10^{76} \text{ atm}^1$

e) $3 Fe_{(s)} + 4 H_2O_{(g)} \rightleftarrows Fe_3O_{4(s)} + 4 H_{2(g)}$

$\Delta H \approx \Delta H° = -1118.4 + 4(0) - 3(0) - 4(-241.818) = -151.128 \text{ kJ/mol}$

$\Delta S \approx \Delta S° = 146.4 + 4(130.684) - 3(27.28) - 4(188.825) = 168.004 \text{ J/mol K}$

$\ln K_{eq} = 5.76$; $K_{eq} = 3.2 \times 10^2$

15.35 a) $2 SO_{2(g)} + O_{2(g)} \rightleftharpoons 2 SO_{3(g)}$

$Q = \dfrac{\left(p_{SO_3}\right)^2}{\left(p_{SO_2}\right)^2\left(p_{O_2}\right)}$, so injecting $CO_{(g)}$ into the system has **no effect** on the equilibrium position.

b) $2 CO_{(g)} + O_{2(g)} \rightleftharpoons 2 CO_{2(g)}$

$Q = \dfrac{\left(p_{CO_2}\right)^2}{\left(p_{CO}\right)^2\left(p_{O_2}\right)}$, so adding $CO_{(g)}$ decreases the value of Q which should cause the reaction to form products from reactants until equilibrium is reestablished. Therefore, **more $CO_{2(g)}$ is produced.**

c) $BaCO_{3(s)} + C_{(s)}\ BaO_{(s)} + 2 CO_{(g)}$

$Q = (p_{CO})^2$, so adding $CO_{(g)}$ makes $Q > K_{eq}$; thus, reaction products will be consumed and reactants will be produced until K_{eq} is reestablished; that is, **$BaCO_{3(s)}$ and $C_{(s)}$ will be produced.**

15.37 $PbCl_{2(s)} \rightleftharpoons Pb^{2+}_{(aq)} + 2 Cl^-_{(aq)}$

a) $Q = [Pb^{2+}][Cl^-]^2$ so adding more $PbCl_{2(s)}$ will not change Q. The system remains at equilibrium and there is no effect on the amount of dissolved $PbCl_{2(s)}$.
b) Adding more H_2O dilutes the solution which lowers $[Pb^{2+}]$ and $[Cl^-]$. Thus, $Q < K_{eq}$ so more $PbCl_{2(s)}$ will dissolve until K_{eq} is reestablished.
c) Adding solid NaCl increases $[Cl^-]$. Thus, $Q > K_{eq}$ so the reaction proceeds to the left and some $PbCl_{2(s)}$ will precipitate.
d) Adding solid KNO_3 has no effect on Q so there is no effect on the amount of dissolved $PbCl_{2(s)}$.

15.39 $SO_{2(g)} + Cl_{2(g)} \rightleftharpoons SO_2Cl_{2(g)} + heat$ $Q = \dfrac{\left(p_{SO_2Cl_2}\right)}{\left(p_{SO_2}\right)\left(p_{Cl_2}\right)}$

Changes that would drive the equilibrium to the left:
1. Heating the system (adding heat).
2. Adding $SO_2Cl_{2(g)}$ to the system.
3. Removing $Cl_{2(g)}$ from the system.
4. Removing $SO_{2(g)}$ from the system.
5. Decreasing the pressure of the system by increasing the volume.

15.41 a) $15\ \text{⚘} \rightarrow 3\ \text{⚘} + 12\ \text{∞} + 12\ \text{⊶}$

$3\ \text{⚘} \rightleftharpoons 12\ \text{∞} + 12\ \text{⊶}$

b) $K_{eq} = \dfrac{(p\ \text{∞})_{eq}\,(p\ \text{⊶})_{eq}}{(p\ \text{⚘})_{eq}} = \dfrac{(12)(12)}{3} = 48$

15.43 Reaction: $\quad$ $H_{2(g)}$ $\quad$ + $\quad$ $F_{2(g)}$ $\rightleftharpoons$ $\quad$ $2\,HF_{(g)}$
$\quad\quad\quad$ Init. (atm) $\quad$ 3.00 $\quad\quad\quad\quad$ 3.00 $\quad\quad\quad\quad$ 0
$\quad\quad\quad$ Change (atm) $\;$ -x $\quad\quad\quad\quad\;$ -x $\quad\quad\quad\quad$ +2x
$\quad\quad\quad$ Equil. (atm) $\;$ 3.00 - x $\quad\quad$ 3.00 - x $\quad\quad$ 2x

$$K_{eq} = \frac{(p_{HF})^2_{eq}}{(p_{H_2})_{eq}(p_{F_2})_{eq}} = 115$$

$$115 = \frac{(2x)^2}{(3.00-x)^2} = \frac{4x^2}{(9.00 - 6.00x + x^2)}$$

$$4x^2 = 115(9.00 - 6.00x + x^2) = 1035 - 690x + 115x^2$$

$$0 = 111x^2 - 690x + 1035$$

$$x = \frac{-b \pm \sqrt{b^2 - 4ac}}{2a} = \frac{690 \pm \sqrt{(690)^2 - 4(111)(1035)}}{2(111)}$$

$$= \frac{690 \pm \sqrt{16560}}{222} = 3.688 \text{ or } 2.528 \text{ atm}$$

x = 2.53 atm because we would get negative partial pressures for $H_{2(g)}$ and $F_{2(g)}$ if we used x = 3.69 atm.

$\Rightarrow$ Equilibrium partial pressures:

$(p_{H_2})_{eq} = 3.00 - 2.53 = 0.47 \text{ atm}$ $\quad\quad\quad\quad$ $(p_{F_2})_{eq} = (p_{H_2})_{eq} = 0.47 \text{ atm}$

$(p_{HF})_{eq} = 2(2.53) = 5.06 \text{ atm}$

15.45 $2\,NO_{2(g)} \rightleftharpoons N_2O_{4(g)}$

at 298 K: $\ln K_{eq} = \dfrac{-\Delta G^\circ}{RT}$

$\Delta G^\circ = \Delta G^\circ(N_2O_4) - 2\,\Delta G^\circ(NO_2)$
$= 97.89 - 2(51.31) = -4.73 \text{ kJ/mol}$

$\ln K_{eq} = \dfrac{-(-4.73)(1000)}{(8.314)(298)} = +1.91$ $\quad\quad$ $K_{eq} = 6.75 \text{ atm}^{-1}$

At 500 K: $\ln K_{eq} = \dfrac{-\Delta H^\circ}{RT} + \dfrac{\Delta S^\circ}{R}$ $\quad$ (Assuming $\Delta H \approx \Delta H^\circ$ and $\Delta S \approx \Delta S^\circ$)

$\Delta H^\circ = \Delta H^\circ(N_2O_4) - 2\,\Delta H^\circ(NO_2) = 9.16 - 2(33.18) = -57.20 \text{ kJ/mol}$

$\Delta S^\circ = \Delta S^\circ(N_2O_4) - 2\,\Delta S^\circ(NO_2) = 304.29 - 2(240.06) = -175.83 \text{ J/mol K}$

$\ln K_{eq} = \dfrac{-(-57.20)(1000)}{(8.314)(500)} + \dfrac{-175.83}{8.314} = -7.39$

$K_{eq} = 6.18 \times 10^{-4}$

15.47 $Ca^{2+}_{(aq)} + 3\,H_2O_{(l)} + CO_{2(g)} \rightleftarrows CaCO_{3(s)} + 2\,H_3O^+_{(aq)}$

$$K_{eq} = \frac{\left[H_3O^+\right]^2_{eq}}{\left[Ca^{2+}\right]_{eq}(p_{CO_2})_{eq}}$$

Concentration units: $Ca^{2+}_{(aq)}$ = M (moles per liter)

$\qquad\qquad\qquad$ $H_2O_{(l)}$ = X (mole fraction)

$\qquad\qquad\qquad$ $CO_{2(g)}$ = atm (partial pressure)

$\qquad\qquad\qquad$ $CaCO_{3(s)}$ = X (mole fraction)

$\qquad\qquad\qquad$ $H_3O^+_{(aq)}$ = M (moles per liter)

$$K_{eq} = \frac{M^2}{M\ atm} = M\ atm^{-1}$$

15.49 $K_{eq} = 25 = \dfrac{[\infty]\,[\clubsuit]}{[\text{oo}]\,[\bullet\bullet]}$

The molecular picture requested will contain 10 of each product molecule and two of each reactant molecule when the system reaches equilibrium.

15.51
Reaction:	$N_{2(g)}$	+	$3\,H_{2(g)}$	$\rightleftarrows$	$2\,NH_{3(g)}$
Initial (atm)	5.0		3.0		0
Change (atm)	- x		-3 x		+ 2x
Equilibrium (atm)	5.0- x		3.0-3x		2x

$$K_{eq} = \frac{\left(p_{NH_3}\right)^2_{eq}}{\left(p_{N_2}\right)_{eq}\left(p_{H_2}\right)^3_{eq}} = 2.81 \times 10^{-5}\ atm^{-2} \qquad T = 472°C = 745\ K$$

$$2.81 \times 10^{-5}\ atm^{-2} = \frac{(2x)^2}{(5.0 - x)(3.0 - 3x)^3}$$

K_{eq} is very small compared to the initial conc. of N_2 and H_2; therefore 5 - x ≈ 5 and 3 - 3x ≈ 3.

$$2.81 \times 10^{-5}\ atm^{-2} = \frac{4x^2}{5.0\,(3.0)^3} = \frac{4x^2}{135};$$

$$x = \sqrt{\frac{(2.81 \times 10^{-5})(135)}{4}} = 3.08 \times 10^{-2}$$

$(p_{NH_3})_{eq} = 2(3.08 \times 10^{-2}) = 6.16 \times 10^{-2}$ atm

$(p_{N_2})_{eq} = 5.0 - 3.08 \times 10^{-2} = 5.0$ atm

$(p_{H_2})_{eq} = 3.0 - 3(3.08 \times 10^{-2}) = 2.9$ atm

15.53 Reaction: $\quad$ $Sn_{(s)}$ $\quad$ + $\quad$ $2\,H_{2(g)}$ $\quad \rightleftharpoons \quad$ $SnH_{4(g)}$

Initial (atm)	2.00×10^2	0
Change (atm)	$-2\,x$	$+x$
Equilibrium (atm)	$2.00 \times 10^2 - 2\,x$	x

$$K_{eq} = \frac{\left(p_{SnH_4}\right)_{eq}}{\left(p_{H_2}\right)_{eq}^2} = 1.07 \times 10^{-33}\ atm^{-1}$$

$1.07 \times 10^{-33} = \dfrac{x}{(2 \times 10^2 - 2x)^2}$ $\qquad$ $x = (1.07 \times 10^{-33})(2 \times 10^2)^2 = 4.28 \times 10^{-29}\ atm$

$(p_{SnH_4})_{eq} = 4.28 \times 10^{-29}\ atm$ $\qquad\qquad$ $PV = nRT$

$$n = \frac{PV}{RT} = \frac{(4.28 \times 10^{-29}\ atm)(10\ L)}{(0.0821\ L\ atm\ /\ K\ mol)(298\ K)} = 1.75 \times 10^{-29}\ mol$$

# of molecules = nN_A = $(1.75 \times 10^{-29}\ mole)(6.022 \times 10^{23}\ molecules/mole)$
= 1.05×10^{-5} molecules of SnH_4

15.55 Reaction: $\quad$ $CCl_{4(g)}$ $\quad \rightleftharpoons \quad$ $2\,Cl_{2(g)}$ $\quad$ + $\quad$ $C_{(g)}$

Initial	0.325	0.35
Change	$-x$	$+2x$
Equilibrium	$0.325 - x$	$0.35 + 2x$

$0.75\ atm = \dfrac{(0.35 + 2x)^2}{(0.325 - x)}$ $\qquad\qquad$ $0.24375 - 0.75\,x = 0.1225 + 1.40 + 4x^2$

$4x^2 + 2.15\,x - 0.12125 = 0$ $\qquad$ $x = \dfrac{-2.15 \pm \sqrt{4.6225 + 1.94}}{8}$ $\qquad$ $x = 0.0515$

$(p_{Cl_2})_{eq} = [0.35 + 2(0.0515)]\ atm = 0.45\ atm$
$(p_{CCl_4})_{eq} = [0.325 - 0.0515]\ atm = 0.27\ atm$

15.57 a) $BeO_{(s)} + H_2O_{(l)} \rightleftharpoons Be^{2+}_{(aq)} + 2\,OH^-_{(aq)}$ $\qquad$ $K_{eq} = [Be^{2+}]_{eq}[OH^-]^2_{eq}\ M^3$

b) $CO_{2(g)} + 2\,H_2O_{(l)} \rightleftharpoons HCO_3^-_{(aq)} + H_3O^+_{(aq)}$, $\quad$ $K_{eq} = \dfrac{\left[HCO_3^-\right]_{eq}\left[H_3O\right]^+_{eq}}{\left(p_{CO_2}\right)_{eq}} M^2\ atm^{-1}$

c) $NH_{3(aq)} + CH_3CO_2H_{(aq)} \rightleftharpoons NH_4^+_{(aq)} + CH_3CO_2^-_{(aq)}$

$$K_{eq} = \frac{\left[NH_4^+\right]_{eq}\left[CH_3CO_2^-\right]_{eq}}{\left[NH_3\right]_{eq}\left[CH_3CO_2H\right]_{eq}}$$

d) $3\,H_2S_{(g)} + 6\,H_2O_{(l)} + 2\,Fe^{3+}_{(aq)} \rightleftharpoons Fe_2S_{3(s)} + 6\,H_3O^+_{(aq)}$

$$K_{eq} = \frac{\left[H_3O^+\right]^6_{eq}}{\left(p_{H_2S}\right)^3_{eq}\left[Fe^{3+}\right]^2} M^4\ atm^{-3}$$

15.59 Reaction: $MgF_{2(s)} + H_2O_{(l)} \rightleftharpoons Mg^{2+}_{(aq)} + 2\,F^-_{(aq)}$

$K_{eq} = K_{sp} = [Mg^{2+}]_{eq}[F^-]^2_{eq}$

$K_{sp} = (1.14 \times 10^{-3}\,M)[2(1.14 \times 10^{-3}\,M)]^2 = 5.93 \times 10^{-9}\,M^3$

15.61 $CO_{2(g)} + 2\,OH^-_{(aq)} \rightleftharpoons CO_3^{2-}_{(aq)} + H_2O_{(l)}$

a) $K_{eq} = \dfrac{\left[CO_3^{2-}\right]_{eq}}{\left(p_{CO_2}\right)_{eq}[OH^-]^2_{eq}}$

b) $Q = \dfrac{\left[CO_3^{2-}\right]}{\left(p_{CO_2}\right)[OH^-]^2}$, so adding Na_2CO_3 makes $Q > K_{eq}$.

When this happens, reaction products are consumed and reactants are produced until K_{eq} is restored. Thus, the pressure of CO_2 in this system will **increase** if $Na_2CO_{3(s)}$ is dissolved in the solution.

c) Bubbling $HCl_{(g)}$ through the solution would cause the concentration of OH^- to decrease, $HCl + OH^- \rightarrow H_2O + Cl^-$; the reaction would shift to the left and p_{CO_2} would increase.

15.63 $Hg_{(g)} + Hg_2Cl_{2(s)} \rightleftharpoons Hg_2Cl_{2(s)}$

a) $\ln K_{eq} = \dfrac{-\Delta G^\circ}{RT}$

$\Delta G^\circ = \Delta G^\circ(Hg_2Cl_{2(s)}) - \Delta G^\circ(Hg) - \Delta G^\circ(HgCl_2) = -210.745 - 31.82 - (-178.6)$
$= -63.97$ kJ/mol

$\ln K_{eq} = \dfrac{-(-63.97)(1000)}{(8.314)(298)} = 25.82;\ K_{eq} = 1.63 \times 10^{11}\,atm^{-1}$

b) $\Delta H^\circ = -265.22 - 61.32 - (-224.3) = -102.24$ kJ/mol
$\Delta S^\circ = 192.5 - 174.96 - 146.0 = -128.46$ J/mol K
(Assuming $\Delta H \approx \Delta H^\circ$ and $\Delta S \approx \Delta S^\circ$)
$\Delta G = \Delta H - T\Delta S = 0$ at equilibrium $\Rightarrow T\Delta S = \Delta H$

$T = \dfrac{\Delta H^\circ}{\Delta S^\circ} = \dfrac{(-102.24)(1000)}{-128.46} = 796$ K

c) $\ln K_{eq} = \dfrac{-\Delta H^\circ}{RT} + \dfrac{\Delta S^\circ}{R}$ $\ln K_{eq} = \dfrac{-(-102.24)(1000)}{(8.314)(1000\,K)} + \dfrac{-128.46}{8.314} = -3.154$

$K_{eq} = 4.269 \times 10^{-2}\,atm^{-1}$

15.65 $C_{(s)} + 2\,H_2O_{(g)} \rightleftharpoons CO_{2(g)} + 2\,H_{2(g)}$

$K_{eq} = \dfrac{\left(p_{CO_2}\right)_{eq}\left(p_{H_2}\right)^2_{eq}}{\left(p_{H_2O}\right)^2_{eq}} = 0.38\,atm$

The research will fail because catalysts do not affect K_{eq} (c.f. p. 745).

15.67 The Haber synthesis is used to produce ammonia that in turn is used either directly or indirectly as much needed fertilizer. In the United States 14.5 million tons of fertilizer are produced via the Haber method. Without these fertilizers, food production would be greatly reduced.

15.69 a) Haber synthesis:

$$CH_{4(g)} + H_2O_{(g)} \underset{800\,°C}{\overset{Ni\ cat.}{\rightleftharpoons}} CO_{(g)} + 3\ H_{2(g)}$$

$$CO_{(g)} + H_2O_{(g)} \underset{250\,°C}{\overset{Fe_2O_3/Cr_2O_3}{\rightleftharpoons}} CO_{2(g)} + H_{2(g)}$$

$$N_{2(g)} + 3\ H_{2(g)} \underset{450\,°C/270\ atm}{\overset{\substack{Fe\ powder\ doped \\ with\ K_2O/Al_2O_3}}{\rightleftharpoons}} 2\ NH_{3(g)}$$

b) Ostwald process:

$$4\ NH_{3(g)} + 5\ O_{2(g)} \xrightarrow[1200\ K]{Pt\ gauze} 4\ NO_{(g)} + 6\ H_2O_{(g)}$$

$$2\ NO_{(g)} + O_{2(g)} \rightleftharpoons 2\ NO_{2(g)}$$

$$3\ NO_{2(g)} + H_2O_{(l)} \rightarrow 2\ HNO_{3(aq)} + NO_{(g)}$$

c) Contact process:

$$S_{(s)} + O_{2(g)} \rightarrow SO_{2(g)}$$

$$2\ SO_{2(g)} + O_{2(g)} \underset{high\ T}{\overset{V_2O_5}{\rightleftharpoons}} 2\ SO_{3(g)}$$

$$SO_{3(g)} + H_2SO_{4(l)} \rightarrow H_2S_2O_{7(l)}$$

$$H_2S_2O_{7(l)} + H_2O_{(l)} \rightarrow 2\ H_2SO_{4(l)}$$

15.71 $Br_{2(g)} + I_{2(g)} \rightarrow 2\ IBr_{(g)}$ $K_{eq} = \dfrac{\left(p_{IBr}\right)^2_{eq}}{\left(p_{Br_2}\right)_{eq}\left(p_{I_2}\right)_{eq}} = 322$

$[p_{Br_2}]eq = 0.512$ atm $[p_{I_2}]eq = 0.327$ atm

$$\left(p_{IBr}\right)_{eq} = \sqrt{322\left(p_{Br_2}\right)_{eq}\left(p_{I_2}\right)_{eq}} = \sqrt{322(0.512)(0.327)} = 7.34\ atm$$

15.73 $2\ SO_{2(g)} + O_{2(g)} \rightleftharpoons 2\ SO_{3(g)}$

$$K_{eq} = \frac{\left(p_{SO_3}\right)^2_{eq}}{\left(p_{SO_2}\right)^2_{eq}\left(p_{O_2}\right)_{eq}} = \frac{(5)^2}{(5)^2(4)} = 0.25\ atm^{-1}$$

New conditions: 3 SO_3, 5 SO_2, 4 O_2

$$Q = \frac{\left(p_{SO_3}\right)^2}{\left(p_{SO_2}\right)^2\left(p_{O_2}\right)} = \frac{(3)^2}{(5)^2(4)} = 9.0 \times 10^{-2}\ atm^{-1}$$

$\Rightarrow Q < K_{eq}$ so the reaction consumes reactants and forms products until K_{eq} is reestablished.

15.75 $Q = \dfrac{\left(p_{SO_3}\right)^2}{\left(p_{SO_2}\right)^2\left(p_{O_2}\right)} = \dfrac{(9)^2}{(9)^2(4)} = 0.25$

$\Rightarrow Q = K_{eq}$ so there is no change in the position of the equilibrium.

15.77 $2\ \text{AcOH} \rightleftharpoons (\text{AcOH})_2$

$K_{eq} = \dfrac{\left(p_{(AcOH)_2}\right)_{eq}}{\left(p_{AcOH}\right)_{eq}^2} = 3.72\ \text{atm}^{-1}$

$(p_{AcOH})_{eq} + (p_{(AcOH)_2})_{eq} = 0.75\ \text{atm}$

$(p_{AcOH})_{eq} = 0.75\ \text{atm} - (p_{(AcOH)_2})_{eq}$

$(p_{AcOH})_{eq} = \sqrt{\dfrac{\left(p_{(AcOH)_2}\right)_{eq}}{3.72}} = 0.75 - \left(p_{(AcOH)_2}\right)_{eq}$

$x/3.72 = (0.75 - x)^2 = 0.563 - 1.50x + x^2$

$x = 3.72(x^2 - 1.50x + 0.563) = 3.72x^2 - 5.58x + 2.093$

$0 = 3.72x^2 - 6.58x + 2.093$

$x = \dfrac{6.58 \pm \sqrt{(6.58)^2 - 4(3.72)(2.093)}}{2(3.72)} = \dfrac{6.58 \pm 3.487}{7.44} = 1.35\ \text{or}\ 0.416$

a) $(p_{dimer})_{eq} = 0.416\ \text{atm}$

b) K_{eq} is lower at 200°C

15.79

Reaction:	$2\ \text{Ef}_{(g)}\ +$	$3\ \text{N}_{2(g)}$	$\rightleftharpoons$	$2\ \text{EfN}_{3(g)}$
Initial (atm)	0.75	1.00		0
Change (atm)	-2x	-3x		+2x
Equilibrium (atm)	0.75 - 2x	1.00 - 3x		2x

$(0.75 - 2x) + (1.00 - 3x) + 2x = 0.85 \qquad\qquad 1.75 - 3x = 0.85$

$0.90 = 3x \qquad\qquad x = 0.30\ \text{atm}$

$K_{eq} = \dfrac{\left(p_{EfN_3}\right)_{eq}^2}{\left(p_{Ef}\right)_{eq}^2\left(p_{N_2}\right)_{eq}^3} = \dfrac{(2x)^2}{(0.75 - 2x)^2(1.00 - 3x)^3} = \dfrac{(0.60)^2}{(0.15)^2(0.10)^3}$

$= 1.6 \times 10^4\ \text{atm}^{-3}$

15.80 $\quad \underset{250ml,\ .2M}{} \underset{}{} \underset{350ml\ .3M}{}$

$2\ AgNO_{3(aq)} + Na_2CO_{3(aq)} \rightleftharpoons Ag_2CO_3 \downarrow + 2\ NaNO_{3(aq)}$

$+ 2\ Ag^+_{(aq)} + CO_3^{-2}{}_{(aq)}$

$K_{sp} = 8.2 \times 10^{-12}\ M^3$

146

figure out ksp

CHAPTER 16: AQUEOUS EQUILIBRIA

16.1 a) H_2O, CH_3CO_2H

 b) NH_4^+, Cl^-, H_2O

 c) K^+, Cl^-, H_2O

 d) Na^+, $CH_3CO_2^-$, H_2O

 e) Na^+, OH^-, H_2O

16.3 a) $CH_3CO_2H_{(l)} + H_2O_{(l)} \rightleftharpoons CH_3CO_2^-{}_{(aq)} + H_3O^+{}_{(aq)}$ $K_{eq} = K_a = 1.8 \times 10^{-5}$ M

 b) $NH_4^+ + H_2O_{(l)} \rightleftharpoons NH_{3(aq)} + H_3O^+$ $K_{eq} = K_a = 5.6 \times 10^{-10}$ M

 c) $KCl_{(s)} + H_2O_{(l)} \rightleftharpoons K^+{}_{(aq)} + Cl^-{}_{(aq)}$ $H_2O + H_2O \rightleftharpoons H_3O^+ + OH^-$

                                                         $K_w = 1 \times 10^{-14}$

 d) $CH_3CO_2^- + H_2O_{(l)} \rightleftharpoons CH_3CO_2H_{(aq)} + OH^-{}_{(aq)}$ $K_{eq} = K_w/K_a = 5.6 \times 10^{-10}$ M

 e) $OH^- + H_2O_{(l)} \rightleftharpoons H_2O_{(l)} + OH^-{}_{(aq)}$ $K_{eq} = 1$

 Ranking of K_{eq}'s: (e) > (a) > (b) = (d) > (c)

16.5 a) $AgCl_{(s)} \rightleftharpoons Ag^+{}_{(aq)} + Cl^-{}_{(aq)}$ $K_{sp} = [Ag^+]_{eq}[Cl^-]_{eq}$

 b) $BaSO_{4(s)} \rightleftharpoons Ba^{2+}{}_{(aq)} + SO_4^{2-}{}_{(aq)}$ $K_{sp} = [Ba^{2+}]_{eq}[SO_4^{2-}]_{eq}$

 c) $Fe(OH)_{2(s)} \rightleftharpoons Fe^{2+}{}_{(aq)} + 2\ OH^-{}_{(aq)}$ $K_{sp} = [Fe^{2+}]_{eq}[OH^-]^2_{eq}$

 d) $Ca_3(PO_4)_2 \rightleftharpoons 3\ Ca^{2+}{}_{(aq)} + 2\ PO_4^{3-}{}_{(aq)}$ $K_{sp} = [Ca^{2+}]^3_{eq}[PO_4^{3-}]^2_{eq}$

16.7 $CaC_2O_{4(s)} \rightleftharpoons Ca^{2+}{}_{(aq)} + C_2O_4^{2-}{}_{(aq)}$

 $K_{sp} = [Ca^{2+}]_{eq}[C_2O_4^{2-}]_{eq}$

 $MM_{CaC_2O_4} = 128.10$ g/mol

 $$[CaC_2O_4] = \left(\frac{6.1 \text{ mg}}{1.0 \text{ L}}\right)\left(\frac{1 \text{ g}}{1000 \text{ mg}}\right)\left(\frac{1 \text{ mol}}{128.10 \text{ g}}\right) = 4.762 \times 10^{-5} \text{ M}$$

 $= [Ca^{2+}]_{eq} = [C_2O_4^{2-}]_{eq}$

 $K_{sp} = (4.762 \times 10^{-5} \text{ M})^2 = 2.3 \times 10^{-9} \text{ M}^2$

16.9 $Cu(NO_3)_{2(aq)} \rightleftharpoons Cu^{2+}_{(aq)} + 2 NO_3^-_{(aq)}$

$KOH_{(aq)} \rightleftharpoons K^+_{(aq)} + OH^-_{(aq)}$ Spectator ions: K^+, NO_3^-

	$Cu^{2+}_{(aq)} +$	$OH^-_{(aq)} \rightleftharpoons$	$Cu(OH)_2$
Init. (M)	0.15 M	0.20 M	
Change (M)	-0.10	-0.20	
Compl. (M)	0.05 M	0	
Change to Equil. (M)	+y	+2y	
Equil. (M)	0.05 + y	2y	

$K_{eq} = 1/K_{sp} = 10^{19.66} = 4.6 \times 10^{19}$

$K_{sp} = [Cu^{2+}]_{eq} [OH^-]^2_{eq} = 2.2 \times 10^{-20} \ M^3$

$= (0.05 + y)(2y)^2 \approx (0.05)(2y)^2 = 0.20y^2$

$y = \sqrt{2.2 \times 10^{-20}/0.20} = 3.3 \times 10^{-10} \ M$

$[Cu^{2+}]_{eq} = 0.05 \ M$ $[OH^-]_{eq} = 2y = 6.6 \times 10^{-10} \ M$ $[K^+] = 0.20 \ M$
$[NO_3^-] = 0.30 \ M$

16.11 $pH = - \log [H_3O^+]$
a) $pH = - \log (4.0) = -0.60$
b) $pH = - \log (3.75 \times 10^{-6}) = 5.43$
c) $pH = - \log (0.0048) = 2.32$
d) $pH = - \log (7.45 \times 10^{-12}) = 11.13$

16.13 $pH + pOH = 14.00$ $pH = 14.00 - pOH$
$pH = 14.00 - (-\log [OH^-]) = 14.00 + \log [OH^-]$
a) $pH = 14.00 + \log (2.0) = 14.30$
b) $pH = 14.00 + \log (3.75 \times 10^{-6}) = 8.57$
c) $pH = 14.00 + \log (0.0048) = 11.68$
d) $pH = 14.00 + \log (7.45 \times 10^{-12}) = 2.87$

16.15 a) CH_3CO_2H = weak acid
b) NH_4Cl = salt
c) KCl = salt
d) $NaCH_3CO_2$ = salt
e) $NaOH$ = strong base

16.17 a) NH_3 = weak base
b) $HClO_4$ = strong acid
c) $HClO$ = weak acid
d) $Ba(OH)_2$ = strong base

16.19

weak acids	conjugate base	weak bases	conj. acids
HClO (or HOCl)	ClO$^-$ (or OCl$^-$)	NH_3	NH_4^+
H_2O	OH$^-$	H_2O	H_3O^+
		NH_2OH	NH_3OH^+

16.21 a) $NaOH \underset{\leftarrow}{\rightarrow} Na^+ + OH^-$ $[NaOH] = [Na^+] = [OH^-] = 1.5$ M

pH = 14.00 - pOH = 14.00 - (- log [OH$^-$])

= 14.00 + log [OH$^-$] = 14.00 + log (1.5) = 14.18

b) $C_5H_5N + H_2O \underset{\leftarrow}{\rightarrow} C_5H_5NH^+ + OH^-$

b) $C_5H_5N + H_2O \underset{\leftarrow}{\rightarrow} C_5H_5NH^+ + OH^-$

Init. (M)	1.5 M	0	0
Change (M)	-x	+x	+x
Equil. (M)	1.5-x	x	x

$pK_b = 8.72$ $K_b = 10^{-8.72} = 1.91 \times 10^{-9} = \dfrac{[C_5H_5NH^+]_{eq}[OH^-]_{eq}}{[C_5H_5N]_{eq}}$

$1.91 \times 10^{-9} = \dfrac{x^2}{1.5-x} \approx \dfrac{x^2}{1.5}$ $x^2 = 1.5(1.91 \times 10^{-9}) = 2.86 \times 10^{-9}$

$x = 5.35 \times 10^{-5} = [OH^-]$

pH = 14.00 - pOH = 14.00 - (- log [OH$^-$]) = 14.00 + log (5.35 $\times 10^{-5}$) = 9.73

c) $NH_2OH + H_2O \underset{\leftarrow}{\rightarrow} NH_3OH^+ + OH^-$

Init. (M)	1.5 M	0	0
Change (M)	-x	+x	+x
Equil. (M)	1.5-x	x	x

$pK_b = 7.96$

$K_b = 10^{-7.96} = 1.10 \times 10^{-8} = \dfrac{[NH_3OH^+]_{eq}[OH^-]_{eq}}{[NH_2OH]_{eq}}$

$1.10 \times 10^{-8} = \dfrac{x^2}{1.5-x}$ $x^2 = 1.5(1.10 \times 10^{-8}) = 1.65 \times 10^{-8}$

$x = 1.28 \times 10^{-4}$ M = [OH$^-$]

pH = 14.00 + log (1.28 $\times 10^{-4}$) = 10.11

d) $HCO_2H + H_2O_{(l)} \underset{\leftarrow}{\rightarrow} HCO_2^-{}_{(aq)} + H_3O^+{}_{(aq)}$

Init. (M)	1.5 M	0	0
Change (M)	-x	+x	+x
Equil. (M)	1.5-x	x	x

$K_a = \dfrac{[HCO_2^-]_{eq}[H_3O^+]_{eq}}{[HCO_2H]_{eq}} = 1.82 \times 10^{-4}$

$K_a = \dfrac{(x)(x)}{1.5-x} = \dfrac{x^2}{1.5} = 1.82 \times 10^{-4}$; $x = 1.65 \times 10^{-2}$ M

pH = - log 1.65 $\times 10^{-2}$ = 1.78

16.23 $N(CH_3)_{3(l)} + H_2O_{(l)} \rightleftarrows HN(CH_3)_3^+{}_{(aq)} + OH^-{}_{(aq)};$

$$K_b = \frac{\left[HN(CH_3)_3^+\right]_{eq}[OH^-]_{eq}}{[N(CH_3)_3]_{eq}}$$

$pK_b = 4.19 \qquad K_b = 10^{-4.19} = 6.46 \times 10^{-5}$

a) Major species: H_2O and $N(CH_3)_3$ Minor species: $HN(CH_3)_3^+$ and OH^-

b)

	$NMe_3 + H_2O \rightleftarrows$	$^+HNMe_3$	+	OH^-
Init. (M)	0.350 M	0		0
Change (M)	-x	+x		+x
Equil. (M)	0.350-x	x		x

$$K_b = 6.46 \times 10^{-5} = \frac{(x)(x)}{(0.350 - x)} \approx \frac{x^2}{0.350}$$

$$x = \sqrt{(0.350)(6.46 \times 10^{-5})} = 4.75 \times 10^{-3} \text{ M}$$

$[N(CH_3)_3]_{eq} = 0.345$ M

$[^+HN(CH_3)_3]_{eq} = [OH^-]_{eq} = 4.75 \times 10^{-3}$ M

c) pH = 14.00 - pOH = 14.00 + log [OH] = 11.68

d) Your molecular drawing should look like the molecular drawings in Figure 16-5, the answer to Exercise 16.3.2 or the answer to Problem 16.22 but with the following species:

NMe_3 + H_2O $\rightleftarrows$ $^+HNMe_3$ + OH^-

16.25 $NH_4NO_{3(aq)} \rightleftarrows NH_4^+{}_{(aq)} + NO_3^-{}_{(aq)}$

$NH_4^+{}_{(aq)} + H_2O \rightleftarrows NH_{3(aq)} + H_3O^+{}_{(aq)}$

a) Major species: $NH_4^+{}_{(aq)}$, $NO_3^-{}_{(aq)}$ and H_2O

b) $NH_4^+{}_{(aq)} + H_2O_{(l)} \rightleftarrows NH_{3(aq)} + H_3O^+{}_{(aq)}$

c)

	NH_4^+	+	$H_2O \rightleftarrows$	NH_3	+	H_3O^+
Init. (M)	0.0100 M			0		0
Change (M)	-x			+x		+x
Equil. (M)	0.0100 M-x			x		x

$pK_a = 9.25 \qquad K_a = 10^{-9.25} = 5.6 \times 10^{-10}$ M

$$K_a = \frac{\left[NH_3\right]_{eq}\left[H_3O^+\right]_{eq}}{\left[NH_4^+\right]} = \frac{(x)(x)}{(0.0100 \text{ M} - x)} \approx \frac{x^2}{0.0100} = 5.6 \times 10^{-10}$$

$$x = \sqrt{(0.0100)(5.6 \times 10^{-10})} = 2.37 \times 10^{-6} \text{ M} = [H_3O^+]$$

pH = - log $[H_3O^+]$ = - log (2.37×10^{-6}) = 5.63

16.27 $HPO_4^{2-}(aq) + OH^-(aq) \rightleftarrows PO_4^{3-}(aq) + H_2O(l)$

$HPO_4^{2-}(aq) + H_2O(l) \rightleftarrows PO_4^{3-}(aq) + H_3O^+(aq)$

$$K_{a_3} = 4.8 \times 10^{-13} \text{ M}$$

$$K_{a_3} = \frac{\left[PO_4^{3-}\right]_{eq}\left[H_3O^+\right]_{eq}}{\left[HPO_4^{2-}\right]_{eq}}$$

$H_3O^+(aq) + OH^-(aq) \rightleftarrows 2\,H_2O(l)$

$$\frac{1}{K_w} = \frac{1}{\left[H_3O^+\right]_{eq}\left[OH^-\right]_{eq}}$$

$$\frac{K_{a_3}}{K_w} = \frac{\left[PO_4^{3-}\right]_{eq}\left[H_3O^+\right]_{eq}}{\left[HPO_4^{2-}\right]_{eq}} \; x \; \frac{1}{\left[H_3O^+\right]_{eq}\left[OH^-\right]_{eq}} = \frac{\left[PO_4^{3-}\right]_{eq}}{\left[HPO_4^{2-}\right]_{eq}\left[OH^-\right]_{eq}} = K_{eq}$$

$$K_{eq} = \frac{K_{a_3}}{K_w} = \frac{4.8 \times 10^{-13} \text{ M}}{1.0 \times 10^{-14} \text{ M}^2} = 48 \text{ M}^{-1}$$

16.29 (b) mmoles H_3O^+ = mmoles HCl = (50 mL)(0.25 M HCl) = 12.5 mmoles H_3O^+

mmoles $CH_3CO_2^-$ = mmoles $NaCH_3CO_2$ = (100 mL)(0.25 M $NaCH_3CO_2$)
$$= 25 \text{ mmoles } CH_3CO_2^-$$

Reaction:	$CH_3CO_2^-(aq)$ +	$H_3O^+(aq)$ $\rightleftarrows$	$CH_3CO_2H(aq)$ + $H_2O(l)$
Initial, mmoles	25	12.5	0
Change, mmoles	-12.5	-12.5	+12.5
Completion, mmoles	12.5	~0	12.5

$$pH = pK_a + \log\frac{[CH_3CO_2^-]}{[CH_3CO_2H]} = 4.75 + \log\left(\frac{12.5}{12.5}\right) = 4.75$$

Note: Remember that in the equation you can substitute amounts of acid and base in moles or mmoles.

d) mmol CH_3CO_2H = (100 mL)(0.25 M CH_3CO_2H) = 25 mmol CH_3CO_2H
mmol OH^- = mmol NaOH= (50 mL)(0.25 M NaOH) = 12.5 mmol OH^-

Reaction:	$CH_3CO_2H(aq)$ +	$OH^-(aq)$ $\rightleftarrows$	$CH_3CO_2^-(aq)$ + $H_2O(l)$
Initial, mmol	25	12.5	
Change, mmol	-12.5	-12.5	+12.5
Completion, mmol	12.5	~0	12.5

$$pH = pK_a + \log\frac{[CH_3CO_2^-]}{[CH_3CO_2H]} = 4.75 + \log\left(\frac{12.5}{12.5}\right) = 4.75$$

16.31   b) Addition of base would change pH 0.1 unit from 4.75 to 4.85.

$$4.85 = 4.75 + \log \frac{[CH_3CO_2^-]}{[CH_3CO_2H]} \qquad \log \frac{[CH_3CO_2^-]}{[CH_3CO_2H]} = 0.10 \qquad \frac{[CH_3CO_2^-]}{[CH_3CO_2H]} = 1.26$$

or $\dfrac{\text{mmoles } CH_3CO_2^-/mL}{\text{mmoles } CH_3CO_2H/mL} = \dfrac{\text{mmoles } CH_3CO_2^-}{\text{mmoles } CH_3CO_2H} = 1.26$

mmoles $CH_3CO_2^-$ = 1.26(mmoles CH_3CO_2H)

Reaction:	$CH_3CO_2H_{(aq)}$ +	$OH^-_{(aq)}$	$\rightleftarrows$	$CH_3CO_2^-_{(aq)}$ + $H_2O_{(l)}$
Initial, mmol	12.5			12.5
Change, mmol	-x	+x	+x	(from Prob. 16.29 (b))
Compl., mmol	12.5-x	x	12.5+x	

12.5 + x = 1.26(12.5 mmol - x)

x + 1.26x = 1.26(12.5 mmol) - 12.5 mmol = 1.26(12.5 mmol) - 1(12.5 mmol)

= 0.26(12.5 mmol)

2.26x = 0.26(12.5 mmol)

$$x = \frac{0.26(12.5 \text{ mmol})}{2.26} = 1.4 \text{ mmoles} = OH^-$$

moles of base = 1.4 x 10^{-3} moles

d) The calculation is exactly the same as for part (b) because we have a buffer containing the same species at the same concentration.

16.33   0.50 M NaH_2PO_4 0.20 M Na_2HPO_4

$H_2PO_4^-_{(aq)} + H_2O_{(l)} \rightleftarrows HPO_4^{2-}_{(aq)} + H_3O^+_{(aq)}$ $pK_{a_2} = 7.21$

$pH = pK_a + \log\left\{\dfrac{[A^-]_{initial}}{[HA]_{initial}}\right\}$; $[A^-] = [HPO_4^{2-}] = 0.20$ M;

 $[HA] = [H_2PO_4^-] = 0.50$ M

$$pH = 7.21 + \log\left(\frac{0.20}{0.50}\right) = 6.81$$

16.35 $[H_2PO_4^-]_{init.}$ = 0.50 M $HPO_4^{2-} + H_3O^+ \rightleftarrows H_2PO_4^- + H_2O_{(l)}$

$[HPO_4^{2-}]_{init.}$ = 0.20 M   volume of buffer = 250 mL

pH range = 6.81 ± 0.2 = 6.61

$6.61 = 7.21 + \log\left(\dfrac{0.20 - x}{0.50 + x}\right)$; $\log\left(\dfrac{0.20 - x}{0.50 + x}\right) = 6.61 - 7.21 = -0.60$

$\left(\dfrac{0.20 - x}{0.50 + x}\right) = 10^{-0.60} = 0.251$

0.20 - x = 0.251(0.50 + x) = 0.126 + 0.251x

1.251x = 0.20 - 0.126 = 0.074   x = 0.059 M H_3O^+

moles H_3O^+ = (0.059 mol/L)(250 mL)(1 L/1000 mL) = 1.49 x 10^{-2} moles

16.37 $HCO_3^-(aq) + OH^-(aq) \rightleftarrows CO_3^{2-}(aq) + H_2O(l)$ $\quad K_{a_2} = 5.6 \times 10^{-11}$ M $\quad pK_a = 10.25$

$$pH = pK_a + \log\frac{[Base]_{eq}}{[Acid]_{eq}} \qquad \log\frac{[Base]_{eq}}{[Acid]_{eq}} = pH - pK_a = 10.60 - 10.25 = 0.35$$

$$\frac{[Base]_{eq}}{[Acid]_{eq}} = 10^{0.35} = 2.23 \qquad K_{a_2} = \frac{[CO_3^{2-}]_{eq}}{[HCO_3^-]_{eq}[OH^-]_{eq}}$$

OH^- is a strong base so every OH^- added takes a proton off HCO_3^- to form CO_3^{2-};

$\therefore [OH^-] = [CO_3^{2-}] = $ base.

$[Base]_{eq} = 2.23 [Acid]_{eq}$ $\quad [Acid]_{eq} = [Acid]_{init} - [Base]_{eq}$

or (mmoles base)$_{eq}$ = 2.23 {(mmoles acid)$_{init}$ - (mmoles base)$_{eq}$}

$x = $ (mmoles base)$_{eq}$ $\qquad x = 2.23$ {(0.200 M)(250 mL) - x}

3.23 x = 111.5 mmoles $\quad$ x = 34.5 mmoles base

$$\frac{34.5 \text{ mmoles base}}{1.0 \text{ M NaOH}} = 34.5 \text{ mL of } 1.0 \text{ M NaOH}$$

16.39 a) $\qquad HA(aq) + H_2O(l) \rightleftarrows A^-(aq) + H_3O^+(aq)$

Initial	10^{-2} M		
Change	-x	+x	+x
Equil.	10^{-2} -x	x	x

$$K_a = \frac{[A^-]_{eq}[H_3O^+]_{eq}}{[HA]_{eq}} = 3.0 \times 10^{-4} \text{ M}$$

$$K_a = \frac{(x)(x)}{(10^{-2} - x)} = 3.0 \times 10^{-4} \text{ M}$$

$x^2 = 3.0 \times 10^{-4}(10^{-2} - x) = 3.0 \times 10^{-6} - 3.0 \times 10^{-4} x$

$0 = x^2 + 3.0 \times 10^{-4} x - 3.0 \times 10^{-6}$

$$= \frac{-3.0 \times 10^{-4} \pm \sqrt{(3.0 \times 10^{-4}) - 4(1)(-3.0 \times 10^{-6})}}{2}$$

$x = 1.6 \times 10^{-3}$ or -1.89×10^{-3}

$[H_3O^+] = 1.6 \times 10^{-3}$ M

$pH = -\log[H_3O^+] = -\log(1.6 \times 10^{-3}) = 2.80$

b) At the stoichiometric point the amount of added OH^- = amt. of HA originally present; thus, aspirin is no longer a major species in solution and pH is determined by proton transfer from H_2O to $A^-(aq)$:

$$K_{eq} = \frac{K_w}{K_a} = \frac{1 \times 10^{-14} \text{ M}^2}{3.0 \times 10^{-4} \text{ M}} = 3.3 \times 10^{-11} \text{ M}$$

(continued)

153

(16.39 continued)

$$A^-_{(aq)} + H_2O_{(l)} \rightleftharpoons HA_{(aq)} + OH^-_{(aq)}$$

Initial	10^{-2} M		
Change	$-x$	$+x$	$+x$
Equil.	$10^{-2} - x$	x	x

$$3.3 \times 10^{-11} \approx \frac{x^2}{10^{-2}} \implies x = 5.77 \times 10^{-7} = [OH^-]; \quad pOH = -\log[OH^-]$$

$$pH = 14.00 + \log(5.77 \times 10^{-7}) = 7.76$$

c) $pH = pK_a$ of weak acid at midpoint, so $pH = -\log K_a = -\log(3.0 \times 10^{-4}) = 3.52$

16.41 At the midpoint of a titration, the pH of the solution $= pK_a$ of the weak acid.
$pH_{midpoint} = 3.88 = pK_a$

16.43 $Zn(OH)_{2(s)} \rightleftharpoons Zn^{2+}_{(aq)} + 2\,OH^-_{(aq)}$

$K_{sp} = [Zn^{2+}]_{eq}[OH^-]^2_{eq} = 1.2 \times 10^{-17}$ M^3

Reaction: $Zn(OH)_{2(s)} \rightleftharpoons Zn^{2+}_{(aq)} + 2\,OH^-_{(aq)}$

Initial (M)	0	0
Change (M)	$+x$	$+2x$
Equil. (M)	x	$2x$

$1.2 \times 10^{-17} = x(2x)^2 = 4x^3$

$$x = \sqrt[3]{(1.2 \times 10^{-17})/4} = 1.44 \times 10^{-6} \text{ M} = [Zn^{2+}]_{eq} = [Zn(OH)_2]_{dissolved}$$

$(1.44 \times 10^{-6}$ M$)(99.40$ g/mole$)(1.0$ L$) = 1.43 \times 10^{-4}$ g $Zn(OH)_{2(s)}$

16.45 a) $H_2SO_{4(aq)} + H_2O_{(l)} \rightleftharpoons HSO_4^-{}_{(aq)} + H_3O^+_{(aq)}$

$HSO_4^-{}_{(aq)} + H_2O_{(l)} \rightleftharpoons SO_4^{2-}{}_{(aq)} + H_3O^+_{(aq)}$

Major Species: $H_2O_{(l)}, H_3O^+_{(aq)}, HSO_4^-{}_{(aq)}$

b) $Na_2SO_{4(s)} + H_2O_{(l)} \rightleftharpoons 2\,Na^+_{(aq)} + SO_4^{2-}{}_{(aq)}$

$SO_4^{2-}{}_{(aq)} + H_2O_{(l)} \rightleftharpoons HSO_4^-{}_{(aq)} + OH^-_{(aq)}$

Major Species: $H_2O_{(l)}, Na^+_{(aq)}, SO_4^{2-}{}_{(aq)}$

c) $CO_{2(g)} + H_2O_{(l)} \rightleftharpoons H_2CO_{3(aq)}$

$H_2CO_{3(aq)} + H_2O_{(l)} \rightleftharpoons HCO_3^-{}_{(aq)} + H_3O^+_{(aq)}$

Major species: $CO_{2(aq)} = H_2CO_{3(aq)}$

d) $NH_4Cl_{(s)} + H_2O_{(l)} \rightleftharpoons NH_4^+_{(aq)} + Cl^-$

$NH_4^+_{(aq)} + H_2O_{(l)} \rightleftharpoons NH_{3(aq)} + H_3O^+_{(aq)}$

Major species: $H_2O_{(l)}, NH_4^+_{(aq)}, Cl^-_{(aq)}$

154

16.47 A $H_2PO_4^-/HPO_4^{2-}$ buffer is required.

$NaH_2PO_{4(s)}/1.00$ M NaOH

$NaH_2PO_{4(s)}/Na_2HPO_{4(s)}/H_2O$

$Na_2HPO_{4(s)}/1.00$ M HCl

16.49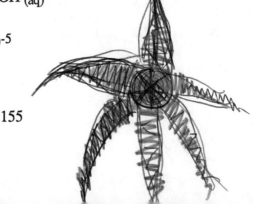

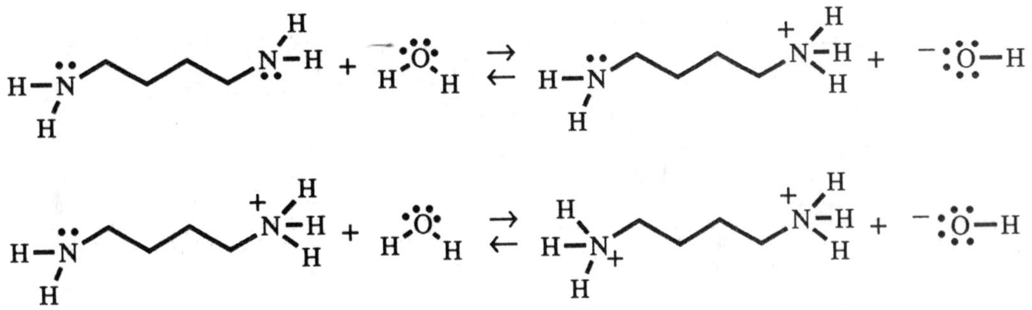

16.51 $HA_{(aq)} + H_2O_{(l)} \rightleftharpoons A^-_{(aq)} + H_3O^+_{(aq)}$

$$K_a = \frac{[A^-]_{eq}[H_3O^+]_{eq}}{[HA]_{eq}}$$

0.060	0	0
-x	+x	+x
0.060-x	x	x

$x = [H_3O^+]_{eq} = 10^{-pH} = 10^{-2.71} = 1.95 \times 10^{-3}$ M

$[HA]_{eq} = 0.060 - 1.95 \times 10^{-3} = 5.81 \times 10^{-2}$ M

$$K_a = \frac{(1.95 \times 10^{-3})^2}{5.81 \times 10^{-2}} = 6.54 \times 10^{-5}$$

$pK_a = -\log K_a = -\log(6.54 \times 10^{-5}) = 4.18$; benzoic acid

16.53 a) $CH_3CO_2H_{(aq)} + H_2O_{(l)} \rightleftharpoons CH_3CO_2^-_{(aq)} + H_3O^+_{(aq)}$

1.00 M	0	0
-x	+x	+x
1.00-x	x	x

$$K_a = 1.8 \times 10^{-5} \text{ M} = \frac{[CH_3CO_2^-]_{eq}[H_3O^+]_{eq}}{[CH_3CO_2H]_{eq}}$$

$1.8 \times 10^{-5} = x^2/1.00\text{-}x \approx x^2/1$; $x = 4.2 \times 10^{-3}$ M $= [H_3O^+]_{eq}$

$pH = -\log [H_3O^+]_{eq} = -\log(4.2 \times 10^{-3}) = 2.38$; acidic

b) $NH_{3(aq)} + H_2O_{(l)} \rightleftharpoons NH_4^+_{(aq)} + OH^-_{(aq)}$

$$K_b = \frac{[NH_4^+]_{eq}[OH^-]_{eq}}{[NH_3]_{eq}} = 1.8 \times 10^{-5}$$

(continued)

155

(16.53 continued)

pOH = 2.38

pH = 14.00 - pOH = 14.00 - 2.38 = 11.62; basic

c) $NH_4Cl_{(s)} \rightleftharpoons NH_4{}^+{}_{(aq)} + Cl^-$ $NH_4{}^+{}_{(aq)} + H_2O_{(l)} \rightleftharpoons NH_3{}^+{}_{(aq)} + H_3O^+{}_{(aq)}$

$$K_a = \frac{[NH_3]_{eq}[H_3O^+]_{eq}}{[NH_4{}^+]_{eq}} = \frac{K_w}{K_b} = \frac{10^{-14}}{1.8 \times 10^{-5}} = 5.6 \times 10^{-10}$$

$$K_a = 5.6 \times 10^{-10} = \frac{x^2}{1.00 - x} \cong x^2; \; x = [H_3O^+] = \sqrt{5.6 \times 10^{-10}} = 2.4 \times 10^{-5}$$

pH = - log(2.4 x 10⁻⁵) = 4.62; acidic

d) $NaCH_3CO_{2(s)} \rightleftharpoons Na^-{}_{(aq)} + CH_3CO_2{}^-{}_{(aq)}$

                        $CH_3CO_{2(aq)} + H_2O_{(l)} \rightleftharpoons CH_3CO_2H_{(aq)} + OH^-{}_{(aq)}$

Init.	1.00 M	0	0
Change	-x	+x	+x
Equil.	1.00-x	x	x

$$K_b = \frac{[CH_3CO_2H]_{eq}[OH^-]_{eq}}{[CH_3CO_2{}^-]_{eq}} = \frac{K_w}{K_a} = \frac{10^{-14}}{1.8 \times 10^{-5}} = 5.6 \times 10^{-10}$$

$$x = [OH^-]_{eq} = \sqrt{5.6 \times 10^{-10}} = 2.4 \times 10^{-5}$$

pOH = 4.62; pH = 14.00 - pOH = 14.00 - 4.62 = 9.38; basic

e) $NH_4CH_3CO_2 \rightleftharpoons NH_4{}^+{}_{(aq)} + CH_3CO_2{}^-{}_{(aq)}$

$NH_4{}^+{}_{(aq)} + H_2O_{(l)} \rightleftharpoons NH_3{}^+{}_{(aq)} + H_3O^+{}_{(aq)}$ $K_a = 5.6 \times 10^{-10}$

$CH_3CO_{2(aq)} + H_2O_{(l)} \rightleftharpoons CH_3CO_2H_{(aq)} + OH^-{}_{(aq)}$ $K_b = 5.6 \times 10^{-10}$; neutral

16.55   a) $H_2CO_{3(aq)} + H_2O_{(l)} \rightleftharpoons HCO_3{}^-{}_{(aq)} + H_3O^+{}_{(aq)}$

        $HCO_3{}^-{}_{(aq)} + H_2O_{(l)} \rightleftharpoons CO_3{}^{2-}{}_{(aq)} + H_3O^+{}_{(aq)}$ acidic

      b) $KHCO_{3(s)} \rightleftharpoons K^+{}_{(aq)} + HCO_3{}^-{}_{(aq)}$

      acidic: $HCO_3{}^-{}_{(aq)} + H_2O_{(l)} \rightleftharpoons CO_3{}^{2-}{}_{(aq)} + H_3O^+{}_{(aq)}$

      basic: $HCO_3{}^-{}_{(aq)} + H_2O_{(l)} \rightleftharpoons H_2CO_{3(s)} + OH^-{}_{(aq)}$ both

      c) $NH_{3(aq)} + H_2O_{(l)} \rightleftharpoons NH_4{}^+{}_{(aq)} + OH^-{}_{(aq)}$ basic

      d) $NaCl_{(s)} \rightleftharpoons Na^+{}_{(aq)} + Cl^-{}_{(aq)}$ neither

16.57 a) $NH_4NO_3 \underset{\leftarrow}{\rightarrow} NH_4^+ + NO_3^-$ (NO_3^-; spectator ion)

$NH_4^+{}_{(aq)} + H_2O_{(l)} \underset{\leftarrow}{\rightarrow} NH_3{}_{(aq)} + H_3O^+{}_{(aq)}$

Major species: $H_2O_{(l)}$, $NH_4^+{}_{(aq)}$

b) $KH_2PO_4 \underset{\leftarrow}{\rightarrow} K^+ + H_2PO_4^-$

$H_2PO_4^-{}_{(aq)} + H_2O_{(l)} \underset{\leftarrow}{\rightarrow} HPO_4^{2-}{}_{(aq)} + H_3O^+{}_{(aq)}$

Major species: $H_2PO_4^-{}_{(aq)}$, $H_2O_{(l)}$

c) $Na_2O + H_2O_{(l)} \underset{\leftarrow}{\rightarrow} 2\, OH^-{}_{(aq)}{}^+ + 2\, Na^+$

Major species: $OH^-{}_{(aq)}$, $H_2O_{(l)}$

d) $C_6H_5CO_2H_{(aq)} + H_2O_{(l)} \underset{\leftarrow}{\rightarrow} C_6H_5CO_2^-{}_{(aq)} + H_3O^+{}_{(aq)}$

Major species: $C_6H_5CO_2H_{(aq)}$, $H_2O_{(l)}$

16.59 $pH = pK_a + \log\dfrac{[A^-]_{init}}{[HA]_{init}} = pK_a + \log\dfrac{[NH_3]}{[NH_4^+]}$

$[NH_3] = 1.00\ M$ $[NH_4^+] = \dfrac{35.0\ g}{53.5\ g/mol}\bigg/ 1\ L = 0.65\ M$

$pH = pK_a + \log\dfrac{1}{0.65}$; $pK_a = \dfrac{pK_w}{pK_b} = 14.00 - 4.75 = 9.25$

$pH = 9.25 + 0.18 = 9.43$ $NH_3{}_{(aq)} + H_3O^+{}_{(aq)} \underset{\leftarrow}{\rightarrow} NH_4^+{}_{(aq)} + H_2O_{(l)}$

$\qquad\qquad\qquad\qquad\qquad$ 1.00 M $\qquad\qquad\qquad$ 0.65 M

$\qquad\qquad\qquad\qquad\qquad$ -x $\qquad\qquad\qquad\qquad$ +x

$\qquad\qquad\qquad\qquad\qquad$ 1.00-x $\qquad\qquad\qquad$ 0.65+x

$pH = 9.25 + \log\dfrac{1.00 - x}{0.65 + x} = 9.43 - 0.05$

$\log\dfrac{1.00 - x}{0.65 + x} = 9.43 - 0.05 - 9.25 = 0.13$

$\dfrac{1.00 - x}{0.65 + x} = 10^{0.13} = 1.35$

$1.00 - x = 1.35(0.65 + x) = 0.88 + 1.35x$

$0.12 = 2.35x$; $x = 0.05$ moles $H_3O^+{}_{(aq)}$

16.61 $HPO_4^{2-}{}_{(aq)} + H_2O_{(l)} \underset{\leftarrow}{\rightarrow} PO_4^{3-}{}_{(aq)} + H_3O^+{}_{(aq)}$

$K_{a_3} = \dfrac{[PO_4^{3-}]_{eq}[H_3O^+]_{eq}}{[HPO_4^{2-}]_{eq}} = 4.8 \times 10^{-13}\ M$

(continued)

(16.61 continued)

$$HPO_4^{2-}{}_{(aq)} + H_2O_{(l)} \rightleftharpoons H_2PO_4^-{}_{(aq)} + OH^-{}_{(aq)}$$

$$K_b = \frac{K_w}{K_{a_2}} = \frac{1 \times 10^{-14}}{6.2 \times 10^{-8}} = 1.6 \times 10^{-7} \text{ M}$$

$K_b > K_{a_3}$ ∴ second reaction dominates

$$HPO_4^{2-}{}_{(aq)} + H_2O_{(l)} \rightleftharpoons H_2PO_4^-{}_{(aq)} + OH^-{}_{(aq)}$$

0.250 M	0	0
-x	+x	+x
0.250 M - x	x	x

$$K_b = 1.6 \times 10^{-7} = \frac{[H_2PO_4^-]_{eq}[OH^-]_{eq}}{[HPO_4^{2-}]_{eq}} = \frac{x^2}{(0.250 - x)} \approx \frac{x^2}{0.250}$$

$x = [OH^-] = 2.0 \times 10^{-4}$ pH = 14.00 - pOH = 14.00 - 3.70 = 10.30

16.63   pH = 8.5   a) methyl orange = yellow
                      b) phenol red = red
                      c) bromocresol green = blue
                      d) thymol blue = yellow to green

16.65 $$PhO^-{}_{(aq)} + H_2O \rightleftharpoons PhOH_{(aq)} + OH^-{}_{(aq)}$$ $$K_b = \frac{[PhOH]_{eq}[OH^-]_{eq}}{[PhO^-]_{eq}}$$

Init. (M)	0.010 M	0	0
Change (M)	-x	+x	+x
Equil. (M)	0.010-x	+x	$x = [OH^-] = 1.00 \times 10^{-3}$

pH = 11 = 14 - pOH; pOH = 3.00

$$K_b = \frac{x^2}{(0.010-x)} = \frac{(1.00 \times 10^{-3})^2}{(0.010 - 1.00 \times 10^{-3})} = 1.11 \times 10^{-4} \text{ M}$$

$$K_a = \frac{K_w}{K_b} = \frac{1.00 \times 10^{-14} \text{ M}^2}{1.11 \times 10^{-4} \text{ M}} = 9.00 \times 10^{-11} \text{ M}$$

$pK_a = -\log K_a = 10.05$

16.67   Reaction: $$H_3BO_{3(aq)} + H_2O_{(l)} \rightleftharpoons H_3O^+{}_{(aq)} + H_2BO_3^-{}_{(aq)}$$

Initial (M)	0.050	0	0
Change (M)	-x	+x	+x
Equil. (M)	0.050-x	x	x

$$K_a = 7.3 \times 10^{-10} \text{ M} = \frac{(x)(x)}{0.050 \text{ M} - x} \cong \frac{x^2}{0.050 \text{ M}}$$

$x = 6.0 \times 10^{-6} = [H_3O^+]_{eq}$ pH = 5.22

16.69 KCN $\rightleftarrows$ K$^+$ + CN$^-$

	CN$^-$(aq)	+ H$_2$O(l) $\rightleftarrows$	HCN(aq)	+	OH$^-$(aq)
Init. (M)	2.00 x 10^{-2} M		0		0
Change (M)	-x		+x		+x
Equil. (M)	2.00 x 10^{-2}-x		x		x

$$K_b = \frac{[HCN]_{eq}[OH^-]_{eq}}{[CN^-]_{eq}} = 10^{-4.69} = 2.04 \times 10^{-5}$$

$$x^2 = (2.04 \times 10^{-5})(2.00 \times 10^{-2} - x) = 4.08 \times 10^{-7} - 2.04 \times 10^{-5}x$$

$$0 = x^2 + 2.04 \times 10^{-5}x - 4.08 \times 10^{-7}$$

$$x = \frac{-2.04 \times 10^{-5} \pm \sqrt{(2.04 \times 10^{-5})^2 - 4(1)(-4.08 \times 10^{-7})}}{2(1)}$$

$$= 6.29 \times 10^{-4} \text{ M} = [OH^-]_{eq} \qquad pOH = 3.20$$

$$pH = 14.00 - pOH = 14.00 - (-\log[OH^-]_{eq}) = 14.00 + \log[OH^-]_{eq} = 10.80$$

16.71 Na$_2$SO$_3$(aq) $\rightleftarrows$ 2 Na$^+$(aq) + SO$_3^{2-}$(aq)

SO$_3^{2-}$(aq) + H$_2$O(l) $\rightleftarrows$ HSO$_3^-$(aq) + OH$^-$(aq)

CH$_3$CO$_2$H(aq) + H$_2$O(l) $\rightleftarrows$ CH$_3$CO$_2^-$(aq) + H$_3$O$^+$(aq)

a) Major species in each solution: SO$_3^{2-}$(aq), Na$^+$(aq), H$_2$O(l)

: CH$_3$CO$_2$H(aq), H$_2$O(l)

b) CH$_3$CO$_2$H(aq) + SO$_3^{2-}$(aq) $\rightleftarrows$ HSO$_3^-$(aq) + CH$_3$CO$_2^-$(aq)

c) acid: CH$_3$CO$_2$H

base: SO$_3^{2-}$

conjugate acid: HSO$_3^-$

conjugate base: CH$_3$CO$_2^-$

16.73 White solid would be Zn(OH)$_2$.

pH = 8.00   [H$_3$O$^+$] = 10^{-8} [OH$^-$] = 10^{-6}

Zn(OH)$_2$(s) $\rightleftarrows$ Zn^{2+}(aq) + 2 OH$^-$(aq)

pK$_{sp}$ = 16.92

K$_{sp}$ = [Zn^{2+}]$_{eq}$[OH$^-$]$^2_{eq}$ = 1.20 x 10^{-17} M

$$\left[Zn^{2+}\right]_{eq} = \frac{K_{sp}}{[OH^-]^2_{eq}} = \frac{1.20 \times 10^{-17} \text{ M}^3}{(10^{-6} \text{ M})^2} = 1.20 \times 10^{-5} \text{ M}$$

(continued)

(16.73 continued)

total mmoles of Zn^{2+} = (100 mL)(0.100 M) = 10.0 mmoles

mmoles of Zn^{2+} in solution = (400 mL)(1.20 x 10^{-5} M) = 4.80 x 10^{-3} mmoles

mmoles of Zn^{2+} precipitated = 10.0 mmoles - 4.80 x 10^{-3} mmoles = 9.9952
$$\cong 10.0 \text{ mmoles}$$

$$\text{mass precipitated} = (10.0 \text{ mmoles } Zn^{2+})\left(\frac{1 \text{ mmole } Zn(OH)_2}{1 \text{ mmole } Zn^{2+}}\right)$$

$$\left(\frac{99.40 \times 10^{-3} \text{ g } Zn(OH)_2}{\text{mmole } Zn(OH)_2}\right) = 0.994 \text{ g } Zn(OH)_2$$

16.75 $$K_{eq} = \frac{[HClO]_{eq}[OH^-]_{eq}}{[ClO^-]_{eq}} = \frac{K_w}{K_a} = \frac{1.00 \times 10^{-14} \text{ M}^2}{3.02 \times 10^{-8} \text{ M}} = 3.31 \times 10^{-7} \text{ M}$$

$$NaClO \rightarrow Na^+_{(aq)} + ClO^-_{(aq)}$$

$$ClO^-_{(aq)} + H_2O_{(l)} \underset{\leftarrow}{\overset{\rightarrow}{}} HClO_{(aq)} + OH^-_{(aq)}$$

Init.	1.00×10^{-2}	0	0
Change	-x	+x	+x
Equil.	1.00×10^{-2}-x	x	x

$$K_{eq} = \frac{x^2}{1.00 \times 10^{-2} - x} \approx \frac{x^2}{1.00 \times 10^{-2}} = 3.31 \times 10^{-7}$$

$$x = \sqrt{(1.00 \times 10^{-2})(3.31 \times 10^{-7})} = 5.75 \times 10^{-5}$$

$$x = [OH^-]_{eq} = 5.75 \times 10^{-5} \text{ M}$$

pOH = 4.24   pH = 14.00 - pOH = 14.00 - 4.24 = 9.76

16.77 $$H_2PO_4^-{}_{(aq)} + H_2O_{(l)} \underset{\leftarrow}{\overset{\rightarrow}{}} HPO_4^{2-}{}_{(aq)} + H_3O^+{}_{(aq)}$$

$K_a = 6.2 \times 10^{-8}$ M

$pK_a = 7.21$

$$7.25 = 7.21 + \log\frac{[HPO_4^{2-}]}{[H_2PO_4^-]}$$

$$0.04 = \log\frac{[HPO_4^{2-}]}{(0.085 \text{ M})}$$

$$\frac{[HPO_4^{2-}]}{0.085 \text{ M}} = 1.1$$

$[HPO_4^{2-}]$ = (1.1)(0.085 M) = 0.094 M

Enzyme loses activity at pH = 7.1

(continued)

$$7.1 = 7.21 + \log\frac{[HPO_4^{2-}]}{[H_2PO_4^-]}$$

$$\log\frac{[HPO_4^{2-}]}{[H_2PO_4^-]} = -0.11 \qquad \frac{[HPO_4^{2-}]}{[H_2PO_4^-]} = 10^{-0.11} = 0.78$$

$[HPO_4^{2-}]_{eq} = 0.78[H_2PO_4^-]_{eq}$
molarity of added H_3O^+ = x $M_{H_3O^+}$
$[HPO_4^{2-}]_{eq}$ = $HPO_4^{2-}]_{init}$ - x $M_{H_3O^+}$
$[H_2PO_4^-]_{eq}$ = $H_2PO_4^-]_{init}$ + x M_{H_3O}
$[HPO_4^{2-}]_{eq}$ = 0.094 M - x $M_{H_3O^+}$
$[H_2PO_4^-]_{eq}$ = 0.085 M + x $M_{H_3O^+}$
0.094 M - x $M_{H_3O^+}$ = (0.78)(0.085 M + x $M_{H_3O^+}$)

0.094 M - 0.066 M = 1.78 x $M_{H_3O^+}$

$$x\ M_{H_3O^+} = \frac{0.028\ M}{1.78} = 0.016\ M$$

$$\text{moles } H_3O^+ = (0.016)\left(\frac{250\ mL}{1000\ mL/L}\right) = 4.0 \times 10^{-3} \text{ moles } H_3O^+$$

Note: This calculation is for the buffer with a pH of 7.25.

16.79

$$NH_{3(aq)} + NH_4^+{}_{(aq)}$$

M: 0.300 M   0.300 M

Vol.: 0.360 L   0.640 L

$$[NH_4^+]_{init.} = \frac{(0.640\ L)(0.300\ M)}{(0.640\ L + 0.360\ L)} = 0.192\ M$$

$$[NH_3]_{init.} = \frac{(0.360)(0.300)}{(0.360 + 0.640)} = 0.108\ M$$

a) $pH = pK_a + \log\left\{\dfrac{[A^-]_{init.}}{[HA]_{init.}}\right\} = (pK_w - pK_b) + \log\dfrac{[NH_3]_{init.}}{[NH_4^+]_{init.}}$

$$= (14.00 - 4.75) + \log\frac{0.108}{0.192} = 9.25 - 0.25 = 9.00$$

b) $NH_{3(aq)} + H_3O^+{}_{(aq)} \rightarrow NH_4^+{}_{(aq)} + H_2O_{(l)}$

c) 0.005 moles of H_3O^+ is added to 0.108 moles NH_3 and 0.192 mole NH_4^+, so a small amount of NH_3 gets protonated.

$[NH_3]$ = (0.108 mol - 0.005 mol)/1 L = 0.103 M

$[NH_4^+]$ = (0.192 + 0.005)/1 L = 0.197 M

pH = 9.25 + log(0.103/0.197) = 8.97

d) As long as the amount of added OH^- is less than the amount of NH_4^+ initially present, the balanced chemical equation is as follows:

$$NH_4^+{}_{(aq)} + OH^-{}_{(aq)} \rightarrow NH_{3(aq)} + H_2O_{(l)}$$

(continued)

(16.79 continued)

e) $\dfrac{14.0 \text{ g KOH}}{56.11 \text{ g/mol}} = 0.250 \text{ mol OH}^-$

Therefore, all of the NH_4^+ is converted to NH_3, with 0.058 M OH^- left over. The excess OH^- will determine the pH of the solution.

pH = 14.00 - (- log[OH^-]) = 14.00 - 1.24 = 12.76

16.81 $MgCO_{3(s)} \rightleftharpoons Mg^{2+}_{(aq)} + CO_3^{2-}_{(aq)}$

$pK_{sp} = 5.00; \ K_{sp} = 1 \times 10^{-5} \ M^2$

$CO_3^{2-}_{(aq)} + H_2O_{(l)} \rightleftharpoons HCO_3^-_{(aq)} + OH^-_{(aq)}$

a) Major species: $Mg^{2+}_{(aq)}$, CO_3^{2-}, $H_2O_{(l)}$

b) $K_{sp} = [Mg^{2+}]_{eq} \ [CO_3^{2-}]_{eq} = 1.00 \times 10^{-5} \ M^2$

$[Mg^{2+}]_{eq} = \sqrt{1.00 \times 10^{-5} \ M^2} = 3.16 \times 10^{-3} \ M$

c) $K_b = \dfrac{[HCO_3^-]_{eq}[OH^-]_{eq}}{[CO_3^{2-}]_{eq}} = \dfrac{K_w}{K_a} = \dfrac{1.00 \times 10^{-14} \ M^2}{5.6 \times 10^{-11} \ M} = 1.79 \times 10^{-4} \ M$

$$CO_3^{2-}_{(aq)} + H_2O_{(l)} \rightleftharpoons HCO_3^-_{(aq)} + OH^-_{(aq)}$$

	CO_3^{2-}	HCO_3^-	OH^-
Init. (M)	3.16×10^{-3} M	0	0
Change	-x	+x	+x
Equil. (M)	3.16×10^{-3}-x	x	x

$K_b = \dfrac{x^2}{3.16 \times 10^{-3} - x} = 1.79 \times 10^{-4};$

$x^2 = 1.79 \times 10^{-4}(3.16 \times 10^{-3} - x) = 5.65 \times 10^{-7} - 1.79 \times 10^{-4}x$

$0 = x^2 + 1.79 \times 10^{-4} - 5.65 \times 10^{-7}$

$x = \dfrac{-1.79 \times 10^{-4} \pm \sqrt{(1.79 \times 10^{-4})^2 - 4(1)(-5.65 \times 10^{-7})}}{2(1)} = 6.67 \times 10^{-4} \ M = [OH^-]_{eq}$

pH = 14.00 - pOH = 14.00 - 3.18 = 10.82 = 10.8

d) $CaCO_{3(s)} \rightleftharpoons Ca^{2+}_{(aq)} + CO_3^{2-}_{(aq)}$

$pK_{sp} = 8.55$

$K_{sp} = [Ca^{2+}]_{eq}[CO_3^{2-}]_{eq} = 2.82 \times 10^{-9} \ M$

$[CO_3^{2-}]_{eq} = 3.16 \times 10^{-3} - 6.67 \times 10^{-4} \ M = 2.50 \times 10^{-3} \ M$

Note: This equilibrium results in a slight adjustment to $[Mg^{2+}]$ in part (b) to ~3.5 x 10^{-3} M.

$[Ca^{2+}]_{eq} = 2.82 \times 10^{-9} \ M^2 / 2.50 \times 10^{-3} \ M = 1.13 \times 10^{-6} \ M$

16.83 [TRIS] = 0.30 M

[TRISH$^+$] = 0.60 M

[H$_3$O$^+$]$_{init.}$ = 12 M

Total vol. = 1.0 L

$$TRIS_{(aq)} + H_3O^+_{(aq)} \rightleftharpoons TRISH^+_{(aq)} + H_2O_{(l)}$$

a) $pH = pKa + \log\left\{\dfrac{[A^-]_{init}}{[HA]_{init.}}\right\}$

$= (14.00 - pK_b) + \log\dfrac{[TRIS]}{[TRISH^+]} = (14.0 - 5.7) + \log\dfrac{0.30}{0.60} = 8.0$

b) $\dfrac{(5.0 \text{ mL H}_3\text{O}^+)}{1000 \text{ mL / L}}12 \text{ M H}_3\text{O}^+ = 6.0 \times 10^{-2}$ moles H$_3$O$^+_{(aq)}$

$[TRIS] = \dfrac{0.30 \text{ mol} - 0.06 \text{ mol}}{1.005 \text{ L}} = 0.239$ M

$[TRISH^+] = \dfrac{0.60 + 0.06}{1.005} = 0.657$ M

$pH = pK_a + \log\dfrac{0.239}{0.657} = 8.3 - 0.439 = 7.86 = 7.9$

16.85 From Problem 16.84: $\dfrac{35 \text{ mL H}_3\text{O}^+}{1000 \text{ mL / L}}12 \text{ M H}_3\text{O}^+ = 0.42$ moles H$_3$O$^+$

$$TRISH^+_{(aq)} + OH^-_{(aq)} \rightleftharpoons TRIS_{(aq)} + H_2O_{(l)}$$

Adding 0.42 moles of OH$^-_{(aq)}$ will restore the buffer to its original pH. Adding 0.63 moles of TRIS would also restore the buffer to its original pH.

16.87 Choose an indicator whose color changes as close to the stoichiometric point as possible: pH$_{stoic}$ = pK$_{indicator}$ ±1

$$CH_3CO_2H_{(aq)} + OH^-_{(aq)} \rightleftharpoons CH_3CO_2^-_{(aq)} + H_2O_{(l)}$$

At the stoichiometric point all CH$_3$CO$_2$H has been converted to CH$_3$CO$_2^-$; the pH is determined by the acid-base equilibrium of CH$_3$CO$_2^-_{(aq)}$:

	CH$_3$CO$_2^-_{(aq)}$ + H$_2$O$_{(l)}$ $\rightleftharpoons$ CH$_3$CO$_2$H$_{(aq)}$	+ OH$^-_{(aq)}$
Init.	0.3 M	0	0
Change	-x	+x	+x
Equil.	0.3-x	x	x

$K_{eq} = \dfrac{K_w}{K_a} = \dfrac{1.00 \times 10^{-14} \text{ M}^2}{1.8 \times 10^{-5} \text{ M}} = 5.6 \times 10^{-10}$ M

(continued)

(16.87 continued)

$$5.6 \times 10^{-10} \, M = \frac{x^2}{0.3 - x} \approx \frac{x^2}{0.3}; \quad x^2 = (0.3)(5.6 \times 10^{-10}) \text{ assuming minimal volume}$$

change during titration.

$$x = 1.3 \times 10^{-5} \, M = [OH^-]_{eq} \qquad pOH = 4.9 \qquad pH = 14 - pOH = 14 - 4.9 = 9.1$$

Possible indicators: thymol blue ($pK_{In} = 8.9$) phenolphthalein ($pK_{In} = 9.4$; best choice), or thymolphthalein ($pK_{In} = 10.0$). The pH at the stoichiometric point would be 9.0 if it is assumed an equal volume of base is added during titration.

16.89 Some salts that would be more soluble in acidic solution than in pure water are: $Ca(OH)_2$, $Cu(OH)_2$, $Fe(OH)_2$, $Fe(OH)_3$, $Ni(OH)_2$.
Salts that would be less soluble in acidic solution than in pure water include: NH_4Cl, NH_4NO_3, NH_4ClO_4, H_2NH_3Cl, CH_3NH_3Cl.
Salts whose solubility does not depend on pH include: KCl, $KClO_4$, $NaClO_4$, $NaCl$, KNO_3, $NaNO_3$.

16.91 a) At the stoichiometric point, the amount of added $OH^- = NH_4^+$ originally present.

$$\text{mol } OH^- = \frac{(0.375 \text{ g } NH_4Cl)}{53.5 \text{ g/mol}} = 7.01 \times 10^{-3} \text{ mol}$$

$$\text{vol. of titrant} = \frac{7.01 \times 10^{-3} \text{ mol } OH^-}{0.08775 \text{ M } OH^-} = 79.9 \text{ mL}$$

vol. final soln. $= 25 + 79.9 = 104.9$ mL

$$NH_4^+ + OH^- \rightarrow NH_3 + H_2O \qquad [NH_3] = \frac{7.01 \times 10^{-3} \text{ mol}}{105 \text{ mL}/1000 \text{ mL / L}} = 6.68 \times 10^{-2} \, M$$

$$K_b = \frac{[NH_4^+]_{eq}[OH^-]_{eq}}{[NH_3]_{eq}} = 10^{-4.75} = 1.75 \times 10^{-5} M$$

$$NH_{3(aq)} + H_2O_{(l)} \overset{\rightarrow}{\underset{\leftarrow}{}} NH_4^+{}_{(aq)} + OH^-{}_{(aq)}$$

Init.	6.68×10^{-2}	0	0
Change	$-x$	$+x$	$+x$
Equil.	$6.68 \times 10^{-2}-x$	x	x

$$K_b = \frac{x^2}{6.68 \times 10^{-2} - x} \approx \frac{x^2}{6.68 \times 10^{-2}} = 1.78 \times 10^{-5};$$

$$x = \sqrt{(6.68 \times 10^{-2})(1.78 \times 10^{-5})} = 1.09 \times 10^{-3} M$$

$$pOH = -\log 1.09 \times 10^{-3} = 2.96 \qquad pH = 14.00 - 2.96 = 11.04 = 11.0$$

b) $NH_{3(aq)} + H_3O^+{}_{(aq)} \rightarrow NH_4^+ + H_2O_{(l)}$
35 mL   x mL
0.15 M   0.537 M
mol added H_3O^+ = mol NH_3 =

$$\left(\frac{0.15 \text{ mol}}{L}\right) 35 \text{ mL} \left(\frac{1 \text{ L}}{1000 \text{ mL}}\right) = 5.25 \times 10^{-3} \text{ moles}$$

(continued)

(16.91 continued)

$$\text{vol } H_3O^+ \text{ added} = \frac{5.25 \times 10^{-3} \text{ mol}}{0.537 \text{ mol} / L}\left(\frac{1000 \text{ mL}}{L}\right) = 9.78 \text{ mL}$$

Total vol. final soln = 35 + 9.78 = 44.8 mL

$[NH_4^+]_{init.} = (0.15)35/44.8 = 0.12 \text{ M}$

$$K_a = \frac{K_w}{K_b} = \frac{10^{-14} \text{ M}^2}{1.78 \times 10^{-5} \text{ M}} = 5.62 \times 10^{-10} \text{ M}$$

$$NH_4^+_{(aq)} + H_2O_{(l)} \rightleftharpoons NH_3_{(aq)} + H_3O^+_{(aq)}$$

Init.	0.12	0	0
Change	-x	+x	+x
Equil.	0.12-x	x	x

$$K_a = \frac{x^2}{0.12-x} \approx \frac{x^2}{0.12} = 5.62 \times 10^{-10}; \quad x^2 = (0.12)(5.62 \times 10^{-10})$$

$$x = [H_3O^+]_{eq} = \sqrt{(5.62 \times 10^{-10})(0.12)} = \sqrt{6.59 \times 10^{-11}} = 8.21 \times 10^{-6} \text{ M}$$

$pH = -\log[H_3O^+]_{eq} = 5.09$

c) $H_3O^+_{(aq)} + OH^-_{(aq)} \rightarrow 2 H_2O_{(l)}$ \qquad pH = 7.0

16.93 $Al(OH)_3 \rightleftharpoons Al^{3+}_{(aq)} + 3 OH^-_{(aq)}$ \qquad $K_{sp} = [Al^{3+}]_{eq}[OH^-]^3_{eq} = 1.8 \times 10^{-33} \text{ M}^4$

a) $K_{sp} = x(3x)^3$ \quad $27x^4 = 1.8 \times 10^{-33}$ \quad $x = \sqrt[4]{\dfrac{1.8 \times 10^{-33}}{27}} = 2.86 \times 10^{-9} \text{ M}$

$[Al^{3+}]_{eq} = x = 2.9 \times 10^{-9} \text{ M}$

b) $[OH^-] = \dfrac{(1200 \text{ L})(0.250 \text{ M})}{1200 \text{ L} + 1300 \text{ L}} = 0.120 \text{ M}$

$[Al^{3+}] = \dfrac{(1300 \text{ L})(0.223 \text{ M})}{1200 \text{ L} + 1300 \text{ L}} = 0.116 \text{ M}$

$1/K_{sp} = 1/1.8 \times 10^{-33} \text{ M}^4 = 5.56 \times 10^{32} \text{ M}^{-4}$

$$Al^{3+}_{(aq)} + 3 OH^-_{(aq)} \rightleftharpoons Al(OH)_{3(s)}$$

Init.	0.116	0.120
Change	-0.040	-0.120
Complete	0.076	0
Change to Equil.	y	3y
Equil.	0.076+y	3y

$$K_{sp} = 1.8 \times 10^{-33} \text{ M}^4 = (0.076 + y)(3y)^3$$

$$y = 9.6 \times 10^{-12}$$

Amount $Al(OH)_3$ that ppts = (0.040 M)(2500 L) = 100 moles

Mass of $Al(OH)_3$ = (100 moles)(78.0 g/mol = 7800 g = 7.80 kg

Residual conc. of Al^{3+} = 0.076 M = 7.6×10^{-2} M

16.95 a) A = HCO_2H,   B = $HCO_2H + HCO_2^-$,   C = HCO_2^-,   D = $OH^- + HCO_2^-$

b) At the stoichiometric point, mol OH^- = mol HCO_2H originally present:

mol OH^- = (0.125 L HCO_2H)(0.135 M HCO_2H) = 1.688 x 10^{-2} moles

or mol OH^- = {29.8 mL/(1000 mL/L)}(0.567 M) = 1.690 x 10^{-2} moles

$$HCO_2H_{(aq)} + OH^-_{(aq)} \rightarrow HCO_2^-_{(aq)} + H_2O_{(l)}$$

Goes to completion to $[HCO_2^-]_{init}$ = $[OH^-]$

$$[HCO_2^-]_{init.} = \frac{1.69 \times 10^{-2} \text{ moles}}{0.125 \text{ L} + 0.0298 \text{ L}} = 0.109 \text{ M}$$

$$K_{eq} = \frac{K_w}{K_a} = \frac{1 \times 10^{-14} \text{ M}^2}{10^{-3.75}} = 5.62 \times 10^{-11} \text{ M}$$

$$HCO_2^-_{(aq)} + H_2O_{(l)} \rightleftarrows HCO_2H_{(aq)} + OH^-_{(aq)}$$

	HCO_2^-	HCO_2H	OH^-
Initial (M)	0.109 M	0	0
Change (M)	-x	+x	+x
Completion (M)	0.109-x	x	x

$$K_{eq} = \frac{x^2}{0.109 - x} \approx \frac{x^2}{0.109} = 5.62 \times 10^{-11} \text{ M}$$

$x = \sqrt{(0.109)(5.62 \times 10^{-11})} = \sqrt{6.14 \times 10^{-12}} = 2.48 \times 10^{-6}$ M

$x = [OH^-]_{eq}$

$pOH = - \log 2.48 \times 10^{-6} = 5.61$

$pH_{stoich.\ pt.} = 14.00 - 5.61 = 8.39$

16.97 $Cd(OH)_{2(s)} + H_2O_{(l)} \rightleftarrows Cd^{2+}_{(aq)} + 2\ OH^-_{(aq)}$ $pK_{sp} = 13.60$

$K_{sp} = 2.51 \times 10^{-14}$ M

$K_{sp} = [Cd^{2+}]_{eq}[OH^-]^2_{eq} = x(2x)^2 = 4x^3 = 2.51 \times 10^{-14}$

$$x = \sqrt[3]{\frac{2.51 \times 10^{-14}}{4}} = \sqrt[3]{6.28 \times 10^{-15}} = 1.84 \times 10^{-5} \text{ M}$$

$[OH^-]_{eq} = 2x = 2(1.84 \times 10^{-5}$ M$) = 3.69 \times 10^{-5}$ M; $pOH = 4.43$

$pH = 14.00 - pOH = 14.00 - 4.43 = 9.57$

16.99 a) $NaOH_{(aq)} \xrightleftharpoons{} Na^+_{(aq)} + OH^-_{(aq)}$ completion

$C_6H_5CO_2H_{(s)} + H_2O_{(l)} \rightleftarrows C_6H_5CO_2^-_{(aq)} + H_3O^+_{(aq)}$ small extent

Sum: $NaOH_{(aq)} + C_6H_5CO_2H_{(s)} + H_2O_{(l)} \rightleftarrows$

$Na^+_{(aq)} + OH^-_{(aq)} + C_6H_5CO_2^-_{(aq)} + H_3O^+_{(aq)}$

Net: $OH^-_{(aq)} + C_6H_5CO_2H_{(s)} \rightarrow C_6H_5CO_2^-_{(aq)} + H_2O_{(l)}$ completion

(continued)

166

(16.99 continued)

b) $(CH_3)_3N_{(aq)} + H_2O_{(l)} \rightleftharpoons (CH_3)_3NH^+_{(aq)} + OH^-_{(aq)}$ small extent

$HNO_{3(aq)} + H_2O_{(l)} \rightleftharpoons H_3O^+_{(aq)} + NO_3^-_{(aq)}$ completion

Sum:

$Me_3N_{(aq)} + HNO_{3(aq)} + 2\,H_2O_{(l)} \rightleftharpoons Me_3NH^+_{(aq)} + NO_3^-_{(aq)} + H_3O^+_{(aq)} + OH^-_{(aq)}$

Net: $(CH_3)_3N_{(aq)} + H_3O^+_{(aq)} \rightarrow (CH_3)_3NH^+_{(aq)} + H_2O_{(l)}$ completion

c) $Na_2SO_{4(aq)} \rightleftharpoons 2\,Na^+_{(aq)} + SO_4^{2-}_{(aq)}$ completion

$CH_3CO_2H_{(aq)} + H_2O_{(l)} \rightleftharpoons CH_3CO_2^-_{(aq)} + H_3O^+_{(aq)}$ small extent

Sum: $Na_2SO_{4(aq)} + CH_3CO_2H_{(aq)} + H_2O_{(l)}$
$$\rightleftharpoons 2\,Na^+_{(aq)} + CH_3CO_2^-_{(aq)} + SO_4^{2-}_{(aq)} + H_3O^+_{(aq)}$$

Net: $SO_4^{2-}_{(aq)} + CH_3CO_2H_{(aq)} \rightleftharpoons CH_3CO_2^-_{(aq)} + HSO_4^-_{(aq)}$ small extent

HSO_4^- $K_a = 1.1 \times 10^{-2}$; CH_3CO_2H $K_a = 1.8 \times 10^{-5}$. HSO_4^- is a stronger acid than CH_3CO_2H; $\therefore$ reaction proceeds to only a small extent.

d) $NH_4Cl_{(aq)} \rightleftharpoons NH_4^+_{(aq)} + Cl^-_{(aq)}$ completion

$NH_4^+ + H_2O_{(l)} \rightleftharpoons NH_{3(aq)} + H_3O^+_{(aq)}$

$Ca(OH)_{2(aq)} \rightarrow Ca^{2+}_{(aq)} + 2\,OH^-_{(aq)}$

$OH^-_{(aq)} + H_3O^+_{(aq)} \rightarrow 2\,H_2O_{(l)}$ completion

Net: $NH_4^+_{(aq)} + OH^-_{(aq)} \rightarrow NH_{3(aq)} + H_2O_{(l)}$ completion

e) $K_2HPO_{4(aq)} \rightleftharpoons 2\,K^+_{(aq)} + HPO_4^{2-}_{(aq)}$ completion

$HPO_4^{2-}_{(aq)} + NH_{3(aq)} \rightleftharpoons NH_4^+_{(aq)} + PO_4^{3-}_{(aq)}$ small extent

$K_{a,\,HPO_4^{2-}} = 4.8 \times 10^{-13}$ $K_{a,\,NH_4^+} = 5.6 \times 10^{-10}$

So, NH_4^+ is a stronger acid than HPO_4^{2-}.

16.101 $pH = pK_a + \log\dfrac{[CH_3CO_2^-]}{[CH_3CO_2H]}$ $4.26 = 4.75 + \log\dfrac{[CH_3CO_2^-]}{[CH_3CO_2H]}$

$\dfrac{[CH_3CO_2^-]}{[CH_3CO_2H]} = 0.32$ or about one $CH_3CO_2^-$ for every three CH_3CO_2H

Using the symbols requested, draw a molecular picture that shows the one to three ratio in this buffer solution.

167

Using the symbols requested, draw a molecular picture that shows the one to three ratio in this buffer solution.

16.103
$$C_{20}H_{25}N_3O + H_2O \rightleftarrows C_{20}H_{26}N_3O^+{}_{(aq)} + OH^-{}_{(aq)}$$

Init.	0.55 M	0	0
Change	-x	+x	+x
Equil.	0.55-x	x	x

$$K_b = \frac{[LSDH^+][OH^-]}{[LSD]} = 10^{-6.12} = 7.59 \times 10^{-7} \text{ M}$$

$$K_b = \frac{x^2}{0.55-x} \cong \frac{x^2}{0.55} = 7.59 \times 10^{-7}$$

$$x = [OH^-]_{eq} = \sqrt{(0.55)(7.59 \times 10^{-7})} = 6.46 \times 10^{-4}$$

$$pOH = -\log[OH^-]_{eq} = 3.19$$

$$pH = 14.00 - pOH = 14.00 - 3.19 = 10.81$$

16.105 Glycine = H_2NCH_2COH

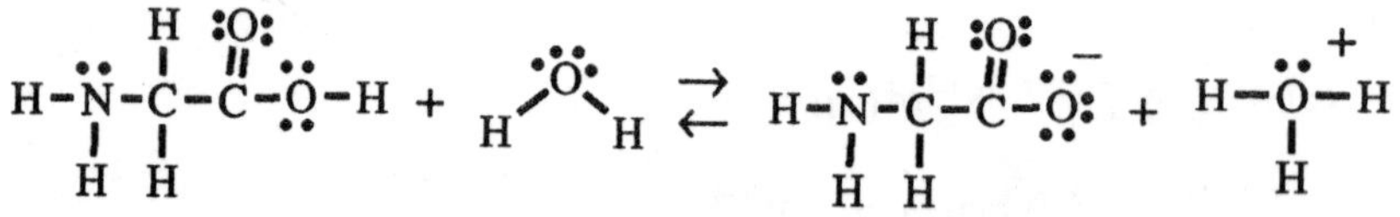

and

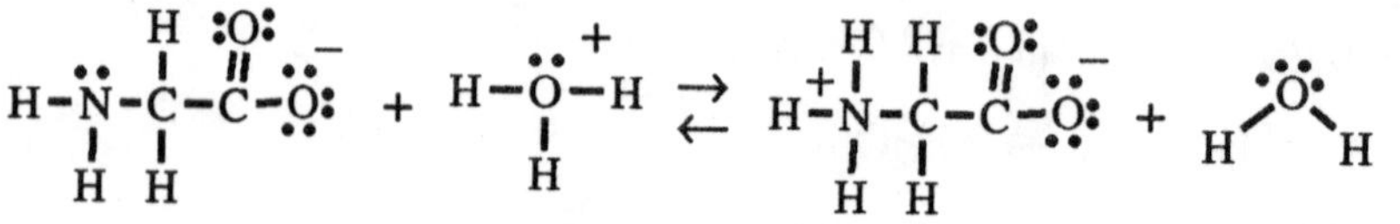

net reaction:

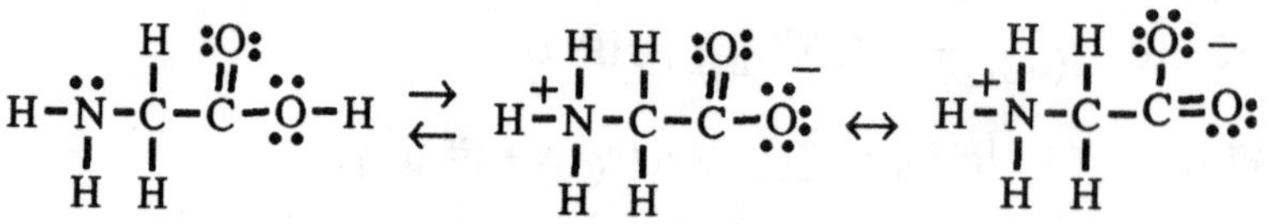

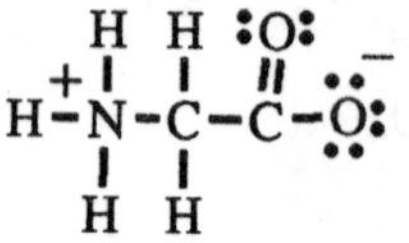

= zwitterion.

168

CHAPTER 17: ELECTRON TRANSFER REACTIONS: REDOX AND ELECTROCHEMISTRY

17.1   a) The oxidation numbers of the atoms in $Fe(OH)_3$ are: $Fe = +3$, $O = -2$, $H = +1$.
       b) The oxidation numbers of the atoms in NH_3 are: $N = -3$ and $H = +1$.
       c) The oxidation numbers of the atoms in PCl_5 are: $P = +5$ and $Cl = -1$.
       d) The oxidation numbers of the atoms in K_2CO_3 are: $K = +1$, $C = +4$, $O = -2$.
       e) The oxidation number of the atoms in P_4 is: $P = 0$.

17.3   (b) is a redox reaction in which Fe^{2+} is oxidized to Fe^{3+} and H_2O_2 is reduced to OH^-.

17.5   a) Cl_2 b) HCl c) $HClO$ d) $HClO_2$ e) $HClO_3$ f) $HClO_4$

17.7   a) Oxidation $Na \rightarrow Na^+$ Reduction $H_2O \rightarrow H_2$

       b) Oxidation $Au \rightarrow AuCl_4^-$ Reduction $HNO_3 \rightarrow NO$

       c) Oxidation $C_2O_4^{2-} \rightarrow CO_2$ Reduction $MnO_4^- \rightarrow Mn^{2+}$

17.9   a) Step 1: $Cu^+ \rightarrow CuO$ (acid solution)
       Step 2: $\mathbf{H_2O} + Cu^+ \rightarrow CuO$
       Step 3: $H_2O + Cu^+ \rightarrow CuO + \mathbf{2\ H^+_{(aq)}}$
       Step 4: $H_2O + Cu^+ \rightarrow CuO + 2\ H^+_{(aq)} + \mathbf{e^-}$

       b) Steps 1 & 2: $S \rightarrow H_2S$ (acid solution)
       Step 3: $\mathbf{2\ H^+_{(aq)}} + S \rightarrow H_2S$
       Step 4: $\mathbf{2\ e^-} + 2\ H^+_{(aq)} + S \rightarrow H_2S$

       c) $AgCl \rightarrow Ag$ (basic solution)
       Steps 1, 2 & 3: $AgCl \rightarrow Ag + \mathbf{Cl^-}$
       Step 4: $\mathbf{e^-} + AgCl \rightarrow Ag + Cl^-$

(continued)

(17.9 continued)

    d) Step 1: $I^- \rightarrow IO_3^-$ (basic solution)

    Step 2: $\mathbf{3\ H_2O} + I^- \rightarrow IO_3^-$

    Step 3: $3\ H_2O + I^- \rightarrow IO_3^- + \mathbf{6\ H^+_{(aq)}}$

    Step 3 a: $3\ H_2O + I^- \rightarrow IO_3^- + 6\ H^+_{(aq)}$

        ✳ $6\ OH^- + 6\ H^+_{(aq)} \rightarrow 6\ H_2O$

        $\mathbf{6\ OH^-} + 3\ H_2O + I^- \rightarrow IO_3^- + \mathbf{6\ H_2O}$

      Cancel H_2O on both sides: $6\ OH^- + I^- \rightarrow IO_3^- + \mathbf{3\ H_2O}$

    Step 4: $6\ OH^- + I^- \rightarrow IO_3^- + 3\ H_2O + \mathbf{6\ e^-}$

    e) Step 1: $IO_3^- \rightarrow IO^-$ (basic solution)

    Step 2: $IO_3^- \rightarrow IO^- + \mathbf{2\ H_2O}$

    Step 3: $\mathbf{4\ H^+_{(aq)}} + IO_3^- \rightarrow IO^- + 2\ H_2O$

    Step 3 a: $4\ H^+_{(aq)} + IO_3^- \rightarrow IO^- + 2\ H_2O$

          $4\ H_2O \rightarrow 4\ H^+_{(aq)} + 4\ OH^-$

          $\mathbf{4\ H_2O} + IO_3^- \rightarrow IO^- + 2\ H_2O + \mathbf{4\ OH^-}$

      Cancel H_2O on both sides: $\mathbf{2\ H_2O} + IO_3^- \rightarrow IO^- + 4\ OH^-$

    Step 4: $\mathbf{4\ e^-} + 2\ H_2O + IO_3^- \rightarrow IO^- + 4\ OH^-$

    f) Step 1: $H_2CO \rightarrow CO_2$ (acid solution)

    Step 2: $\mathbf{H_2O} + H_2CO \rightarrow CO_2$

    Step 3: $H_2O + H_2CO \rightarrow CO_2 + \mathbf{4\ H^+_{(aq)}}$

    Step 4: $H_2O + H_2CO \rightarrow CO_2 + 4\ H^+_{(aq)} + \mathbf{4\ e^-}$

17.11 a) Oxidation: $H_2O + Cu^+ \rightarrow CuO + 2\ H^+_{(aq)} + e^-$

      Reduction: $2\ e^- + 2\ H^+_{(aq)} + S \rightarrow H_2S$

Multiply the oxidation half-reaction by 2:

    $2\ H_2O + 2\ Cu^+ \rightarrow 2\ CuO + 4\ H^+_{(aq)} + 2\ e^-$

Combine the half reactions:

    $2\ H_2O + 2\ Cu^+ + 2\ e^- + 2\ H^+_{(aq)} + S \rightarrow 2\ CuO + 4\ H^+_{(aq)} + 2\ e^- + H_2S$

Cancel duplicate species:

    $2\ H_2O + 2\ Cu^+ + S \rightarrow 2\ CuO + 2\ H^+_{(aq)} + H_2S$

    b) Oxidation: $6\ OH^- + I^- \rightarrow IO_3^- + 3\ H_2O + 6\ e^-$

      Reduction: $e^- + AgCl \rightarrow Ag + Cl^-$

Multiply the reduction half-reaction by 6:

    $6\ e^- + 6\ AgCl \rightarrow 6\ Ag + 6\ Cl^-$

Combine the half reactions:

    $6\ OH^- + I^- + 6\ e^- + 6\ AgCl \rightarrow IO_3^- + 3\ H_2O + 6\ e^- + 6\ Ag + 6\ Cl^-$

(continued)

170

(17.11 continued)

Cancel duplicate species:

$$6 \text{ OH}^- + \text{I}^- + 6 \text{ AgCl} \rightarrow \text{IO}_3^- + 3 \text{ H}_2\text{O} + 6 \text{ Ag} + 6 \text{ Cl}^-$$

c) Oxidation: $6 \text{ OH}^- + \text{I}^- \rightarrow \text{IO}_3^- + 3 \text{ H}_2\text{O} + 6 \text{ e}^-$

Reduction: $4 \text{ e}^- + 2 \text{ H}_2\text{O} + \text{IO}_3^- \rightarrow \text{IO}^- + 4 \text{ OH}^-$

Multiply the oxidation half-reaction by 2:

$$12 \text{ OH}^- + 2 \text{ I}^- \rightarrow 2 \text{ IO}_3^- + 6 \text{ H}_2\text{O} + 12 \text{ e}^-$$

Multiply the reduction half-reaction by 3:

$$12 \text{ e}^- + 6 \text{ H}_2\text{O} + 3 \text{ IO}_3^- \rightarrow 3 \text{ IO}^- + 12 \text{ OH}^-$$

Combine the half-reactions:

$$12 \text{ OH}^- + 2 \text{ I}^- + 12 \text{ e}^- + 6 \text{ H}_2\text{O} + 3 \text{ IO}_3^- \rightarrow$$
$$2 \text{ IO}_3^- + 6 \text{ H}_2\text{O} + 12 \text{ e}^- + 3 \text{ IO}^- + 12 \text{ OH}^-$$

Cancel duplicate species: $2 \text{ I}^- + \text{IO}_3^- \rightarrow 3 \text{ IO}^-$

d) Oxidation: $\text{H}_2\text{CO} + \text{H}_2\text{O} \rightarrow \text{CO}_2 + 4 \text{ H}^+_{(aq)} + 4 \text{ e}^-$

Reduction: $2 \text{ e}^- + 2 \text{ H}^+_{(aq)} + \text{S} \rightarrow \text{H}_2\text{S}$

Multiply the reduction half-reaction by 2:

$$4 \text{ e}^- + 4 \text{ H}^+_{(aq)} + 2 \text{ S} \rightarrow 2 \text{ H}_2\text{S}$$

Combine the half-reactions:

$$\text{H}_2\text{CO} + \text{H}_2\text{O} + 4 \text{ e}^- + 4 \text{ H}^+_{(aq)} + 2 \text{ S} \rightarrow \text{CO}_2 + 4 \text{ H}^+_{(aq)} + 4 \text{ e}^- + 2 \text{ H}_2\text{S}$$

Cancel duplicate species:

$$\text{H}_2\text{CO} + \text{H}_2\text{O} + 2 \text{ S} \rightarrow \text{CO}_2 + 2 \text{ H}_2\text{S}$$

17.13 a) $\text{PbO} + \text{Co(NH}_3)_6^{3+} \rightarrow \text{PbO}_2 + \text{Co(NH}_3)_6^{2+}$ (basic)

I and II. Oxidation:

Step 1: $\text{PbO} \rightarrow \text{PbO}_2$

Step 2: $\mathbf{H_2O} + \text{PbO} \rightarrow \text{PbO}_2$

Step 3: $\text{H}_2\text{O} + \text{PbO} \rightarrow \text{PbO}_2 + \mathbf{2 \text{ H}^+_{(aq)}}$

Step 3 a: $\text{H}_2\text{O} + \text{PbO} \rightarrow \text{PbO}_2 + 2 \text{ H}^+_{(aq)}$

$2 \text{ H}^+_{(aq)} + 2 \text{ OH}^- \rightarrow 2 \text{ H}_2\text{O}$

$\mathbf{2 \text{ OH}^-} + \text{H}_2\text{O} + \text{PbO} \rightarrow \text{PbO}_2 + \mathbf{2 \text{ H}_2\text{O}}$

Cancel duplicate species: $2 \text{ OH}^- + \text{PbO} \rightarrow \text{PbO}_2 + \mathbf{H_2O}$

Step 4: $2 \text{ OH}^- + \text{PbO} \rightarrow \text{PbO}_2 + \text{H}_2\text{O} + \mathbf{2 \text{ e}^-}$

Reduction:

Steps 1 to 3: $\text{Co(NH}_3)_6^{3+} \rightarrow \text{Co(NH}_3)_6^{2+}$

Step 4: $\mathbf{e^-} + \text{Co(NH}_3)_6^{3+} \rightarrow \text{Co(NH}_3)_6^{2+}$

III. Multiply the red. half-reaction by 2: $2 \text{ e}^- + 2 \text{ Co(NH}_3)_6^{3+} \rightarrow 2 \text{ Co(NH}_3)_6^{2+}$

IV. Combine the half-reactions: $2 \text{ OH}^- + \text{PbO} + 2 \text{ e}^- + 2 \text{ Co(NH}_3)_6^{3+} \rightarrow$
$$\text{PbO}_2 + \text{H}_2\text{O} + 2 \text{ e}^- + \text{Co(NH}_3)_6^{2+}$$

(continued)

171

(17.13 continued)

Cancel duplicate species:

$$2 \text{ OH}^- + \text{PbO} + 2 \text{ Co(NH}_3)_6^{3+} \rightarrow \text{PbO}_2 + \text{H}_2\text{O} + \text{Co(NH}_3)_6^{2+}$$

b) $\text{O}_2 + \text{As} \rightarrow \text{HAsO}_2 + \text{H}_2\text{O}$ (acidic)

I and II. Oxidation

Step 1: $\text{As} \rightarrow \text{HAsO}_2$

Step 2: $\mathbf{2 \text{ H}_2\text{O}} + \text{As} \rightarrow \text{HAsO}_2$

Step 3: $2 \text{ H}_2\text{O} + \text{As} \rightarrow \text{HAsO}_2 + \mathbf{3 \text{ H}^+_{(aq)}}$

Step 4: $2 \text{ H}_2\text{O} + \text{As} \rightarrow \text{HAsO}_2 + 3 \text{ H}^+_{(aq)} + \mathbf{3 \text{ e}^-}$

          Reduction:

Step 1: $\text{O}_2 \rightarrow \text{H}_2\text{O}$

Step 2: $\text{O}_2 \rightarrow \mathbf{2 \text{ H}_2\text{O}}$

Step 3: $\mathbf{4 \text{ H}^+_{(aq)}} + \text{O}_2 \rightarrow 2 \text{ H}_2\text{O}$

Step 4: $\mathbf{4 \text{ e}^-} + 4 \text{ H}^+_{(aq)} + \text{O}_2 \rightarrow 2 \text{ H}_2\text{O}$

III. Multiply the ox. half-rxn. by 4: $8 \text{ H}_2\text{O} + 4 \text{ As} \rightarrow 4 \text{ HAsO}_2 + 12 \text{ H}^+_{(aq)} + 12 \text{ e}^-$

Multiply the reduction half-reaction by 3: $12 \text{ e}^- + 12 \text{ H}^+_{(aq)} + 3 \text{ O}_2 \rightarrow 6 \text{ H}_2\text{O}$

IV. Combine the half-reactions: $8 \text{ H}_2\text{O} + 4 \text{ As} + 12 \text{ e}^- + 12 \text{ H}^+_{(aq)} + 3 \text{ O}_2 \rightarrow$
$$4 \text{ HAsO}_2 + 12 \text{ H}^+_{(aq)} + 12 \text{ e}^- + 6 \text{ H}_2\text{O}$$

Cancel duplicate species: $2 \text{ H}_2\text{O} + 4 \text{ As} + 3 \text{ O}_2 \rightarrow 4 \text{ HAsO}_2$

c) $\text{Br}^- + \text{MnO}_4^- \rightarrow \text{MnO}_2 + \text{BrO}_3^-$ (basic)

I and II. Oxidation:

Step 1: $\text{Br}^- \rightarrow \text{BrO}_3^-$

Step 2: $\mathbf{3 \text{ H}_2\text{O}} + \text{Br}^- \rightarrow \text{BrO}_3^-$

Step 3: $3 \text{ H}_2\text{O} + \text{Br}^- \rightarrow \text{BrO}_3^- + \mathbf{6 \text{ H}^+_{(aq)}}$

Step 3 a: $3 \text{H}_2\text{O} + \text{Br}^- \rightarrow \text{BrO}_3^- + 6 \text{ H}^+_{(aq)}$

          $6 \text{ H}^+_{(aq)} + 6 \text{ OH}^- \rightarrow 6 \text{ H}_2\text{O}$

          $\mathbf{6 \text{ OH}^-} + 3 \text{ H}_2\text{O} + \text{Br}^- \rightarrow \text{BrO}_3^- + \mathbf{6 \text{ H}_2\text{O}}$

     Cancel duplicate species: $6 \text{ OH}^- + \text{Br}^- \rightarrow \text{BrO}_3^- + 3 \text{ H}_2\text{O}$

Step 4: $6 \text{ OH}^- + \text{Br}^- \rightarrow \text{BrO}_3^- + 3 \text{ H}_2\text{O} + \mathbf{6 \text{ e}^-}$

          Reduction:

Step 1: $\text{MnO}_4^- \rightarrow \text{MnO}_2$

Step 2: $\text{MnO}_4^- \rightarrow \text{MnO}_2 + \mathbf{2 \text{ H}_2\text{O}}$

Step 3: $\mathbf{4 \text{ H}^+_{(aq)}} + \text{MnO}_4^- \rightarrow \text{MnO}_2 + 2 \text{ H}_2\text{O}$

Step 3 a: $4 \text{ H}^+_{(aq)} + \text{MnO}_4^- \rightarrow \text{MnO}_2 + 2 \text{ H}_2\text{O}$

          $4 \text{ H}_2\text{O} \rightarrow 4 \text{ H}^+_{(aq)} + 4 \text{ OH}^-$

          $\mathbf{4 \text{ H}_2\text{O}} + \text{MnO}_4^- \rightarrow \text{MnO}_2 + 2 \text{ H}_2\text{O} + \mathbf{4 \text{ OH}^-}$

(continued)

172

(17.13 continued)

Cancel duplicate species: $2 H_2O + MnO_4^- \rightarrow MnO_2 + 4 OH^-$

Step 4: $3 e^- + 2 H_2O + MnO_4^- \rightarrow MnO_2 + 4 OH^-$

III. Multiply the red. half-rxn. by 2: $6 e^- + 4 H_2O + 2 MnO_4^- \rightarrow 2 MnO_2 + 8 OH^-$

IV. Combine the half-reactions: $6 OH^- + Br^- + 6 e^- + 4 H_2O + 2 MnO_4^- \rightarrow$
$$BrO_3^- + 3 H_2O + 6 e^- + 2 MnO_2 + 8 OH^-$$

Cancel duplicate species: $Br^- + H_2O + 2 MnO_4^- \rightarrow BrO_3^- + 2 MnO_2 + 2 OH^-$

d) $NO_2 \rightarrow NO_3^- + NO$ (acidic)
I and II. Oxidation

Step 1: $NO_2 \rightarrow NO_3^-$
Step 2: $H_2O + NO_2 \rightarrow NO_3^-$
Step 3: $H_2O + NO_2 \rightarrow NO_3^- + 2 H^+_{(aq)}$
Step 4: $H_2O + NO_2 \rightarrow NO_3^- + 2 H^+_{(aq)} + e^-$

    Reduction:

Step 1: $NO_2 \rightarrow NO$
Step 2: $NO_2 \rightarrow NO + H_2O$
Step 3: $2 H^+_{(aq)} + NO_2 \rightarrow NO + H_2O$
Step 4: $2 e^- + 2 H^+_{(aq)} + NO_2 \rightarrow NO + H_2O$

III. Multiply the ox. half-rxn. by 2: $2 H_2O + 2 NO_2 \rightarrow 2 NO_3^- + 4 H^+_{(aq)} + 2 e^-$

IV. Combine the half-reactions: $2 H_2O + 2 NO_2 + 2 e^- + 2 H^+_{(aq)} + NO_2 \rightarrow$
$$2 NO_3^- + 4 H^+_{(aq)} + 2 e^- + NO + H_2O$$

Cancel duplicate species: $H_2O + 3 NO_2 \rightarrow 2 NO_3^- + 2 H^+_{(aq)} + NO$

e) $ClO_4^- + Cl^- \rightarrow ClO^- + Cl_2$ _____ (acidic)

I and II. Oxidation: $Cl^- \rightarrow Cl_2$
Step 1: $2 Cl^- \rightarrow Cl_2$
Steps 2, 3, 4: $2 Cl^- \rightarrow Cl_2 + 2 e^-$
    Reduction:
Step 1: $ClO_4^- \rightarrow ClO^-$
Step 2: $ClO_4^- \rightarrow ClO^- + 3 H_2O$
Step 3: $6 H^+_{(aq)} + ClO_4^- \rightarrow ClO^- + 3 H_2O$
Step 4: $6 e^- + 6 H^+_{(aq)} + ClO_4^- \rightarrow ClO^- + 3 H_2O$
III. Multiply the oxidation half-reaction by 3: $6 Cl^- \rightarrow 3 Cl_2 + 6 e^-$
IV. Combine the half-reactions:
$$6 Cl^- + 6 e^- + 6 H^+_{(aq)} + ClO_4^- \rightarrow 3 Cl_2 + ClO^- + 3 H_2O + 6 e^-$$

(continued)

173

(17.13 continued)

        Cancel duplicate series:

$$6\ Cl^- + 6\ H^+_{(aq)} + ClO_4^- \rightarrow 3\ Cl_2 + 6\ e^- + ClO^- + 3\ H_2O$$

        Note: The oxidation and reduction half-reactions chosen could also be:

           Oxidation: $Cl^- \rightarrow ClO^-$

           Reduction: $ClO_4^- \rightarrow Cl_2$

        The balanced reaction would then be:

$$7\ Cl^- + 2\ H^+_{(aq)} + 2\ ClO_4^- \rightarrow 7\ ClO^- + Cl_2 + H_2O$$

        f) $AlH_4^- + H_2CO \rightarrow Al^{3+} + CH_3OH$ (basic)

        I and II. Oxidation

        Steps 1 & 2: $AlH_4^- \rightarrow Al^{3+}$

        Step 3: $AlH_4^- \rightarrow Al^{3+} + \mathbf{4\ H^+_{(aq)}}$

        Step 3 a: $AlH_4^- \rightarrow Al^{3+} + 4\ H^+_{(aq)}$

             $4\ H^+_{(aq)} + 4\ OH^- \rightarrow 4\ H_2O$

             $\mathbf{4\ OH^-} + AlH_4^- \rightarrow Al^{3+} + \mathbf{4\ H_2O}$

        Step 4: $4\ OH^- + AlH_4^- \rightarrow Al^{3+} + 4\ H_2O + \mathbf{8\ e^-}$

           Reduction:

        Steps 1 & 2: $H_2CO \rightarrow CH_3OH$

        Step 3: $\mathbf{2\ H^+_{(aq)}} + H_2CO \rightarrow CH_3OH$

        Step 3 a: $2\ H^+_{(aq)} + H_2CO \rightarrow CH_3OH$

             $2\ H_2O \rightarrow 2\ H^+_{(aq)} + 2\ OH^-$

             $\mathbf{2\ H_2O} + H_2CO \rightarrow CH_3OH + \mathbf{2\ OH^-}$

        Step 4: $\mathbf{2\ e^-} + 2\ H_2O + H_2CO \rightarrow CH_3OH + 2\ OH^-$

        III. Multiply the red. half-rxn. by 4: $8\ e^- + 8\ H_2O + 4\ H_2CO \rightarrow 4\ CH_3OH + 8\ OH^-$

        IV. Combine the half-reactions: $4\ OH^- + AlH_4^- + 8\ e^- + 8\ H_2O + 4\ H_2CO \rightarrow$
                                             $Al^{3+} + 4\ H_2O + 8\ e^- + 4\ CH_3OH + 8\ OH^-$

        Cancel duplicate species: $AlH_4^- + 4\ H_2O + 4\ H_2CO \rightarrow Al^{3+} + 4\ CH_3OH + 4\ OH^-$

17.15 a) $O_2 + Cu \rightarrow 2\ CuO$

      $\Delta G^\circ_{rxn} = 2\ mol\left(\Delta G^\circ_{f\ CuO}\right) = 2\ mol\ (-129.7\ kJ/mol) = -259.4\ kJ$; spontaneous

      b) $O_2 + 2\ Hg \rightarrow 2\ HgO$

      $\Delta G^\circ_{rxn} = 2\ mol\left(\Delta G^\circ_{f\ HgO}\right) = 2\ mol\ (-58.539\ kJ/mol) = -117.078\ kJ$; spontaneous

      c) $CuS + O_2 \rightarrow Cu + SO_2$ $\Delta G^\circ_{rxn} = 1\ mol\left(\Delta G^\circ_{f\ SO_2}\right) - 1\ mol\left(\Delta G^\circ_{f\ CuS}\right)$

            $= 1\ mol(-300.2\ kJ/mol) - 1\ mol(-53.6\ kJ/mol) = -246.6\ kJ$; spontaneous

(continued)

174

(17.15 continued)

d) $FeS + O_2 \rightarrow Fe + SO_2$ $\qquad \Delta G^\circ_{rxn} = 1\,mol\left(\Delta G^\circ_{f\,SO_2}\right) - 1\,mol\left(\Delta G^\circ_{f\,FeS}\right)$

$= 1\,mol(-300.2\,kJ/mol) - 1\,mol(-100.4\,kJ/mol) = -199.8\,kJ$; spontaneous

Note: value for $\Delta G^\circ_{f\,FeS}$ is from CRC HANDBOOK OF CHEMISTRY AND PHYSICS.

17.17 Your sketch should illustrate an electrode similar to the electrode shown on the left in the answer given at the end of the chapter for Exercise 17.3.1. Your sketch needs to show solid AgCl and solid Ag in contact with each other, electrons flowing into the solid AgCl from an inert electrode, and AgCl becoming solid Ag and aqueous chloride ions. The aqueous chloride ions move away into the aqueous medium surrounding the electrode and solids.

17.19 Draw a sketch that uses two Pt electrodes like the one shown on the right in Figure 17-7 or on the left in the drawing in Figure 17-13. One side would need to be the $H^+_{(aq)}/H_2{_{(g)}}$ shown in Figure 17-13. The other compartment would need to be $Cl_2{_{(g)}}/Cl^-_{(aq)}$ in place of $H^+_{(aq)}/H_2{_{(g)}}$. A porous plate will need to separate the two compartments.

17.21 $Pb + PbO_2 + 2\,HSO_4^- + 2\,H^+ \rightarrow 2\,PbSO_4 + 2\,H_2O$

$$15\,s \times \frac{5.9\,C}{s} \times \frac{1\,mol\,e^-}{96,485\,C} \times \frac{1\,mole\,Pb}{2\,mol\,e^-} \times \frac{207.2\,g\,Pb}{mole\,Pb} = 0.095\,g\,Pb$$

$$15\,s \times \frac{5.9\,C}{s} \times \frac{1\,mol\,e^-}{96,485\,C} \times \frac{1\,mole\,PbO_2}{2\,mol\,e^-} \times \frac{239.2\,g\,PbO_2}{mole\,PbO_2} = 0.11\,g\,PbO_2$$

17.23 $1.750\,amp - 1.350\,amp = 0.400\,amp$

$$0.850\,g\,PbSO_4 \times \frac{mol\,PbSO_4}{303.3\,g\,PbSO_4} \times \frac{2\,mol\,e^-}{mol\,PbSO_4} \times \frac{96,485\,C}{mol\,e^-}$$

$$\times \frac{s}{0.400\,C} \times \frac{min}{60\,s} = 22.5\,min$$

17.25 a) $\qquad Cr_2O_7^{2-} + C_2H_4O \rightarrow CH_3CO_2H + Cr^{3+}$ (acidic)

I and II. Oxidation:

Step 1: $\qquad C_2H_4O \rightarrow CH_3CO_2H$

Step 2: $\qquad \mathbf{H_2O} + C_2H_4O \rightarrow CH_3CO_2H$

Step 3: $\qquad H_2O + C_2H_4O \rightarrow CH_3CO_2H + \mathbf{2\,H^+}_{(aq)}$

Step 4: $\qquad H_2O + C_2H_4O \rightarrow CH_3CO_2H + 2\,H^+_{(aq)} + \mathbf{2\,e^-}$

(continued)

175

(17.25 continued)

Reduction: $Cr_2O_7^{2-} \rightarrow Cr^{3+}$

Step 1: $\quad Cr_2O_7^{2-} \rightarrow 2\ Cr^{3+}$

Step 2: $\quad Cr_2O_7^{2-} \rightarrow 2Cr^{3+} + \textbf{7 H}_2\textbf{O}$

Step 3: $\quad \textbf{14 H}^+_{(aq)} + Cr_2O_7^{2-} \rightarrow 2\ Cr^{3+} + 7\ H_2O$

Step 4: $\quad \textbf{6 e}^- + 14\ H^+_{(aq)} + Cr_2O_7^{2-} \rightarrow 2\ Cr^{3+} + 7\ H_2O$

III. Multiply the oxidation half-reaction by 3:

$$3\ H_2O + 3\ C_2H_4O \rightarrow 3\ CH_3CO_2H + 6\ H^+_{(aq)} + 6\ e^-$$

IV. Combine the half-reactions: $3\ H_2O + 3\ C_2H_4O + 6\ e^- + 14\ H^+_{(aq)} +$

$$Cr_2O_7^{2-} \rightarrow 3\ CH_3CO_2H + 6\ H^+_{(aq)} + 6\ e^- + 2\ Cr^{3+} + 7\ H_2O$$

Cancel duplicate species:

$$3\ C_2H_4O + 8\ H^+_{(aq)} + Cr_2O_7^{2-} \rightarrow 3\ CH_3CO_2H + 2\ Cr^{3+} + 4\ H_2O$$

b) $1.00\ g\ C_2H_4O \left(\dfrac{1\ mol\ C_2H_4O}{44.0\ g\ C_2H_4O} \right) \left(\dfrac{2\ mol\ e^-}{1\ mol\ C_2H_4O} \right) = 0.0455\ mol\ e^-$

17.27 a) $Co(NH_3)_6^{3+} + e^- \rightarrow Co(NH_3)_6^{2+}$ $\qquad E° = 0.108\ V$

$PbO_2 + H_2O + 2\ e^- \rightarrow PbO + 2\ OH^-$ $\quad E° = 0.247\ V$ (from CRC)
$\qquad E° = 0.108\ V - 0.247\ V = -0.139\ V$

b) $O_2 + 4\ H^+_{(aq)} + 4\ e^- \rightarrow 2\ H_2O$ $\qquad E° = 1.229\ V$

$HAsO_2 + 3\ H^+_{(aq)} + 3\ e^- \rightarrow As + 2\ H_2O$ $\qquad E° = 0.248\ V$
$\qquad E° = 1.229\ V - 0.248\ V = 0.981\ V$

c) $BrO_3^- + 3\ H_2O + 6\ e^- \rightarrow Br^- + 6\ OH^-$ $\qquad E° = 0.61\ V$ (from CRC)

$MnO_4^- + 2\ H_2O + 3\ e^- \rightarrow MnO_2 + 4\ OH^-$ $\qquad E° = 0.595\ V$ (from CRC)
$\qquad E° = 0.595\ V - 0.61\ V = -0.02\ V$

17.29 Draw a sketch that includes two Pt electrodes, compartments, etc., like the compartment on the right in Figure 17-7 or the left in Figure 17-13. One side would need to be the $H^+_{(aq)}/H_{2(g)}$ as shown in Figures 17-7 and 17-13. The other compartment would need to be $F_{2(g)}/F^-_{(aq)}$ in place of $H^+_{(aq)}/H_{2(g)}$. An external circuit and a porous plate to connect the compartments are needed.

$F_{2\ (g)} + 2\ e^- \rightarrow 2\ F^-_{(aq)}$ $\qquad\qquad E°_{F_2/F^-} = ?$

$2\ H^+_{(aq)} + 2\ e^- \rightarrow H_{2(g)}$ $\qquad\qquad E°_{H^+/H_2} = 0.000\ V$

$E°_{cell} = E°_{F_2/F^-} - E°_{H^+/H_2}$ $\qquad E°_{cell} = E°_{F_2/F^-}$

The anode would be the standard hydrogen electrode.

17.31 $E°_{cell} = E°_{cathode} - E°_{anode}$

anode: $Ru \rightarrow Ru^{3+} + 3\ e^-$

cathode: $PbSO_4 + 2\ e^- \rightarrow Pb + SO_4^{2-}$ $\qquad E° = -0.3588\ V$

(continued)

(17.31 continued)
$$E^{\circ}_{cell} = 0.745 \text{ V}$$

$$E^{\circ}_{cell} = E^{\circ}_{cathode} - E^{\circ}_{anode}$$

$$E^{\circ}_{anode} = E^{\circ}_{cathode} - E^{\circ}_{cell} = -0.3588 \text{ V} - 0.745 \text{ V} = -1.104 \text{ V}$$

If the Pb electrode were the anode instead of the cathode

$$E^{\circ}_{cell} = E^{\circ}_{cathode} - E^{\circ}_{anode}$$

$E^{\circ}_{cathode} = 0.745 \text{ V} + (-0.3588 \text{ V}) = 0.386 \text{ V}$ (more logical value as reduction potentials of different ions of Ru are all positive)

17.33 The standard potentials for a - c were determined in Problem 17.27.

a) $\Delta G^{\circ} = -nFE^{\circ} = -(2 \text{ mol})(9.6485 \times 10^4 \text{ C/mol})(-0.139 \text{ V}) = +2.68 \times 10^4 \text{ J}$
    or $+26.8$ kJ

b) $\Delta G^{\circ} = -nFE^{\circ} = -(12 \text{ mol})(9.6485 \times 10^4 \text{ C/mol})(0.981 \text{ V}) = -1.14 \times 10^6 \text{ J}$
    or -1.14×10^3 kJ

c) $\Delta G^{\circ} = -nFE^{\circ} = -(6 \text{ mol})(9.6485 \times 10^4 \text{ C/mol})(-0.02 \text{ V}) = 1 \times 10^4$ J or 10 kJ

d) $NO_2 + 2 H^+_{(aq)} + 2 e^- \rightarrow NO + H_2O$ $E^{\circ} = 1.03$ V

$NO_3^- + 2 H^+_{(aq)} + e^- \rightarrow NO_2 + H_2O$ $E^{\circ} = 0.81$ V (calculated from CRC)
     $E^{\circ} = 1.03 \text{ V} - (0.81 \text{ V}) = 0.22 \text{ V}$

$\Delta G^{\circ} = -nFE^{\circ} = -(2 \text{ mol})(9.6485 \times 10^4 \text{ C/mol})(0.22 \text{ V}) = -4.2 \times 10^4$ J or -42 kJ

e) $Cl_2 + 2 e^- \rightarrow 2 Cl^-$ $E^{\circ} = 1.35827$ V

$ClO_4^- + 6 H^+_{(aq)} + 6 e^- \rightarrow ClO^- + 3 H_2O$ $E^{\circ} = 1.349$ V (calculated from other

$E = 1.349 \text{ V} - (1.358 \text{ V}) = -0.009 \text{ V}$ E° in App. H)

$\Delta G^{\circ} = -nFE^{\circ} = -(6 \text{ mol})(9.6485 \times 10^4 \text{ C/mol})(-0.009 \text{ V}) = +5 \times 10^3$ J or 5 kJ

17.35 $Ca^{2+} + 2 e^- \rightarrow Ca$ $E^{\circ} = -2.868$ V

$Ca(OH)_2 + 2 e^- \rightarrow Ca + 2 OH^-$ $E^{\circ} = -3.02$ V

$Ca(OH)_2 + 2 e^- \rightarrow Ca + 2 OH^-$

         $Ca \rightarrow Ca^{2+} + 2 e^-$

       $Ca(OH)_2 \rightarrow Ca^{2+} + 2 OH^-$ $K_{sp} = [Ca^{2+}][OH^-]^2$

$E^{\circ} = -3.02 \text{ V} - (-2.868 \text{ V}) = -0.15 \text{ V}$

$\Delta G^{\circ} = -nFE^{\circ} = -RT \ln K$ $\ln K = nFE^{\circ}/RT$

$$\log K = \frac{nE^{\circ}}{0.05916 \text{ V}} = \frac{2(-0.15 \text{ V})}{0.05916 \text{ V}} = -5.07 \quad K = K_{sp} = 8.5 \times 10^{-6}$$

Table 17-1 contains sufficient data only for the calculation for $Ca(OH)_2$. Other possible compounds are either missing data or are not metal hydroxides.

17.37 $E° = 1.35$ V $\quad 2NiO(OH)_{(s)} + 2 H_2O_{(l)} + Cd_{(s)} \rightarrow 2 Ni(OH)_{2(s)} + Cd(OH)_{2(s)}$

$$E = E° - \frac{RT}{nF} \log Q \qquad Q = 1 \qquad E = 1.35 \text{ V}$$

The concentration of OH^- does not appear in Q. If one looks at the individual half-cells and their potentials, one finds that the concentration of OH^- affects each half-cell equally. Therefore, upon subtraction of the half-cell potentials, any changes due to changes in the concentration of OH^- will cancel.

17.39 Net reaction for the standard dry cell:

$$Zn_{(s)} + 2 NH_4{}^+{}_{(aq)} \rightleftharpoons Zn^{2+}{}_{(aq)} + 2 NH_{3(g)} + H_{2(g)}$$

$$E = E° - \frac{0.05916 \text{ V}}{2} \log \frac{[Zn^{2+}]\left(p_{NH_3}\right)^2 \left(p_{H_2}\right)}{[NH_4{}^+]^2}$$

As the reaction proceeds, the numerator (products) of the log term increases and the denominator (reactants) decreases, the log value increases, and E decreases as the cell is used.

17.41 $Cr^{2+} + 2 e^- \rightarrow Cr \qquad\qquad E° = -0.913$ V

$Fe^{2+} + 2 e^- \rightarrow Fe \qquad\qquad E° = -0.447$ V

Chromium will corrode because the potential for the oxidation of Cr is more positive than the potential for the oxidation of Fe.

17.43 The presence of $NaHCO_3$ makes the water (solution) slightly basic.

$Ag_2S + 2 e^- \rightarrow 2 Ag + S^{2-} \qquad$ reduction $\qquad E° = -0.691$ V

$3 Ag_2S + 6 e^- \rightarrow 6 Ag + 3 S^{2-}$

$Al + 4 OH^- \rightarrow Al(OH)_4{}^- + 3 e^- \qquad$ oxidation $\qquad E° = +2.33$ V

$2 Al + 8 OH^- \rightarrow 2 Al(OH)_4{}^- + 6 e^-$

$3 Ag_2S + 2 Al + 8 OH^- \rightarrow 6 Ag + 2 Al(OH)_4{}^- + 3 S^{2-} \qquad$ overall

$$E° = +1.64 \text{ V}$$

17.45 $\underline{2 Cl^-} \rightarrow \underline{|Cl_2} + \underline{2 e^-}$

$$(200 \text{ min})\left(\frac{60 \text{ s}}{\text{min}}\right)\left(\frac{4.50 \text{ C}}{\text{s}}\right)\left(\frac{\text{mol e}^-}{9.6485 \times 10^4 \text{ C}}\right)\left(\frac{1 \text{ mol Cl}_2}{2 \text{ mol e}^-}\right)\left(\frac{70.9 \text{ g Cl}_2}{\text{mol Cl}_2}\right)$$

$$= 19.8 \text{ g Cl}_2$$

17.47 $Cu^{2+} + 2 e^- \rightarrow Cu$
$\qquad\qquad E° = 0.3419$ V

$$(0.250 \text{ L})\left(\frac{0.245 \text{ mol CuSO}_4}{\text{L}}\right)\left(\frac{2 \text{ mol e}^-}{\text{mol CuSO}_4}\right)\left(\frac{9.6485 \times 10^4 \text{ C}}{\text{mol e}^-}\right)\left(\frac{\text{s}}{2.45 \text{ C}}\right)\left(\frac{\text{min}}{60 \text{ s}}\right)$$

$$= 80.4 \text{ min}$$

17.49 $Zn^{2+} + 2 e^- \rightarrow Zn$

$$(7.55 \text{ g Zn})\left(\frac{1 \text{ mol Zn}}{65.39 \text{ g Zn}}\right)\left(\frac{2 \text{ mol e}^-}{1 \text{ mol Zn}}\right)\left(\frac{9.6485 \times 10^4 \text{ C}}{\text{mol e}^-}\right) = 2.23 \times 10^4 \text{ C}$$

17.51 a) $2 \text{ CuFeS}_{2(s)} + 3 \text{ O}_{2(g)} \xrightarrow{\Delta} 2 \text{ FeO}_{(s)} + 2 \text{ CuS}_{(s)} + 2 \text{ SO}_{2(g)}$
  or $2 \text{ CuFeS}_{2(s)} + 4 \text{ O}_{2(g)} \xrightarrow{\Delta} 2 \text{ FeO}_{(s)} + \text{Cu}_2\text{S}_{(s)} + 3 \text{ SO}_{2(g)}$

  b) $\text{Al(O)OH}_{(s)} + \text{NaOH}_{(aq)} + \text{H}_2\text{O}_{(l)} \rightarrow \text{Na[Al(OH)}_4]_{(aq)}$

  c) $\text{Si}_{(g)} + \text{O}_{2(g)} \rightarrow \text{SiO}_{2(s)} + \text{CaO}_{(s)} \rightarrow \text{CaSiO}_{3(l)}$

  d) $\text{TiCl}_{4(l)} + 4 \text{ Na}_{(s)} \rightarrow \text{Ti}_{(s)} + 4 \text{ NaCl}_{(s)}$

17.53 $$(5.60 \times 10^4 \text{ kg ore})\left(\frac{10^3 \text{ g}}{\text{kg}}\right)\left(\frac{2.37 \text{ g Cu}_2\text{S}}{100 \text{ g ore}}\right)\left(\frac{\text{mol Cu}_2\text{S}}{159.2 \text{ g Cu}_2\text{S}}\right)\left(\frac{2 \text{ mol Cu}}{\text{mol Cu}_2\text{S}}\right)$$

$$\left(\frac{63.55 \text{ g Cu}}{\text{mol Cu}}\right) = 1.06 \times 10^6 \text{ g Cu (or } 1.06 \times 10^3 \text{ kg Cu)}$$

$$(5.60 \times 10^4 \text{ kg ore})\left(\frac{10^3 \text{ g}}{\text{kg}}\right)\left(\frac{2.37 \text{ g Cu}_2\text{S}}{100 \text{ g ore}}\right)\left(\frac{\text{mol Cu}_2\text{S}}{159.2 \text{ g Cu}_2\text{S}}\right)\left(\frac{1 \text{ mol SO}_2}{1 \text{ mol Cu}_2\text{S}}\right)$$

$$= 8.34 \times 10^3 \text{ mol SO}_2$$

$PV = nRT$ $\qquad V_{SO_2} = \dfrac{nRT}{P} = \dfrac{(8.34 \times 10^3 \text{ mol})\left(0.0821\dfrac{\text{L atm}}{\text{K mol}}\right)(296.6 \text{ K})}{(755 \text{ torr} / 760 \text{ torr atm}^{-1})}$

$\qquad\qquad\qquad V_{SO_2} = 2.04 \times 10^5 \text{ L}$

17.55 $\text{ZnO}_{(s)} + \text{C}_{(s)} \rightarrow \text{Zn}_{(s)} + \text{CO}_{(g)}$

$\Delta G^\circ_{rxn} = 1 \text{ mol}\left(\Delta G^\circ_{f_{CO}}\right) - 1 \text{ mol}\left(\Delta G^\circ_{f_{ZnO}}\right)$
$= 1 \text{ mol}(-137.168 \text{ kJ/mol}) - 1 \text{ mol}(-318.30 \text{ kJ/mol}) = 181.13 \text{ kJ}$

$\text{ZnO}_{(s)} + \text{CO}_{(g)} \rightarrow \text{Zn}_{(s)} + \text{CO}_{2(g)}$

$\Delta G^\circ_{rxn} = 1 \text{ mol}\left(\Delta G^\circ_{f_{CO_2}}\right) - \left[1 \text{ mol}\left(\Delta G^\circ_{f_{CO}}\right) + 1 \text{ mol}\left(\Delta G^\circ_{f_{ZnO}}\right)\right]$

$\Delta G^\circ_{rxn} = 1 \text{ mol}(-394.359 \text{ kJ} / \text{mol})$

$-[1 \text{ mol}(-137.168 \text{ kJ} / \text{mol}) + 1 \text{ mol}(-318.30 \text{ kJ} / \text{mol})] = 61.11 \text{ kJ}$

17.57 $SO_4^{2-}{}_{(aq)} + 4 H^+{}_{(aq)} + 2 e^- \rightarrow SO_{2(g)} + 2 H_2O$

$H_2SO_{3(aq)} + H_2O \rightarrow 2 e^- + 4 H^+{}_{(aq)} + SO_4^{2-}{}_{(aq)}$

$H_2SO_{3(aq)} \rightarrow SO_{2(g)} + H_2O$

$E° = 0.20\ V - 0.17\ V = 0.03\ V$ $\qquad \Delta G° = -nFE° = -RT \ln K$

$\ln K = \dfrac{nFE°}{RT} = \dfrac{2(9.6485 \times 10^4\ C\,/\,mol)(0.03\ V)}{(8.314\ J/K \cdot mol)(298\ K)} = 2.3$ $\qquad C = J/V$

$K = 10 \qquad\qquad K = \dfrac{p_{SO_2}}{[H_2SO_3]} \qquad\qquad [H_2SO_3] = 1.00\ M \qquad p_{SO_2} = 10\ atm$

17.59 $I_2 + 2 e^- \rightleftarrows 2 I^-$ $\qquad\qquad E° = 0.5355\ V$

$2 IO_3^- + 12 H^+{}_{(aq)} + 10 e^- \rightleftarrows I_2 + 6 H_2O \qquad E° = 1.195\ V$

$10 I^- + 2 IO_3^- + 12 H^+{}_{(aq)} \rightleftarrows 6 I_2 + 6 H_2O$

Divide through by 2:

$5 I^-{}_{(aq)} + IO_3^-{}_{(aq)} + 6 H^+{}_{(aq)} \rightleftarrows 3 I_{2(s)} + 3 H_2O_{(l)}$

$E° = 1.195\ V - 0.5355\ V = 0.660\ V$

a) pH = 2.00 $\qquad$ $[H^+] = 0.010\ M$

$E = E° - \dfrac{0.05916\ V}{n} \log Q = E° - \dfrac{0.05916\ V}{n} \log \dfrac{1}{[I^-]^5[IO_3^-][H^+]^6}$

$\qquad = 0.660\ V - \dfrac{0.05916\ V}{5} \log \dfrac{1}{(0.100)^5(0.100)(0.010)^6}$

$\qquad = 0.660\ V - \dfrac{0.05916\ V}{5} \log 1.0 \times 10^{18}$

$\qquad = 0.660\ V - 0.213\ V = 0.45\ V$ (spontaneous as written)

b) pH = 11.00 $\quad$ $[H^+] = 1.0 \times 10^{-11}$ $\qquad$ Using $E° = 0.660\ V$ (acidic solution)

$E = 0.660\ V - \dfrac{0.05916\ V}{5} \log \dfrac{1}{(0.100)^5(0.100)(1.0 \times 10^{-11})^6}$

$E = 0.660\ V - \dfrac{0.05916\ V}{5} \log 1.0 \times 10^{72}$

$E = 0.660\ V - 85\ V = -0.19\ V$ (spontaneous in the reverse direction from written)

c) At equilibrium, E = 0 $\quad$ $E° = \dfrac{0.05916\ V}{n} \log K$

$\log K = \dfrac{nE°}{0.05916\ V} = \dfrac{5(0.660\ V)}{0.05916\ V} = 55.8$

$K = 6 \times 10^{55}$ $\qquad\qquad K = 1/[I^-]^5[IO_3^-][H^+]^6$

$[H^+] = \left(\dfrac{1}{[I^-]^5[IO_3^-]K} \right)^{1/6}$ $\qquad [H^+] = \left(\dfrac{1}{(0.100)^5(0.100)(6 \times 10^{55})} \right)^{1/6}$

$[H^+] = 5 \times 10^{-9}$ $\qquad\qquad$ pH = 8.3

17.61 $AgCl + e^- \rightarrow Ag + Cl^-$ $E° = 0.22233$ V

$E_{cathode} = E°_{cathode} - (0.05916$ V/1$)\log[Cl^-] = 0.22233$ V $- 0.05916$ V $\log(0.500)$
   $= 0.22233$ V $+ 0.0178$ V $= 0.240$ V

$Mg^{2+} + 2 e^- \rightarrow Mg$ $E° = -2.372$ V
net reaction: $2 AgCl + Mg \rightarrow 2 Ag + Mg^{2+} + 2 Cl^-$

b) $E_{cell} = E_{cathode} - E_{anode} = 0.240$ V $- (-2.372$ V$) = 2.612$ V

or $E_{cell} = 0.22233$ V $- (-2.372$ V$) - \dfrac{0.05916}{2}\log[Mg^{2+}][Cl^-]^2 = 2.612$ V

c) Your molecular picture should depict an electrochemical cell in which one vessel is like the left compartment in the drawing in Figure 17-5 except that the zinc is replaced with magnesium (both the atoms and the ions) and the SO_4^- is replaced with Cl^-. The other vessel needs to be a drawing of an electrode like the iron electrode shown in the answer given on page 881 for Section Exercise 17.3.1. The Fe would need to be replaced with Ag and the Fe_2O_3 replaced by AgCl. In this vessel the anion would need to be the chloride ion just as it is in the other vessel. The two vessels need to be connected by a porous separator. The chloride ions will flow from the vessel containing the silver electrode to the vessel containing the magnesium electrode.

17.63 $Fe^{2+} + 2 e^- \rightarrow Fe$ $E° = -0.447$ V
$O_2 + 4 H^+_{(aq)} + 4 e^- \rightarrow 2 H_2O$ $E° = 1.229$ V
$O_2 + 2 H_2O + 4 e^- \rightarrow 4 OH^-$ $E° = 0.401$ V

pH = 10 $2 Fe + O_2 + 2 H_2O \rightarrow 2 Fe^{2+} + 4 OH^-$
      $E° = 0.401$ V $- (-0.447$ V$) = 0.848$ V

pH = 3 $2 Fe + O_2 + 4 H^+_{(aq)} \rightarrow 2 Fe^{2+} + 2 H_2O$
      $E° = 1.229$ V $- (0.447$ V$) = 1.676$ V

Iron is more likely to rust at pH = 3. Similar results will be obtained if one looks at Fe^{3+} instead of Fe^{2+} or if one calculates adjustments in the values for nonstandard concentrations of H^+ or OH^-.

17.65 From Table 17-1
$F_2 + 2 e^- \rightarrow 2 F^-$ $E° = 2.866$ V   cathode

$Mg^{2+} + 2 e^- \rightarrow Mg$ $E° = -2.372$ V   anode
      $E°_{cell} = 2.866$ V $- (-2.372$ V$) = 5.238$ V

$Na^+ + e^- \rightarrow Na$ $E° = -2.714$ V   anode
      $E°_{cell} = 2.866$ V $- (-2.714$ V$) = 5.580$ V

(continued)

181

(17.65 continued)

$$Ca^{2+} + 2 e^- \rightarrow Ca \qquad\qquad E° = -2.868 \text{ V} \qquad \text{anode}$$

$$E°_{cell} = 2.866 \text{ V} - (-2.868 \text{ V}) = 5.734 \text{ V}$$

$$Ca(OH)_2 + 2 e^- \rightarrow Ca + 2 \text{ OH}^- \qquad E° = -3.02 \text{ V} \qquad \text{anode}$$

$$E°_{cell} = 2.866 \text{ V} - (-3.02 \text{ V}) = 5.89 \text{ V}$$

$$Li^+ + e^- \rightarrow Li \qquad\qquad E° = -3.0401 \text{ V} \qquad \text{anode}$$

$$E°_{cell} = 2.866 \text{ V} - (-3.0401 \text{ V}) = 5.906 \text{ V}$$

The reaction rate and battery life must be considered as well as $E°$. It would be difficult to contain F_2 and very dangerous if the container should fail and free F_2.

17.67 $\quad E° = \dfrac{0.05916 \text{ V}}{n} \log K = \dfrac{0.05916 \text{ V}}{2} \log(2.69 \times 10^{12}) = 0.368 \text{ V}$

17.69 The better oxidizing agent will have the more positive (or less negative) potential for its reduction reaction.
a) MnO_4^- is a better oxidizing agent than $Cr_2O_7^{2-}$ (in acidic solution).
b) H_2O_2 is a better oxidizing agent than O_2 (in basic solution).
c) Sn^{2+} is a better oxidizing agent than Fe^{2+}.

17.71 $\quad H_2O_{2(aq)} + 2 H^+_{(aq)} + 2 e^- \rightarrow 2 H_2O_{(l)} \qquad E° = 1.776 \text{ V (from CRC)}$
$\qquad O_{2(g)} + 2 H^+_{(aq)} + 2 e^- \rightarrow H_2O_{2(aq)} \qquad E° = +0.695 \text{ V}$

$\qquad 2 H_2O_{2(aq)} \rightarrow 2 H_2O_{(l)} + O_{2(g)} \qquad E° = 1.776 \text{ V} - 0.695 \text{ V} = 1.081 \text{ V}$

$$\ln K_{eq} = \frac{nFE°}{RT} = \frac{nE°}{0.05916 \text{ V}} = \frac{(2)(1.081 \text{ V})}{(0.05916 \text{ V})} = 36.54 \qquad K_{eq} = 7.4 \times 10^{15} \text{ atm M}^{-2}$$

Solutions of hydrogen peroxide are unstable. The equilibrium lies far to the right. As the hydrogen peroxide decomposes, the oxygen gas is given off and the reaction continues to completion.

17.73 $\quad MnO_4^- + H_2SO_3 \rightarrow Mn^{2+} + HSO_4^- \qquad$ (acidic)
$\qquad\qquad$ Oxidation:

$H_2SO_3 \rightarrow HSO_4^-$

$\textbf{H}_2\textbf{O} + H_2SO_3 \rightarrow HSO_4^-$

$H_2O + H_2SO_3 \rightarrow HSO_4^- + \textbf{3 H}^+_{\textbf{(aq)}}$

$H_2O + H_2SO_3 \rightarrow HSO_4^- + 3 H^+_{(aq)} + \textbf{2 e}^-$
$\qquad\qquad$ Reduction:

$MnO_4^- \rightarrow Mn^{2+}$

$MnO_4^- \rightarrow Mn^{2+} + \textbf{4 H}_2\textbf{O}$

$\textbf{8 H}^+_{\textbf{(aq)}} + MnO_4^- \rightarrow Mn^{2+} + 4 H_2O$

$\textbf{5 e}^- + \textbf{8 H}^+_{\textbf{(aq)}} + MnO_4^- \rightarrow Mn^{2+} + 4 H_2O$

(continued)

(17.73 continued)

Multiply the oxid. half-reaction by 5 and the red. half-reaction by 2:

$5 H_2O + 5 H_2SO_3 \rightarrow 5 HSO_4^- + 15 H^+_{(aq)} + 10 e^-$

$10 e^- + 16 H^+_{(aq)} + 2 MnO_4^- \rightarrow 2 Mn^{2+} + 8 H_2O$

Combine the half-reactions:

$5 H_2O + 5 H_2SO_3 + 10 e^- + 16 H^+_{(aq)} + 2 MnO_4^- \rightarrow$
$$5 HSO_4^- + 15 H^+_{(aq)} + 10 e^- + 2 Mn^{2+} + 8 H_2O$$

Cancel duplicate species:

$5 H_2SO_3 + H^+_{(aq)} + 2 MnO_4^- \rightarrow 5 HSO_4^- + 2 Mn^{2+} + 3 H_2O$

$MnO_4^- + SO_2 \rightarrow Mn^{2+} + HSO_4^-$ (acidic)

Oxidation:

$SO_2 \rightarrow HSO_4^-$

$\mathbf{2 H_2O} + SO_2 \rightarrow HSO_4^-$

$2 H_2O + SO_2 \rightarrow HSO_4^- + \mathbf{3 H^+_{(aq)}}$

$2 H_2O + SO_2 \rightarrow HSO_4^- + 3 H^+_{(aq)} + \mathbf{2 e^-}$

Reduction (same as part a):

$5 e^- + 8 H^+_{(aq)} + MnO_4^- \rightarrow Mn^{2+} + 4 H_2O$

Multiply the oxid. half-reaction by 5 and the red. half-reaction by 2:

$10 H_2O + 5 SO_2 \rightarrow 5 HSO_4^- + 15 H^+_{(aq)} + 10 e^-$

$10 e^- + 16 H^+_{(aq)} + 2 MnO_4^- \rightarrow 2 Mn^{2+} + 8 H_2O$

Combine the half-reactions:

$10 H_2O + 5 SO_2 + 10 e^- + 16 H^+_{(aq)} + 2 MnO_4^- \rightarrow$
$$5 HSO_4^- + 15 H^+_{(aq)} + 10 e^- + 2 Mn^{2+} + 8 H_2O$$

Cancel duplicate species:

$2 H_2O + 5 SO_2 + H^+_{(aq)} + 2 MnO_4^- \rightarrow 5 HSO_4^- + 2 Mn^{2+}$

$MnO_4^- + H_2S \rightarrow Mn^{2+} + HSO_4^-$ (acidic)

Oxidation:

$H_2S \rightarrow HSO_4^-$

$\mathbf{4 H_2O} + H_2S \rightarrow HSO_4^-$

$4 H_2O + H_2S \rightarrow HSO_4^- + \mathbf{9 H^+_{(aq)}}$

$4 H_2O + H_2S \rightarrow HSO_4^- + 9 H^+_{(aq)} + \mathbf{8 e^-}$

Reduction (same as part a):

$5 e^- + 8 H^+_{(aq)} + MnO_4^- \rightarrow Mn^{2+} + 4 H_2O$

Multiply the oxid. half-reaction by 5 and the red. half-reaction by 8:

$20 H_2O + 5 H_2S \rightarrow 5 HSO_4^- + 45 H^+_{(aq)} + 40 e^-$

$40 e^- + 64 H^+_{(aq)} + 8 MnO_4^- \rightarrow 8 Mn^{2+} + 32 H_2O$

(continued)

183

(17.73 continued)

Combine the half-reactions:

$$20 \ H_2O + 5 \ H_2S + 40 \ e^- + 64 \ H^+_{(aq)} + 8 \ MnO_4^- \rightarrow$$
$$5 \ HSO_4^- + 45 \ H^+_{(aq)} + 40 \ e^- + 8 \ Mn^{2+} + 32 \ H_2O$$

Cancel duplicate species:

$$5 \ H_2S + 19 \ H^+_{(aq)} + 8 \ MnO_4^- \rightarrow 5 \ HSO_4^- + 8 \ Mn^{2+} + 12 \ H_2O$$

$$MnO_4^- + H_2S_2O_3 \rightarrow Mn^{2+} + HSO_4^- \qquad \text{(acidic)}$$

Oxidation:

$$H_2S_2O_3 \rightarrow HSO_4^-$$
$$H_2S_2O_3 \rightarrow \mathbf{2} \ HSO_4^-$$
$$\mathbf{5 \ H_2O} + H_2S_2O_3 \rightarrow 2 \ HSO_4^-$$
$$5 \ H_2O + H_2S_2O_3 \rightarrow 2 \ HSO_4^- + \mathbf{10 \ H^+_{(aq)}}$$
$$5 \ H_2O + H_2S_2O_3 \rightarrow 2 \ HSO_4^- + 10 \ H^+_{(aq)} + \mathbf{8 \ e^-}$$

Reduction (same as part a):

$$5 \ e^- + 8 \ H^+_{(aq)} + MnO_4^- \rightarrow Mn^{2+} + 4 \ H_2O$$

Multiply the oxid. half-reaction by 5 and the red. half-reaction by 8:

$$25 \ H_2O + 5 \ H_2S_2O_3 \rightarrow 10 \ HSO_4^- + 50 \ H^+_{(aq)} + 40 \ e^-$$
$$40 \ e^- + 64 \ H^+_{(aq)} + 8 \ MnO_4^- \rightarrow 8 \ Mn^{2+} + 32 \ H_2O$$

Combine the half-reactions:

$$25 \ H_2O + 5 \ H_2S_2O_3 + 40 \ e^- + 64 \ H^+_{(aq)} + 8 \ MnO_4^- \rightarrow$$
$$10 \ HSO_4^- + 50 \ H^+_{(aq)} + 40 \ e^- + 8 \ Mn^{2+} + 32 \ H_2O$$

Cancel duplicate species:

$$5 \ H_2S_2O_3 + 14 \ H^+_{(aq)} + 8 \ MnO_4^- \rightarrow 10 \ HSO_4^- + 8 \ Mn^{2+} + 7 \ H_2O$$

17.75 $HgS_{(s)} + O_{2(g)} \xrightarrow{\Delta} Hg_{(l)} + SO_{2(g)}$

$PbS_{(s)} + O_{2(g)} \xrightarrow{\Delta} Pb_{(l)} + SO_{2(g)}$

$ZnO_{(s)} + C_{(s)} \xrightarrow{\Delta} Zn_{(l)} + CO_{(g)}$

$TiCl_{4(l)} + 2 \ Mg_{(s)} \xrightarrow{\Delta} Ti_{(s)} + 2 \ MgCl_{2(g)}$

$MgCl_{2(l)} \xrightarrow{\text{electrolysis}} Mg_{(s)} + Cl_{2(g)}$

$2 \ NaCl_{(l)} \xrightarrow{\text{electrolysis}} 2 \ Na_{(l)} + Cl_{2(g)}$

$Cr_2O_{3(s)} + 2 \ Al_{(s)} \xrightarrow{\Delta} Al_2O_{3(s)} + 2 \ Cr_{(s)}$

17.77 $Cr(OH)_3 + H_2 \rightarrow Cr + H_2O$ (basic solution)

a) Oxidation:

$2\,OH^- + H_2 \rightarrow 2\,H_2O + 2\,e^-$

Reduction:

$3\,e^- + Cr(OH)_3 \rightarrow Cr + 3\,OH^-$

Multiply the oxid. half-reaction by 3 and the red. half-reaction by 2:

$6\,OH^- + 3\,H_2 \rightarrow 6\,H_2O + 6\,e^-$

$6\,e^- + 2\,Cr(OH)_3 \rightarrow 2\,Cr + 6\,OH^-$

Combine the half-reactions:

$6\,OH^- + 3\,H_2 + 6\,e^- + 2\,Cr(OH)_3 \rightarrow 6\,H_2O + 6\,e^- + 2\,Cr + 6\,OH^-$

Cancel duplicate species:

$3\,H_2 + 2\,Cr(OH)_3 \rightarrow 6\,H_2O + 2\,Cr$

b) $Cr(OH)_3 + 3\,e^- \rightarrow Cr + OH^-$ $E° = -1.48\ V$

$2\,H_2O + 2\,e^- \rightarrow H_2 + 2\,OH^-$ $E° = -0.828\ V$

$E° = -1.48 - (-0.828\ V) = -0.65\ V$

c) $\Delta G° = -nFE° = -(6\ mol)\left(\dfrac{9.6485 \times 10^4\ C}{mol}\right)(-0.65\ V)$

$$= 3.8 \times 10^5\ J = 3.8 \times 10^2\ kJ$$

17.79 a) Your molecular picture should include a drawing like that in Figure 17-16 except the copper electrodes are replaced with zinc electrodes. The Cu^{2+} and SO_4^- in the right compartment are replaced with 1.25 M Zn^{2+} and 2.5 M NO_3^-. The Cu^{2+} and SO_4^- in the left compartment are replaced with 0.250 M Zn^{2+} and 0.500 M Cl^-. The nitrate ions will flow from the right vessel to the left one. Your molecular picture needs to show the spontaneous production of zinc ions and electrons from zinc metal in the left compartment while zinc ions and electrons form zinc metal at the right electrode.

b) Cathode: $2\,e^- + Zn^{2+} \rightarrow Zn$

Anode: $Zn \rightarrow Zn^{2+} + 2\,e^-$

$E = E° - \dfrac{0.05916\ V}{n}\log Q$ $E° = 0$

$E = -\dfrac{0.05916\ V}{n}\log Q = -\dfrac{0.05916\ V}{n}\log \dfrac{[Zn^{2+}]_{dilute}}{[Zn^{2+}]_{concentrated}}$

$E = -\dfrac{0.05916\ V}{2}\log \dfrac{(0.250\ M)}{(1.25\ M)} = 0.0207\ V$

17.81 $\text{Tl}^+ + e^- \rightarrow \text{Tl}$ $\qquad\qquad$ $E° = -0.34$ V

$\qquad$ $2 \text{ H}^+ + 2 e^- \rightarrow \text{H}_2$ $\qquad\qquad$ $E° = 0.00$ V

$\qquad$ $2 \text{ H}^+_{(aq)} + 2 \text{ Tl}_{(s)} \rightarrow \text{H}_{2(g)} + 2 \text{ Tl}^+_{(aq)}$

$\qquad$ $E° = 0.00$ V - (-0.34 V) = 0.34 V

$$E = E° - \frac{0.05916 \text{ V}}{n} \log Q = E° - \frac{0.05916 \text{ V}}{n} \log \frac{p_{\text{H}_2}[\text{Tl}^+]^2}{[\text{H}^+]^2}$$

$$= 0.34 \text{ V} - \frac{0.05916 \text{ V}}{2} \log \frac{(0.90)(0.050)^2}{(0.50)^2} = 0.34 \text{ V} - (-0.061 \text{ V}) = 0.40 \text{ V}$$

CHAPTER 18: THE CHEMISTRY OF LEWIS ACIDS AND BASES

18.1   a) Ni + 4 CO → [Ni(CO)$_4$]   Lewis acid: Ni   Lewis base: CO

       b) SbCl$_3$ + 2 Cl$^-$ → SbCl$_5^{2-}$ Lewis acid: SbCl$_3$ Lewis base: Cl$^-$

       c) (CH$_3$)$_3$P + AlBr$_3$ → (CH$_3$)$_3$P–AlBr$_3$
                                  Lewis acid: AlBr$_3$ Lewis base: (CH$_3$)$_3$P

       d) BF$_3$ + ClF$_3$ → ClF$_2^+$ + BF$_4^-$ Lewis acid: BF$_3$ Lewis base: ClF$_3$

       e) I$_2$ + I$^-$ → I$_3^-$ Lewis acid: I$_2$ Lewis base: I$^-$

18.3   BF$_3$ and AlF$_3$ are electron-deficient with an empty orbital that accepts a pair of
       electrons to form a fourth bond. The central atoms of SiF$_4$ and PF$_5$ have empty
       valence d orbitals that allow them to act as Lewis acids by accepting a pair of
       electrons. All of the valence orbitals of the C in CF$_4$ are used to form bonds with F.
       The valence orbitals of the N in NF$_3$ are used forming the 3 bonds to F and holding
       the non-bonding pair of electrons. The S in SF$_6$ uses sp^{3}d^2 hybridized orbitals to
       form 6 bonds to F that occupy the space around the S making it difficult for any
       further bonds to be formed.

18.5   a) Au = [Xe]6s^{1}5d^{10} Au$^+$ = [Xe]5d^{10} Au^{3+} = [Xe]5d^8

       b) Ni = [Ar]4s^{2}3d^8 Ni^{2+} = [Ar]3d^8 Ni^{3+} = [Ar]3d^7

       c) Mn$^-$ = [Ar]4s^{2}3d^6 Mn = [Ar]4s^{2}3d^5 Mn$^+$ = [Ar]4s^{1}3d^5

18.7   a) [Ru(NH$_3$)$_6$]Cl$_2$ = hexaammineruthenium(II) chloride

       b) [Cr(en)$_2$I$_2$]I = bis(ethylenediamine)diiodochromium(III) iodide

       c) cis-[PdCl$_2$(P(CH$_3$)$_3$)$_2$] = cis-dichlorobis(trimethylphosphine)palladium(II)

       d) fac-[Ir(NH$_3$)$_3$Cl$_3$] = fac-triamminetrichloroiridium(III)

       e) [Ni(CO)$_4$] = tetracarbonylnickel(0)

18.9   a) [Rh(en)$_3$]Cl$_3$ Rh(III), 4d^6 b) cis-[Mo(CO)$_4$Br$_2$]   Mo(II), 4d^4

       c) Na$_3$[IrCl$_6$]   Ir(III), 5d^6 d) mer-[Ir(NH$_3$)$_3$Cl$_3$]   Ir(III), 5d^6

       e) [Mn(CO)$_5$Cl]   Mn(I), 3d^6

18.11 a)

$$\left[\begin{array}{c} NH_2 \\ NH_2\cdots Rh \cdots NH_2 \\ NH_2 \quad NH_2 \\ NH_2 \end{array} \right]^{3+}, \; 3\;Cl^-$$

b)

$$\begin{array}{c} CO \\ Br\cdots Mo\cdots CO \\ Br \quad CO \\ CO \end{array}$$

c)

$$3\;Na^+, \; \left[\begin{array}{c} Cl \\ Cl\cdots Ir \cdots Cl \\ Cl \quad Cl \\ Cl \end{array} \right]^{3-}$$

d)

$$\begin{array}{c} NH_3 \\ Cl\cdots Ir \cdots NH_3 \\ H_3N \quad Cl \\ Cl \end{array}$$

e)

$$\begin{array}{c} CO \\ OC\cdots Mn \cdots CO \\ OC \quad CO \\ Cl \end{array}$$

18.13 a) 4 NH_3, Cl^-, NO_2^-, Co^{3+} *cis*-[Co(NH$_3$)$_4$Cl(NO$_2$)]$^+$

      b) NH_3, 3 Cl^-, Pt^{2+} [Pt(NH$_3$)Cl$_3$]$^-$

      c) 2 H_2O, 2 en, Cu^{2+} *trans*-[Cu(en)$_2$(H$_2$O)$_2$]$^{2+}$

      d) 4 Cl^-, Fe^{3+} [FeCl$_4$]$^-$

18.15 [Ni(CO)$_4$]   geometry = tetrahedral   color = colorless

Ni(CO)$_4$ is poisonous because it is so volatile. Once inhaled, some of the CO ligands dissociate from the metal and dissolve in the blood stream, replacing O_2 in hemoglobin. Suffocation results.

18.17 $E = \dfrac{hc}{\lambda} = \dfrac{(6.626 \times 10^{-34} \text{ Js})(2.998 \times 10^8 \text{ m/s})}{(430.5 \text{ nm})(10^{-9} \text{ m/nm})} \dfrac{(6.022 \times 10^{23})}{1000 \text{ J/kJ}} = 278 \text{ kJ/mol}$

$$= \dfrac{(6.626 \times 10^{-34} \text{ Js})(2.998 \times 10^8 \text{ m/s})(6.022 \times 10^{23})}{(611.5 \text{ nm})(10^{-9} \text{ m/nm})(1000 \text{ J/kJ})} = 196 \text{ kJ/mol}$$

a) $\Delta = 278$ kJ/mol and $\Delta = 196$ kJ/mol

b) green/blue green

Note: The wavelengths around 510 nm are the only ones in the visible region that are transmitted. This area is the wavelength of blue-green to green colored light.

c) tetraaquadichlorochromium(III) ion

d)

$$\left[\begin{array}{c} OH_2 \\ H_2O\cdots Cr \cdots Cl \\ H_2O \quad Cl \\ OH_2 \end{array} \right]^+ \qquad \left[\begin{array}{c} Cl \\ H_2O\cdots Cr \cdots OH_2 \\ H_2O \quad OH_2 \\ Cl \end{array} \right]^+$$

            cis   *trans*

        yellow   blue-green

(continued)

(18.17 continued)

e)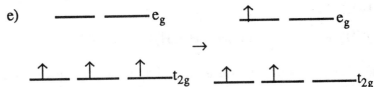

18.19 Zr^{2+} complexes are dark purple; Zr^{2+} is d^2

Zr^{4+} complexes are colorless; Zr^{4+} is d^0

Since Zr^{4+} complexes have no d electrons to undergo transition, they are colorless.

18.21 $[Cr(H_2O)_6]^{3+}$; $[Cr(NH_3)_6]^{3+}$
Both are Cr^{3+}, d^3, but NH_3 gives complexes with larger splitting than H_2O. According to Table 18-2, the larger Δ should be associated with the orange complex. Therefore, $[Cr(NH_3)_6]^{3+}$ is orange; $[Cr(H_2O)_6]^{3+}$ is violet.

18.23 a) $BF_3 > BCl_3 > AlCl_3$ Boron is above aluminum in the periodic table; therefore, BF_3 and BCl_3 are harder than $AlCl_3$. BF_3 has F that is more electronegative than the Cl in BCl_3. BF_3 is harder than BCl_3.

b) $Al^{3+} > Tl^{3+} > Tl^+$ Al^{3+} and Tl^{3+} are harder than Tl^+ because they have a greater charge. Al^{3+} is harder than Tl^{3+} because it is above Tl^{3+} in the periodic table.

c) $AlCl_3 > AlBr_3 > AlI_3$ The electronegativities of the anions increase Cl > Br > I. Therefore, the hardness of the compounds decrease: $AlCl_3 > AlBr_3 > AlI_3$

18.25 SO_3 is a harder Lewis acid than SO_2 because the oxidation state of sulfur is +6 in SO_3 but only +4 in SO_2. The greater positive charge on S in SO_3 exerts a stronger pull on the electron cloud thus making it less polarizable and a harder Lewis acid.

18.27 $(H_3C)_2N-PF_2$

$$H_3C-\overset{..}{\underset{CH_3}{N}}-\overset{\overset{F}{|}}{\underset{..}{P}}-F$$

BF_3 is a harder Lewis acid than BH_3; the phosphorus end of $(H_3C)_2N-PF_2$ is probably a harder Lewis base than the nitrogen end because the fluoride atoms are so electronegative. Therefore, the phosphorus end forms an adduct with BF_3 and the nitrogen end forms an adduct with BH_3.

$$\underset{H_3C\quad CH_3}{\overset{\overset{\displaystyle BH_3}{|}}{N}\text{IIII}\cdots PF_2} \qquad \text{and} \qquad \underset{F\quad F}{\overset{\overset{\displaystyle BF_3}{|}}{P}\text{IIII}\cdots N(CH_3)_2}$$

18.29 a) $NBr_3 + GaCl_3 \rightarrow Br_3N-GaCl_3$

b) $Al(CH_3)_3 + LiCH_3 \rightarrow$ no reaction

(continued)

189

(18.29 continued)

c) $SiF_4 + 2\,LiF \rightarrow Li_2[SiF_6]^{2-}$

d) $4\,LiCH_2CH_2CH_2CH_3 + SnCl_4 \rightarrow Sn(CH_2CH_2CH_2CH_3)_4 + 4\,LiCl$

18.31 The bulkiness of the CH_3 groups hinders the ability of approaching Lewis bases to form bonds to the boron atom.

18.33 $PCl_3 + (H_3C)_2NH \rightarrow Cl_3P\!-\!NH(CH_3)_2 \xrightarrow{-\,HCl} Cl_2P\!-\!N(CH_3)_2$

$\xrightarrow{+\,HN(CH_3)_2} ((H_3C)_2N)Cl_2P\!-\!NH(CH_3)_2 \xrightarrow{-\,HCl} ((H_3C)_2N)_2ClP$

$\xrightarrow{+\,HN(CH_3)_2} ((H_3C)_2N)_2ClP\!-\!NH(CH_3)_2 \xrightarrow{-\,HCl} P(N(CH_3)_2)_3$

18.35 Large atoms with low oxidation states, such as Hg^{2+}, are highly polarizable so they are soft Lewis acids. Zn, on the other hand, is a relatively smaller atom in the +2 oxidation state. As a result, Zn^{2+} is less polarizable and a harder Lewis acid than Hg^{2+}. Because Hg^{2+} is a soft Lewis acid it does not form compounds with hard Lewis bases. Zinc, a hard Lewis acid, forms compounds with hard Lewis bases such as oxides, carbonates, and silicates.

18.37 Transition metals have vacant orbitals so they are Lewis acids. They will form bonds with ligands that have lone pairs of electrons (Lewis bases). Very few, if any, Lewis bases having lone pairs of electrons are cations; they are anions or neutral molecules.

18.39

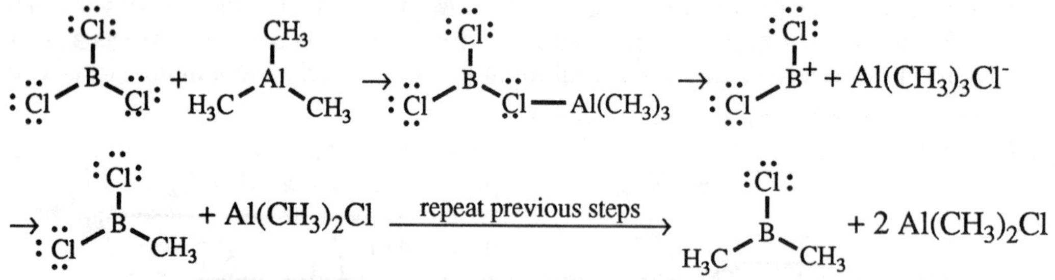

a) The Lewis acid is $Al(CH_3)_3$ because it forms an adduct by reacting with an electron pair from a Cl of BCl_3. Further, it cannot be a Lewis base because it has no lone pair of electrons on the molecule.

b) As $Al(CH_3)_2Cl$ molecules are formed, they dimerize to form unreactive:

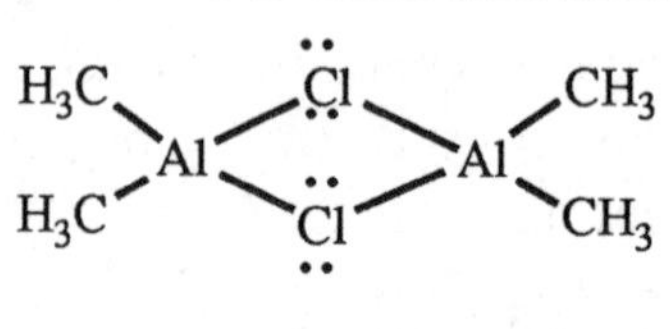

190

18.41 $CaO + SiO_2 \rightarrow CaSiO_3$

In this reaction, an O^{2-} anion adds to the SiO_2 structure to form an SiO_3^{2-} anion. Thus, O^{2-} acts as a Lewis base, donating a pair of electons to form an Si—O bond, and SiO_2 acts as a Lewis acid.

18.43 $W(CO)_6$

Since the complex is colorless, it must be because the energy difference corresponds to photon energies in the ultraviolet.

18.45 Sulfur is a soft Lewis base. Covalent sulfur–sulfur single bonds are a major determinant of protein structure. Enzymes are proteins. Sulfur is in many enzymes. If the soft metal ions (lead or mercury) are present, they bind with the sulfur atoms of the enzymes. This changes the enzymes' shape and the enzymes can no longer perform their function of catalyzing essential biochemical reactions in the body. This results in illness or even death.

18.47 Strong adducts form when the donor and acceptor can approach each other closely. In dimethyltetrahydrofuran, the electron clouds around the two methyl groups repel an approaching BF_3 molecule, lengthening the O—B bond and making the adduct weaker.

18.49 Because F^- is at the small splitting end of the spectrochemical series, $[FeF_6]^{3-}$ will absorb light of longer wavelengths than $[Fe(H_2O)_6]^{3+}$. This places its absorption in the infrared; so the complex is colorless.

18.51 The $[Ni(CN)_4]^{2-}$ complex is diamagnetic meaning that all eight valence electrons of Ni^{2+} are paired. This can only occur if the complex is square planar (See Figure 18-16). The $[NiCl_4]^{2-}$ complex is paramagnetic because it has unpaired spins. This requires tetrahedral geometry (See Fig. 18-17).

square planar

tetrahedral

18.53 Rust removers containing oxalate anion take advantage of the chelate effect. The formation constant for the reaction of Fe^{2+} or Fe^{3+} with the bidentate oxalate must be larger than it is for monodentate O^{2-}. Thus, the iron cations in rust will preferentially react with oxalate. Once the iron is tied up with oxalate ligands, the water soluble complex can be washed away with water.

CHAPTER 19: NUCLEAR CHEMISTRY AND RADIOCHEMISTRY

19.1 a) $Z = 3$, $A = 6$, $N = 3$ b) $Z = 20$, $A = 43$, $N = 23$

c) $Z = 92$, $A = 238$, $N = 146$ d) $Z = 52$, $A = 130$, $N = 78$

e) $Z = 10$, $A = 20$, $N = 10$ f) $Z = 82$, $A = 205$, $N = 123$

19.3 a) $^{4}_{2}He$ b) $^{184}_{74}W$ c) $^{26}_{12}Mg$ d) $^{60}_{28}Ni$

19.5 a) unstable (neutron-poor, $Z > N$)   b) unstable ($Z > 83$)

c) stable   d) unstable (Z-N, odd-odd; Tc has no stable nuclides)

19.7 $\Delta m = 1.00$ met ton $= 1.00(1000 \text{ kg}) = 1.00(10^3 \text{ kg})(10^3 \text{ g/kg}) = 1.00 \ 10^6$ g

$\Delta E = \Delta m(8.988 \times 10^{10} \text{ kJ/g}) = (1.00 \times 10^6 \text{ g})(8.988 \times 10^{10} \text{ kJ/g}) = 8.99 \times 10^{16}$ kJ

19.9 $MM_{Cs} = 132.905$ g / mol $^{133}_{55}Cs$

protons: $55(1.007276 \text{ g/mol}) = 55.400180$ g/mol
neutrons: $78(1.008665 \text{ g/mol}) = 78.675870$ g/mol
electrons: $55(0.0005486 \text{ g/mol}) = \underline{0.0301730 \text{ g/mol}}$
134.106223 g/mol

$\Delta m = 132.905$ g/mol - 134.106 g/mol = -1.201 g/mol

$\Delta E = (-1.201 \text{ g/mol})(8.988 \times 10^{10} \text{ kJ/g}) = -1.079 \times 10^{11}$ kJ/mol

ΔE per nucleon $= \dfrac{-1.079 \times 10^{11} \text{ kJ / mol}}{133 \text{ nucleons}} = -8.11 \times 10^8$ kJ / mol nucleons

19.11

	symbol	name
a) a photon (high energy)	γ	Gamma-ray
b) positive particle with mass number 4	$\propto$	alpha-particle
c) positron:	$\beta+$	positron

19.13 a) $^{125}_{52}Te \rightarrow {}^{125}_{52}Te + \gamma$ b) $^{123}_{52}Te + {}^{0}_{-1}e \rightarrow {}^{123}_{51}Sb$

c) $^{127}_{52}Te \rightarrow {}^{0}_{-1}\beta + {}^{127m}_{53}I$

$^{127m}_{53}I \rightarrow {}^{127}_{53}I + \gamma$

19.15 $^{232}_{90}Th \rightarrow {}^4_2\alpha + {}^{228}_{88}Ra$ $\qquad$ $^{228}_{88}Ra \rightarrow {}^0_{-1}\beta + {}^{228}_{89}Ac$

$\qquad$ $^{228}_{89}Ac \rightarrow {}^0_{-1}\beta + {}^{228}_{90}Th$ $\qquad$ $^{228}_{90}Th \rightarrow {}^4_2\alpha + {}^{224}_{88}Ra$

$\qquad$ $^{224}_{88}Ra \rightarrow {}^4_2\alpha + {}^{220}_{86}Rn$ $\qquad$ $^{220}_{86}Rn \rightarrow {}^4_2\alpha + {}^{216}_{84}Po$

$\qquad$ $^{216}_{84}Po \rightarrow {}^0_{-1}\beta + {}^{216}_{85}At$ $\qquad$ $^{216}_{85}At \rightarrow {}^4_2\alpha + {}^{212}_{83}Bi$

$\qquad$ $^{212}_{83}Bi \rightarrow {}^0_{-1}\beta + {}^{212}_{84}Po$ $\qquad$ $^{212}_{84}Po \rightarrow {}^4_2\alpha + {}^{208}_{82}Pb$

19.17 a) $^{12}_6C + {}^1_0n \rightarrow \left({}^{13m}_{6}C\right) \rightarrow {}^{12}_5B + {}^1_1p$

$\qquad$ b) $^{16}_8O + {}^4_2\alpha \rightarrow \left({}^{20m}_{10}Ne\right) \rightarrow {}^{20}_{10}Ne + \gamma$

$\qquad$ c) $^{247}_{96}Cm + {}^{11}_5B \rightarrow \left({}^{258m}_{101}Md\right) \rightarrow {}^{255}_{101}Md + 3\,{}^1_0n$

19.19 $^{14}_7N + {}^4_2\alpha \rightarrow {}^{18}_8O + {}^0_{+1}\beta+$

19.21 The most likely products have A = 95 and A = 138.
Element with stable isotope with A = 95: Mo
Elements with stable isotopes with A = 138: Ba, La, Ce
Other elements with stable isotopes with A = 90-100: Zr, Nb, Ru, Rh
Other elements with stable isotopes with A = 132-143: Xe, Cs, Pr, Nd

19.23 It is important that heat be transferred from the core to the turbines without transfer of matter because the core generates lethal radioactive products that cannot be allowed to escape or transfer radioactivity. Liquid sodium metal or high pressure water are the two most used substances for circulation around the core to absorb the heat produced by the nuclear fission. This hot fluid then passes through a steam generator, transfers its heat energy to water resulting in the water being vaporized to steam. This steam is used to produce electricity by driving a conventional steam turbine. Sodium metal or water are good choices to carry the heat from the core to the steam generator because if they were to form radioactive products while passing through the core, these products have short half-lifes, decay by β emission (which can be shielded) and produce stable products.

19.25 mass of ^3_1H = 3.01605 g/mol

$\qquad$ Assuming ^3_1H and ^2_1H both have a radius of 1.4×10^{-3} pm

$$E = \frac{(1.389 \times 10^5 \text{ kJ pm/mol})(Z_1)(Z_2)}{d}$$

(continued)

(19.25 continued)

$$E = \frac{(1.389 \times 10^5 \text{ kJ pm/mol}) (1)(1)}{(2 \times 1.4 \times 10^{-3} \text{ pm})} = 5.0 \times 10^7 \text{ kJ/mol}$$

$$E = 1/2mv^2$$

$$v = \sqrt{\frac{E}{1/2 \, m}} = \sqrt{\frac{(5.0 \times 10^7 \text{ kJ/mol}) (10^3 \text{ J/kJ}) \left(\dfrac{\text{kg} \cdot \text{m}^2}{\text{s}^2 \cdot \text{J}} \right)}{\left(\dfrac{1}{2} \right) (3.01605 \text{ g/mol}) \left(\dfrac{\text{kg}}{10^3 \text{ g}} \right)}} = 5.8 \times 10^6 \text{ m/s}$$

19.27   Presently fusion can occur only on a very large scale. It occurs only at very high
temperatures that require high-intensity power sources. Confinement of the reaction
requires high-intensity fields that are beyond current technology as the temperatures
are too high for physical containment. If a sustained fusion reaction is achieved,
some way must be found to convert its energy output (consisting of photons) into
electricity. This will require huge amounts of money.

19.29

Second-generation star	Temperature	Composition
hydrogen-burning stage	probably lower than first generation as ^{12}C catalyzed	$^1_1$H, $^{12}_6$C, $^4_2$He, $\left(^{13}_6\text{C}\right)$, $^0_{-1}$e, $\left(^{14}_6\text{C}\right)$, $\left(^{15}_7\text{N}\right)$ and other nuclides up to Z = 26
helium-burning stage	10^8 K	$^{13}_6$C, $^4_2$He, $^{16}_8$O, $^1_0$n, $^{56}_{26}$Fe, $^{57}_{27}$Co and many more reactions because $^1_0$n are generated; all possible stable nuclides can form
carbon-burning stage	10^9 K	May become unstable, explode, and produce still heavier nuclides

A second-generation star is a combination of interstellar hydrogen and matter from
an exploded first-generation star. Therefore, a second-generation star will contain
higher-Z nuclides because a first-generation star does not become unstable and
explode (supernova) until it reaches the third stage at which time nuclides with Z
and A values up to iron-56 are being produced. A greater range of reactions takes
place in a second-generation star because the fusion of $^4_2$He and $^{13}_6$C generates
neutrons which can be sequentially captured resulting in the production of all
possible stable nuclides. An explosion of a second-generation star produces not
only heavier nuclide debris, but also large numbers of neutrons that can be captured
and for still more heavy nuclides.

19.31 Iron (Z = 26) is the most stable nuclide. Fusion reactions do not produce nuclides with Z greater than 26. Higher-Z nuclides can be produced by sequential capture of neutrons and β-emission, but first-generation stars do not have sources of neutrons and, therefore, do not produce nuclides with Z > 26.

19.33 $N = (1.50 \text{ mL})\left(\dfrac{1 \text{ L}}{10^3 \text{ mL}}\right)\left(\dfrac{2.50 \times 10^{-9} \text{ mol Tc}}{\text{L}}\right)\left(\dfrac{6.022 \times 10^{23} \text{ atoms Tc}}{\text{mol Tc}}\right)$

$$= 2.26 \times 10^{13} \text{ atoms Tc}$$

$$\frac{\Delta N}{\Delta t} = \frac{-N \ln 2}{t_{1/2}} = \frac{-(2.26 \times 10^{13} \text{ Tc})(0.693)}{(6.0 \text{ hr})(60 \text{ min / hr})(60 \text{ s / min})} = -7.3 \times 10^8 \text{ Tc/s}$$

7.3×10^8 Tc decays/s = 7.3×10^8 γ-rays emitted per second.

19.35 After the foods are packaged, the package is exposed to enough gamma radiation to kill microorganisms. Moderate doses retard spoilage and high doses can prevent spoilage completely for as long as the package remains intact. There is fear that irradiation might generate unhealthy, or even carcinogenic, byproducts by causing chemical reactions.

19.37 $N_0 = N_{Sr} + N_{Rb}$

$$\frac{N_0}{N} = \frac{N_{Sr} + N_{Rb}}{N_{Rb}} = 1 + \frac{N_{Sr}}{N_{Rb}} = 1 + 0.0050 = 1.0050$$

$$t = \frac{t_{1/2}}{\ln 2} \ln\left(\frac{N_0}{N}\right) = \frac{(4.9 \times 10^{11} \text{ yr})}{(0.693)} \ln(1.0050) = 3.5 \times 10^9 \text{ yr}$$

19.39 Advantageous properties of the ^{99m}Te for use as a medical imaging isotope are a half-life of 6 hours (long enough to use but short enough to minimize long-term effects); emits γ-rays of moderate energy (easy to detect but not highly damaging to living tissue); element can be bound to chemicals that the body recognizes and processes (can be used to image the thyroid gland, brain, lungs, heart, liver, stomach, kidneys and bones); and can readily be extracted in pure form from natural molybdenum that has been exposed to neutron bombardment.

19.41 $^{209}_{83}$Bi

Protons:	83(1.007276 g/mol)	= 83.603908 g/mol
Neutrons:	126(1.008665 g/mol)	= 127.091790 g/mol
Electrons:	83(0.0005486 g/mol) =	<u>0.0455338 g/mol</u>
		210.741232 g/mol

Δm = 208.980 g/mol - 210.741232 g/mol = -1.7612 g/mol

ΔE = (-1.7612 g/mol)(8.988 x 10^10 kJ/g) = -1.583 x 10^11 kJ/mol

(continued)

(19.41 continued)

binding energy per mole nucleons =

$$\frac{\Delta E}{209 \text{ nucleons}} = \frac{(-1.583 \times 10^{11} \text{ kJ/mol})}{209 \text{ nucleons}} = -7.574 \times 10^8 \text{ kJ/mol nucleons}$$

binding energy per nucleon =

$$\frac{(-7.574 \times 10^8 \text{ kJ/mol nucleons})}{6.022 \times 10^{23} \text{ nucleons / mol nucleons}} = 1.258 \times 10^{-15} \text{ kJ / nucleon}$$

19.43 $(10^5 \text{ neutrons/s})(30 \text{ s}) = 3.0 \times 10^6 \text{ neutrons}$

$$\ln\left(\frac{N_0}{N}\right) = \frac{t \ln 2}{t_{1/2}} = \ln\left(\frac{3.0 \times 10^6 \text{ neutrons}}{N}\right) = \left(\frac{(\ln 2)(1 \text{ hr})\left(\frac{3600 \text{ s}}{\text{hr}}\right)}{1100 \text{ s}}\right) = 2.268$$

$$\frac{3.0 \times 10^6 \text{ neutrons}}{N} = e^{2.268} = 9.66$$

$$N = \frac{3.0 \times 10^6 \text{ neutrons}}{9.66} = 3.1 \times 10^5 \text{ neutrons (if } t_{1/2} = 1100 \text{ s)}$$

$$\ln\left(\frac{3.0 \times 10^6 \text{ neutrons}}{N}\right) = \left(\frac{(\ln 2)(1 \text{ hr})\left(\frac{3600 \text{ s}}{\text{hr}}\right)}{876 \text{ s}}\right) = 2.849$$

$$N = \frac{3.0 \times 10^6 \text{ neutrons}}{e^{2.849}} = 1.7 \times 10^5 \text{ neutrons (if } t_{1/2} = 876 \text{ s)}$$

19.45 $^1_0n \rightarrow {}^1_1p + {}^0_{-1}\beta$ β-ray (an electron) is the other product.

See Sample Problem 19-2:

$\Delta m = (1.007276 \text{ g/mol}) + 0.0005486 \text{ g/mol} - 1.008665 \text{ g/mol} = -0.0008404 \text{ g/mol}$

$$\Delta E_{\text{per electron}} = \frac{(8.988 \times 10^{10} \text{ kJ / g})(-0.0008404 \text{ g / mol})}{(6.022 \times 10^{23} \text{ electrons / mol})}$$

$$= -1.25 \times 10^{-16} \text{ kJ / electron} = -1.25 \times 10^{-13} \text{ J / electron}$$

19.47 $^0_{+1}\beta + {}^0_{-1}e \rightarrow 2 \, {}^0_0\gamma$

$\Delta m = 0 - 2(0.0005486 \text{ g/mol}) = -0.0010972 \text{ g/mol}$

(continued)

196

(19.47 continued)

energy per mole of γ-rays =

$$\frac{\Delta E}{2} = \frac{(8.988 \times 10^{10} \text{ kJ / g})(-0.0010972 \text{ g / mol})}{2} = -4.931 \times 10^7 \text{ kJ / mol}$$

$$\text{energy per }\gamma\text{-ray} = \frac{-4.931 \times 10^7 \text{ kJ /mol}}{6.022 \times 10^{23} \text{ / mol}} = -8.188 \times 10^{-17} \text{ kJ or } -8.19 \times 10^{-14} \text{ J}$$

19.49 $\ln\left(\dfrac{N_0}{N}\right) = \dfrac{t \ln 2}{t_{1/2}} = \dfrac{(1 \text{ yr}) \ln 2}{103 \text{ yr}} = 0.00673$ $\dfrac{N_0}{N} = e^{0.00673} = 1.0068$

$$\frac{N}{N_0} = 0.993$$

Assume one mole of ^{209}Po and one mole of ^{210}Po at beginning:

for ^{209}Po: $\ln\left(\dfrac{1 \text{ mol}}{N}\right) = \dfrac{(10 \text{ yr})(\ln 2)}{103 \text{ yr}} = 0.0673$

$\left(\dfrac{1 \text{ mol}}{N}\right) = e^{0.0673} = 1.070$ $N = 0.935 \text{ mol } ^{209}$Po

for ^{210}Po: $\ln\left(\dfrac{1 \text{ mol}}{N}\right) = \dfrac{(10 \text{ yr})(\ln 2)}{(138.4 \text{ days})(1 \text{ yr / 365.25 days})} = 18.29$

$\left(\dfrac{1 \text{ mol}}{N}\right) = e^{18.29} = 8.77 \times 10^7$ $N = 1.14 \times 10^{-8} \text{ mol } ^{210}$Po

^{210}Po/^{209}Po = $(1.14 \times 10^{-8} \text{ mol } ^{210}Po)/(0.935 \text{ mol } ^{209}Po) = 1.2 \times 10^{-8}$

19.51 $^{11}_{6}$C   N / Z = 5/6 = 0.833

$^{15}_{8}$O   N / Z = 7/8 = 0.875

Both ^{11}C and ^{15}O are located below the "belt of stability" (N/Z < 1).

$^{11}_{6}$C → $^{0}_{+1}\beta$ + $^{11}_{5}$B

$^{15}_{8}$O → $^{0}_{+1}\beta$ + $^{15}_{7}$N

19.53 $\ln\left(\dfrac{15.3 \text{ counts / g min}}{0.03 \text{ counts / g min}}\right) = \dfrac{t \ln 2}{5730 \text{ yr}}$

$t = \ln\left(\dfrac{15.3}{0.03}\right)\dfrac{5730 \text{ yr}}{\ln 2} = 5.2 \times 10^4 \text{ yrs}$

19.55 a) $^{64}_{29}Cu \rightarrow \ ^{0}_{-1}\beta + \ ^{64}_{30}Zn$ and $^{64}_{29}Cu \rightarrow \ ^{0}_{+1}\beta + \ ^{64}_{28}Ni$

b) $\Delta E = (8.988 \times 10^{10} \text{ kJ/g})(\Delta m)$

$$\Delta m = \frac{\Delta E}{(8.988 \times 10^{10} \text{ kJ / g})(10^3 \text{ J / kJ})}$$

for $^{0}_{-1}\beta$: $\Delta m = \dfrac{-(9.3 \times 10^{-14} \text{ J / atom})(6.022 \times 10^{23} \text{ atom / mol})}{(8.988 \times 10^{10} \text{ kJ / g})(10^3 \text{ J / kJ})}$

$= -0.000623 \text{ g/mol}$

for $^{0}_{+1}\beta$: $\Delta m = \dfrac{-(1.04 \times 10^{-13} \text{ J / atom})(6.022 \times 10^{23} \text{ atom / mol})}{(8.988 \times 10^{10} \text{ kJ / g})(10^3 \text{ J / kJ})}$

$= -0.000697 \text{ g / mol}$

for $^{0}_{-1}\beta$: $\Delta m = -0.000623 \text{ g / mol} = m_e + m_{Zn} - m_{Cu}$

$m_{Zn} = -0.000623 \text{ g/mol} + 63.92976 \text{ g/mol} - 0.0005486 \text{ g/mol} = 63.92859 \text{ g / mol}$

for $^{0}_{+1}\beta$: $\Delta m = -0.000697 \text{ g / mol} = m_e + m_{Ni} - m_{Cu}$

$m_{Ni} = -0.000697 \text{ g/mol} - 0.0005486 \text{ g/mol} + 63.92976 \text{ g/mol} = 63.92851 \text{ g/mol}$

19.57 $^{232}_{90}Th$ $^{238}_{92}U$ $^{206,207,208}_{82}Pb$

α-emission decreases A by 4 units and Z by 2 units.

β-emission increases Z by 1 unit and leave A unchanged.

γ-emission changes neither A nor Z.

Difference of A for Th and A for each Pb isotope:

^{206}Pb: 232 - 206 = 26 ^{207}Pb: 232 - 207 = 25 ^{208}Pb: 232 - 208 = 24

Difference of A for U and A for each Pb isotope:

^{206}Pb: 238 - 206 = 32 ^{207}Pb: 238 - 207 = 31 ^{208}Pb: 238 - 208 = 30

For radioactive decay consisting entirely of α-, β- and γ-emissions, only α-emissions change the mass number (A) and then only by 4 units. Therefore, an isotope decaying by this scheme can only produce isotopes of new elements which differ in mass numbers by a multiple of 4.

Therefore, neither ^{232}Th nor ^{238}U can decay by this scheme and produce ^{207}Pb (differences of A are 25 and 31, respectively). The decay of ^{232}Th can produce ^{208}Pb (difference of A = 24 = 4 x 6) but not ^{206}Pb (difference of A = 26). The decay of ^{238}U can produce ^{206}Pb (difference of A = 32 = 4 x 8) but not ^{208}Pb (difference of A = 30).

The decay of $^{232}_{90}Th$ to $^{208}_{82}Pb$ requires 6 α-emissions and 4 β-emissions to counteract the Z change.

The decay of $^{238}_{92}U$ to $^{206}_{82}Pb$ requires 8 α-emissions and 6 β-emissions to counteract the Z change.

The lead from ^{238}U decay has a lower atomic mass than lead from ^{232}Th decay.

19.59 $^{10}_{5}B + ^{1}_{0}n \rightarrow ^{4}_{2}\alpha + ^{7}_{3}Li$

This does not present a health hazard because it is more difficult to shield for $^{1}_{0}n$ than for $^{4}_{2}\alpha$. The shielding necessary to protect against neutrons easily shields against $^{4}_{2}\alpha$.

19.61 1% of ^{226}Ra disappeared; $N = 0.99\, N_0$

$$t = \frac{(1622\ yr)\ln\left(N_0/0.99\, N_0\right)}{\ln 2} = 24\ yr$$

1% of ^{226}Ra remains; $N = 0.01\, N_0$

$$t = \frac{(1622\ yr)\ln\left(N_0/0.01\, N_0\right)}{\ln 2} = 1 \times 10^4\ yr$$

19.63 $PV = nRT$; $n = \dfrac{PV}{RT} = \dfrac{(1\ atm)(6.0 \times 10^{-5}\ cm^3)(1\ mL/1\ cm^3)(L/10^3\ mL)}{(0.0821\ L\ atm\ /\ mol\ K)(298\ K)}$

$$= 2.45 \times 10^{-9}\ mol\ ^{4}_{2}He$$

$$n_U = \frac{1.3 \times 10^{-7}\ g\ U}{238\ g\ U\ /\ mol\ U} = 5.46 \times 10^{-10}\ mol\ U; \quad N = 5.46 \times 10^{-10}\ mol\ U$$

$$N_0 = 5.46 \times 10^{-10}\ mol\ U + \left(2.45 \times 10^{-9}\ mol\ ^{4}_{2}He\right)\left(1\ mol\ U/8\ mol\ ^{4}_{2}He\right)$$

$$= 8.52 \times 10^{-10}\ mol\ U$$

$$\ln\left(\frac{N_0}{N}\right) = \frac{t\ \ln 2}{t_{1/2}}$$

$$t = \frac{t_{1/2}\ \ln\left(N_0/N\right)}{\ln 2} = \frac{(4.51 \times 10^9\ yr)\left(\ln\left(\dfrac{8.52 \times 10^{-10}\ mol\ U}{5.46 \times 10^{-10}\ mol\ U}\right)\right)}{0.6931} = 2.9 \times 10^9\ yr$$

19.65 $\ln(N_0/N) = \dfrac{(3250\ yr)\ \ln 2}{(5730\ yr)} = 0.3931$; $\dfrac{N_0}{N} = e^{0.3931} = 1.482$; $N_0 = 1.482\ N$;

if $N = 1$, $N_0 = 1.48$,

but if N_0 was 20% greater than thought, $N_0 = (1.20)(1.482) = 1.778$

$$\ln\left(\frac{1.78}{1}\right) = \frac{t\ \ln 2}{(5730\ yr)}; \quad t = \frac{(5730\ yr)\left(\ln(1.778/1)\right)}{\ln 2} = 4.76 \times 10^3\ yr$$

19.67 $^{197}_{79}$Au; molar mass = 196.967 g/mol

79 protons: (79)(1.007276 g/mol) =   79.574804 g/mol
79 electrons: (79)(0.0005486 g/mol) =   0.0433394 g/mol
118 neutrons: (118)(1.007276 g/mol) = 119.022470 g/mol
                                                       198.640613 g/mol

Δm = 196.967 g/mol - 198.6406 g/mol = -1.6736 g/mol

ΔE = (8.988 x 10^{10} kJ/g)(-1.6736 g/mol) = -1.504 x 10^{11} kJ/mol

$$\text{binding energy per nucleus of gold} = \frac{\Delta E}{6.023 \times 10^{23} \text{ nuclei/mol}}$$

$$= \frac{-1.504 \times 10^{11} \text{ kJ/mol}}{6.023 \times 10^{23} \text{ nuclei/mol}} = -2.498 \times 10^{-13} \text{ kJ/nucleus}$$

$$\text{binding energy per nucleon} = \frac{-2.498 \times 10^{-13} \text{ kJ/nucleus}}{197 \text{ nucleons/nucleus}}$$

$$= -1.268 \times 10^{-15} \text{ kJ/nucleon} = -1.268 \times 10^{-12} \text{ J/nucleon}$$

19.69 $^{27}_{13}$Al + $^{4}_{2}\alpha \rightarrow \left(^{31m}_{15}\text{P}\right) \rightarrow ^{1}_{0}\text{n} + ^{30}_{15}\text{P}$

The $^{30}_{15}$P will not be stable because it has an odd number of protons and an odd number of neutrons and is not one of the five such isotopes that are stable. It has an equal number of protons and neutrons which gives a N/Z ratio of 1. Thus, it will probably decay by positron-emission that would increase N by 1 and decrease Z by 1 becoming $^{30}_{14}$Si. $^{30}_{14}$Si would have a N/Z ratio greater than 1 and is stable.

19.71 $\ln\left(\dfrac{N_0}{N}\right) = \dfrac{t \ln 2}{t_{1/2}};$ $t = \dfrac{t_{1/2} \ln(N_0/N)}{\ln 2} = \dfrac{(15.0 \text{ hr})\ln(25 \text{ μg}/1 \text{ μg})}{\ln 2} = 70 \text{ hr}$

19.73 World's supply of ^{235}U =

$$(10 \times 10^6 \text{met ton U})\left(\frac{1000 \text{ kg}}{\text{met ton}}\right)\left(\frac{0.7 \text{ kg } ^{235}\text{U}}{100 \text{ kg U}}\right)\left(\frac{10^3 \text{ g}}{\text{kg}}\right) = 7 \times 10^{10} \text{ g } ^{235}\text{U}$$

atoms of ^{235}U =

$$\frac{(7 \times 10^{10} \text{ g } ^{235}\text{U})(6.022 \times 10^{23} \text{ atoms/mol})}{(235 \text{ g/mol})} = 1.8 \times 10^{32} \text{ atoms } ^{235}\text{U}$$

energy supplied by ^{235}U =

$$(1.8 \times 10^{32} \text{ atoms } ^{235}\text{U})(1 \text{ nucleus/atom})\left(2.9 \times 10^{-11} \text{ J/nucleus}\right) = 5.2 \times 10^{21} \text{ J}$$

$$\text{year's supply} = \frac{5.2 \times 10^{21} \text{ J}}{(2 \times 10^{17} \text{ kJ/yr})(10^3 \text{ J/kJ})} = 26 \text{ years}$$

Enough ^{235}U to supply world's energy for 30 years (1 significant figure).

19.75 The simplest way would be to prepare (or buy) alcohol containing some ^{18}O. The condensation reaction with a carboxylic acid could then be performed. If the oxygen in the water molecule comes from the carboxylic acid, there would not be any ^{18}O in the water. If the oxygen in the water molecule comes from the alcohol, there would be ^{18}O in the water. The experiment could also be performed using a carboxylic acid containing ^{18}O and water containing normal ^{16}O but preparation and results are more complicated because the carboxylic acid contains 2 oxygen atoms. Experiments have shown that the oxygen in the water does come from the carboxylic acid. The ^{18}O isotope is not unstable (radioactive); it just has more mass than ^{16}O. Its presence is identified by using mass spectrometry.

19.77 $\ln\left(\dfrac{N_0}{N}\right) = \dfrac{t \ln 2}{t_{1/2}}$;

$\ln\left(\dfrac{125\ mg}{N}\right) = \dfrac{(4\ days)\ln 2}{(14.3\ days)} = 0.1939$

$\dfrac{125\ mg}{N} = e^{0.1939} = 1.21$;

$N = \dfrac{125\ mg}{1.21} = 103\ mg$